SERVICE AUTOMATION AND DYNAMIC PROVISIONING TECHNIQUES IN IP/MPLS ENVIRONMENTS

WILEY SERIES IN COMMUNICATIONS NETWORKING & DISTRIBUTED SYSTEMS

The 'Wiley Series in Communications Networking & Distributed Systems' is a series of expert-level, technically detailed books covering cutting-edge research, and brand new developments as well as tutorial-style treatments in networking, middleware and software technologies for communications and distributed systems. The books will provide timely and reliable information about the state-of-the-art to researchers, advanced students and development engineers in the Telecommunications and the Computing sectors.

Other titles in the series:

Wright: *Voice over Packet Networks* 0-471-49516-6 (February 2001)
Jepsen: *Java for Telecommunications* 0-471-49826-2 (July 2001)
Sutton: *Secure Communications* 0-471-49904-8 (December 2001)
Stajano: *Security for Ubiquitous Computing* 0-470-84493-0 (February 2002)
Martin-Flatin: *Web-Based Management of IP Networks and Systems*, 0-471-48702-3 (September 2002)
Berman, Fox, Hey: *Grid Computing. Making the Global Infrastructure a Reality*, 0-470-85319-0 (March 2003)
Turner, Magill, Marples: *Service Provision. Technologies for Next Generation Communications* 0-470-85066-3 (April 2004)
Welzl: *Network Congestion Control: Managing Internet Traffic* 0-470-02528-X (July 2005)
Raz, Juhola, Serrat-Fernandez, Galis: *Fast and Efficient Context-Aware Services* 0-470-01668-X (April 2006)
Heckmann: *The Competitive Internet Service Provider* 0-470-01293-5 (April 2006)
Dressler: *Self-Organization in Sensor and Actor Networks* 0-470-02820-3 (November 2007)
Berndt: *Towards 4G Technologies: Services with Initiative* 978-0-470-01031-0 (March 2008)

SERVICE AUTOMATION AND DYNAMIC PROVISIONING TECHNIQUES IN IP/MPLS ENVIRONMENTS

Christian Jacquenet, Gilles Bourdon and Mohamed Boucadair

All at
France Telecom, France

John Wiley & Sons, Ltd

Telephone (+44) 1243 779777

Email (for orders and customer service enquiries): cs-books@wiley.co.uk
Visit our Home Page on www.wiley.com

Other Wiley Editorial Offices

John Wiley & Sons Inc., 111 River Street, Hoboken, NJ 07030, USA

Jossey-Bass, 989 Market Street, San Francisco, CA 94103-1741, USA

Wiley-VCH Verlag GmbH, Boschstr. 12, D-69469 Weinheim, Germany

John Wiley & Sons Australia Ltd, 42 McDougall Street, Milton, Queensland 4064, Australia

John Wiley & Sons (Asia) Pte Ltd, 2 Clementi Loop #02-01, Jin Xing Distripark, Singapore 129809

John Wiley & Sons Canada Ltd, 6045 Freemont Blvd, Mississauga, ONT, L5R 4J3, Canada

Wiley also publishes its books in a variety of electronic formats. Some content that appears in print may not be available in electronic books.

Library of Congress Cataloging-in-Publication Data

Jacquenet, Christian.
Service automation and dynamic provisioning techniques in IP/MPLS environments / Christian Jacquenet, Gilles Bourdon and Mohamed Boucadair.
p. cm.
Includes index.
ISBN 978-0-470-01829-3 (cloth : alk. paper)
1. MPLS standard. 2. TCP/IP (Computer network protocol) I. Bourdon, Gilles. II. Boucadalr, Mohamed. III. Title.
TK5105.573.J33 2008
004.6'2–dc22

2007043741

British Library Cataloguing in Publication Data

A catalogue record for this book is available from the British Library

ISBN 978-0-470-01829-3 (H/B)

Typeset in 10/12 pt Times by Thomson Digital, India.
Printed and bound in Great Britain by Antony Rowe Ltd, Chippenham, England.

Contents

Preface

Just remember the set of services offered by the Internet a few years ago – emails, web services, sometimes experimental voice services, over what used to be referred to as a 'high-speed' connection of a few hundred kbits/second! The Internet has gone through a profound transformation and has been evolving at an unprecedented rate compared with other industries, thus becoming the central component of all forms of communication: data (emails, web services, search engines, peer-to-peer, e-commerce, stock trading, etc.), voice but also video (TV broadcasting, videoconferencing).

New innovative applications and services will undoubtedly continue to emerge, and we are still at an early stage of what the Internet will be able to provide in the near future. With no doubt, the impact of the Internet on how people communicate around the world and access to information will continue to increase rapidly. New forms of communication will arise such as tele-presence, ubiquitous services and distributed gaming, and the Internet will ineluctably extend its reach to 'objects', which is sometimes referred to the 'Internet of things', with billions of objects interconnected with each other and new forms of machine-to-machine communication. This new era of services will lead to endless possibilities and opportunities in a variety of domains.

The offering of a wide range of new services has required the design of networking technologies in the form of sophisticated protocols and mechanisms based on open standards driven by the Internet Engineering Task Force (IETF). The non-proprietary nature of the Internet Protocol (IP) led to interoperable solutions, thus making the Internet a unique platform of innovation.

As a direct implication of the Internet becoming critical to our personal and professional lives, user expectation has become very high in terms of reliability, quality of service (QoS) and security. A network failure of a few minutes is now considered as unacceptable! Fast network failure detection and traffic reroute mechanisms have been designed to find alternate paths in the network within the timeframe of a few milliseconds while maintaining path quality.

Fine granularity in terms of QoS is now a must: although some applications are inherently delay tolerant (e.g. asynchronous communications such as emails), other traffic types impose bounded delays, jitters and reliability constraints that require complex configuration tasks to engineer the network. QoS guarantees imply traffic classification at the edge of the network, sophisticated local forwarding techniques (multipriority scheduling and traffic discard) and traffic engineering.

The ability to effectively engineer the traffic within the network is now of the utmost importance and is known as a fairly difficult task for service providers considering the high volume of varying traffic. Furthermore, service providers have to engineer the network carefully in order to meet the quality of services imposed by demanding applications while having to deal with resource constraints. Security has become a central component: user identification and authentication and protection against attacks of different forms, including denial of service (DoS) attacks, require the configuration of complex networking technologies. Last but not least, the ability to efficiently manage and monitor the network is an absolute requirement to check service level agreements, enforce policies, detect network faults and perform network troubleshooting to increase the network availability.

A considerable amount of attention has been paid to service automation, network provisioning and policy enforcement. Network technology designers have been actively working on various tools to effectively provision, configure and monitor the network with sophisticated network components so as to ensure the toll quality that the Internet is now delivering, far from the 'best effort' service of the early days of the Internet. These tasks are increasingly crucial and complex, considering the diversity of the set of services provided by the Internet and the scale at which such tasks must be performed, with hundreds of millions of end-users, hundreds of services and a very significant traffic growth.

This is the right book at the right time, and the authors are known for their deep level of expertise in this domain. The organization of the book is particularly well suited to the topic. The first part examines the protocols and architecture required for network provisioning and policy enforcement in IP/MPLS networks. However, a book on this key subject would not be thorough without a strong emphasis on issues of a practical nature, and this is what the second part of the book is about. A number of highly relevant examples are provided on QoS, traffic engineering and virtual private networks, ideally complementing the theory expounded in the first part of the book.

JP Vasseur

Cisco Distinguished Engineer

Chair of the IETF Path Computation Element Working Group

Acknowledgements

Christian	To my wife Béatrice and my sons Pierre and Paul, with all my love
Gilles	To my wife and my son
Mohamed	To my parents and my wife, with all my love

Part I

Architectures and Protocols for Service Automation

Part I

1

Introduction

1.1 To Begin With

The Internet has become a privileged playground for the deployment of a wide range of value-added IP service offerings. These services rely upon the combination of complex yet advanced capabilities to forward the corresponding traffic with the desired level of quality, as per a set of policies (in terms of forwarding, routing, security, etc.) that have been defined by the service provider, and sometimes negotiated with the customers.

This is a book about techniques that allow the dynamic enforcement of such policies.

Before discussing the motivation for such a book and detailing its organization, this chapter begins with an introductory reminder about the basics of IP networks. A 30 000 ft overview of the Internet as we know it.

1.1.1 On IP Networks in General, and Routers in Particular

An IP network is a set of transmission and switching resources that process IP traffic. The IP traffic is composed of protocol data units (PDUs) (RFC 791 [1]), which are called datagrams. The transmission resources of an IP network rely upon various link-level transport technologies, such as asynchronous transfer mode (ATM), synchronous digital hierarchy (SDH), etc.

The switching resources of an IP network are called 'routers'. IP routers are in charge of processing each IP datagram, as per the following chronology:

- Upon receipt of a datagram, the router analyzes the contents of the destination address field of the datagram. This allows the router to identify the output interface through which the IP datagram will be forwarded, according to the contents of the forwarding information base, or FIB. An FIB of an IP router is typically composed of a set of {next hop; IP network} associations. The first member of these associations corresponds to the interface identifier of the next router capable of processing the datagram whose

Service Automation and Dynamic Provisioning Techniques in IP/MPLS Environments C. Jacquenet, G. Bourdon and M. Boucadair

destination address field corresponds to the IP network (expressed as an IP address) which is the second member of the pair.

- The analysis of the FIB allows the router to perform the switching features that will direct the datagram to the appropriate output interface through which the next hop router's interface identified in the aforementioned pair can be reached.
- Then the router performs the forwarding task which will actually transmit the datagram over the selected output interface.

Thus the forwarding of an IP datagram relies upon the hop-by-hop paradigm owing to the systematic identification of the next router on the path towards the final destination [2–4]. Note also that Postel [1] also mentions the *source routing* mode, where the path to be followed by IP datagrams can either be partially ('loose source routing') or fully ('strict source routing') defined by the source that sends the IP datagram.

An FIB of an IP router is fed by information that comes from the use of a routing process, which can be either static or dynamic. In the case of static routing, the set of paths towards destination prefixes is manually configured on every router of the network.

In the case of a dynamic routing process, the FIB is dynamically fed by information that is stored and maintained in a specific table – the routing information base (RIB). There are at least as many RIB databases as routing protocols activated on the IP router.

The IP routers, which are operated by a globally unique administrative entity within the Internet community, form an autonomous system (AS) (see Figure 1.1) or border gateway protocol (BGP) domain (RFC 4271 [5]). From a typological standpoint, an AS is composed of a set of routers, thus yielding the distinction between the inner of an AS and the outer of an AS. The outer of an AS is the rest of the Internet.

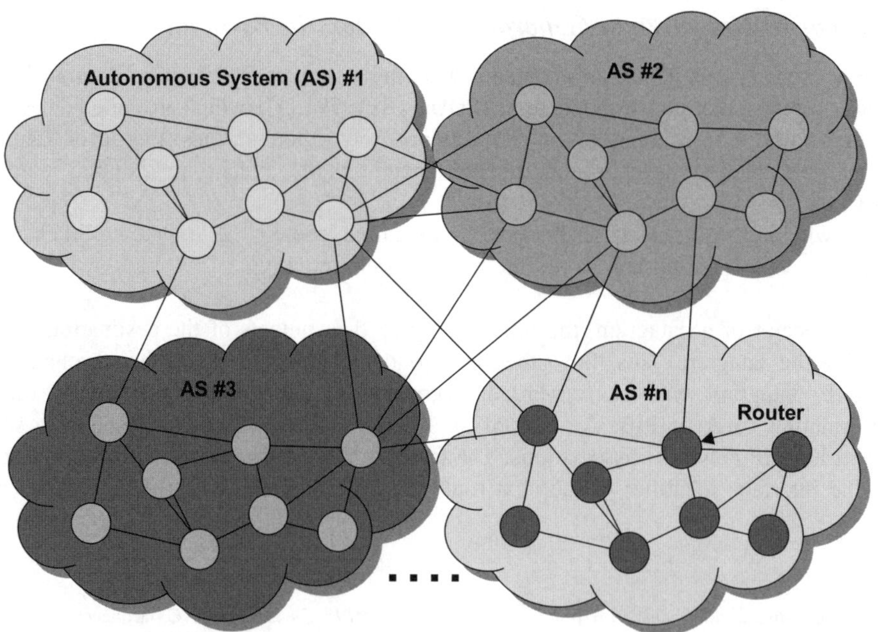

Figure 1.1 The Internet organized into autonomous systems

1.1.2 On the Usefulness of Dynamic Routing Protocols in IP Networks

The deployment of IP networks of large scale (such as those that compose today's Internet) has rapidly led to the necessity of using dynamic routing protocols, so that routers might determine as efficiently as possible (that is, as fast as possible) the best route to reach a given destination (such an efficiency can be qualified in terms of convergence time).

Protocol convergence can be defined as the time it takes for a routing protocol to compute, select, install and disseminate the routing information [that is, the required information to reach a (set of) destination prefix(es)] at the scale of a region, be it an OSPF area or a BGP domain. That is, for a given destination prefix, a converged state is reached when information regarding this prefix has been added/modified or withdrawn in all relevant databases of the routers in the region. Traffic for a 'converged' prefix should be forwarded consistently inside the region.

As a matter of fact, static routing reveals itself as being incompatible with the number of IP networks that currently compose the Internet, because the static feeding of the FIB databases (which may therefore contain tens of thousands of entries, as per http://bgp.potaroo.net/) is a tedious task that may obviously impact upon the forwarding efficiency of such IP networks, because of network failures or congestion occurrences. Indeed, static routing leads to 'frozen' network architectures, which cannot adapt easily to the aforementioned events, unlike dynamic routing.

Dynamic routing protocols therefore allow routers to *dynamically* exchange network reachability information. Such information is stored in the RIB bases of these routers (as mentioned above) and is dynamically refreshed. The organization of the Internet into multiple autonomous systems yields the following routing protocol classification:

- dynamic routing protocols making it possible to exchange reachability information about networks that are part of the autonomous system: such protocols are called interior gateway protocols, or IGP;
- dynamic routing protocols making it possible to exchange reachability information about networks that are outside the autonomous system: such protocols are called exterior gateway protocols, or EGP.

Figure 1.2 depicts such a classification. Note that the white arrow of the figure should not be understood as a limitation of EGP exchanges that would be restricted to inter-AS communications. As a matter of fact, there are also BGP exchanges within domains.

These dynamic routing protocols use a specific algorithm whose calculation process takes into account one or several parameters which are often called *metrics*. These metrics are used by the routing algorithm to enforce a routing policy when the administrator of an IP network has the ability to actually define (and possibly modify) the values of such metrics.

Among the most commonly used metrics, one can cite:

- the number of routers (*hop count metric*) to cross before reaching a given destination [the fewer the routers, the better will be the route, whatever the characteristics of the links (in terms of speed, among others) that interconnect the routers];
- the *cost metric*, the meaning of which is broader than the previous hop count metric, and which generally reflects a weight assigned to an interface, a transmission link, the crossing of an autonomous system or a combination of these components.

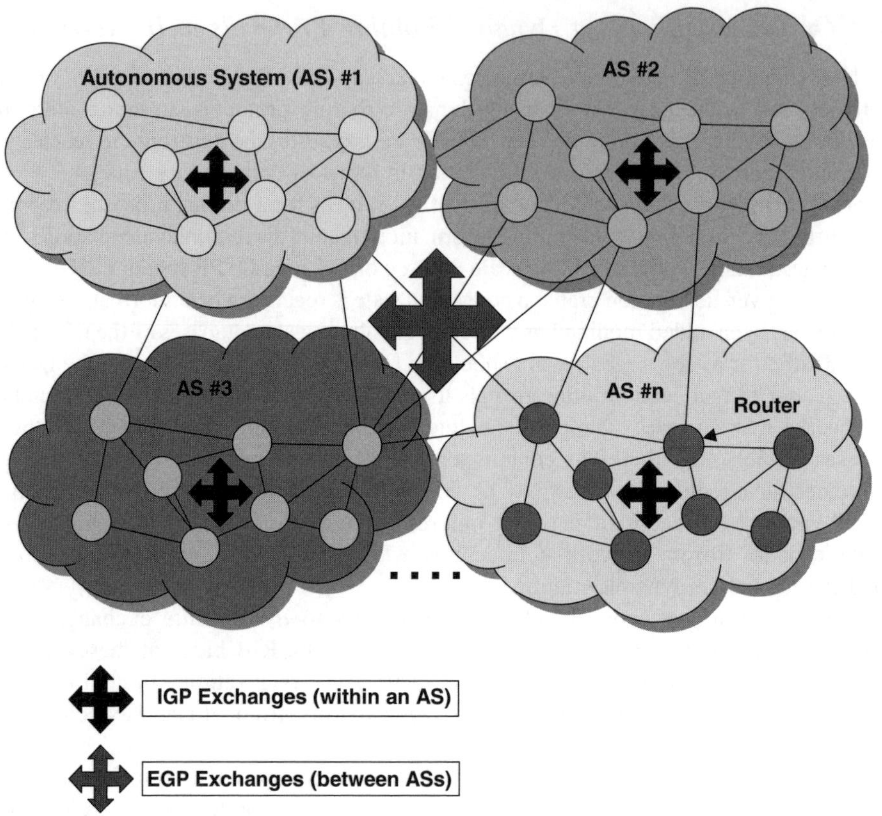

Figure 1.2 Two kinds of dynamic routing protocol (IGP and EGP)

The nature of the routing algorithms yields another typological effort, which consists in distinguishing the following:

- Routing protocols using algorithms based upon distance-vector calculation. Such an algorithm is generally inspired by the Bellman–Ford probabilistic calculation.
- Routing protocols using algorithms that take into account the state of the links interconnecting the routers. Such routing protocols are called 'link-state' routing protocols, and their algorithms are generally based upon the use of the Dijkstra probabilistic calculation.

Table 1.1 provides a summary of the principal IGP-specific characteristics of both distance-vector and link-state routing algorithms.

The very first IGP to be specified, standardized, developed and implemented by router vendors was the routing information protocol (RIP) (RFC 1058 [6], RFC 2453 [7]) back in 1984. The route selection process of RIP relies upon the use of a distance-vector calculation, directly inspired from the Bellman–Ford algorithm.

Table 1.1 Comparison between distance-vector and link-state routing protocols

Distance-vector Routing Protocols	Link-state Routing Protocols
Each router (periodically but also spontaneously) sends reachability information (routes to destination prefixes) to its directly-connected neighbors.	Each router (periodically but also spontaneously) sends reachability information to all the routers of the domain to which it belongs (a domain corresponds either to the autonomous system to which the router belongs to or part of the autonomous system).
The reachability information is composed of a cost estimation (generally expressed in terms of hop count, that is, the number of routers that need to be crossed to reach a given destination) of each of the paths that make it possible to reach all the networks (destination prefixes) of which the router is aware.	The reachability information is composed of the cost of the paths (generally expressed as a combination of metrics that reflect the cost of each path better than the hop count metric used by distance-vector routing protocols; link-state protocols use metrics that reflect the link bandwidth associated with a given interface, for example) towards adjacent networks. Thus, routers of a given domain acquire a more accurate knowledge of the domain's topology, and hence a better estimation of the shortest path to reach a given destination within the domain.

An example of a link-state routing protocol is the open shortest path first (OSPF) protocol (RFC 2328 [8]), which is supported by most of the routers on the market.

1.1.3 On the Inability of an IGP to Address Interdomain Communication Needs

The organization of the Internet into autonomous systems does not necessarily justify the aforementioned IGP/EGP typology, since the network reachability information exchange between autonomous systems is primarily based upon the use of a dynamic routing protocol, whatever this protocol might be (static routing between ASs is not an option, for the reasons mentioned in Section 1.1.2).

Therefore, why not use an IGP protocol to exchange network reachability information between autonomous systems? Here is a couple of reasons:

1. A router that activates a *distance-vector routing protocol* advertizes to its neighbors the whole set of networks it can reach. This information is displayed as a vector list that includes the cost of the path associated with each network. Each router of the network builds its own RIB database according to the information contained in these vector lists,

but this information does not provide any clue concerning the identity of the routers and the networks that have to be crossed before reaching a given destination. This may present some difficulty when exchanging such reachability information between autonomous systems:

- The distance-vector routing protocol states that all the routers running it have a common understanding of the metric that allows them to select a next hop rather than another. This common understanding may not be the case for routers belonging to different autonomous systems.
- The routing policy that has been defined within an autonomous system might be such that communication with specific autonomous systems is forbidden (e.g. for exchanging specific network reachability information). A distance-vector routing protocol has no means to reflect such filtering capabilities in the vector lists it can propagate.

2. A router that activates a *link-state routing protocol* advertizes network reachability information which is partly composed of the costs associated with the links that connect the router to adjacent networks, so that each of these routers has the ability to build up a complete image of the network topology. This advertisement mechanism relies upon the use of a flooding capability, which may encounter some scalability issues when considering communication between autonomous systems:
   - The autonomous systems do not necessarily have a common understanding of the metrics that are used to compute a shortest path, so that the topological information that is maintained by the routers may be dramatically different from one autonomous system to another.
   - The aforementioned flooding capability of a link-state protocol can rapidly become incompatible with networks of large scale (in terms of the number of routers composing a given domain), especially when considering the traffic volume associated with the broadcasting of network reachability information.

The basic motivation that yielded the specification, the standardization and the development of routing protocols of the EGP type was based upon the following information: since the metrics used by IGP routing protocols can be understood differently by routers belonging to different autonomous systems, the network reachability information to be exchanged between autonomous systems should rely upon other metrics.

Thus, a router belonging to autonomous system A would advertize to autonomous systems B, C, etc., the networks it can reach, including the autonomous systems that have to be crossed to reach such networks. This very basic concept is used by EGP routing protocols, and it is called ‘path-vector routing’.

An EGP routing protocol has the following characteristics:

- The information exchanged between routers that belong to different autonomous systems does not contain any clues about the use of a specific metric, or the value of any cost.
- The information exchanged between routers that belong to different autonomous systems describe a set of routes towards a set of destination prefixes. The description of such routes includes (but is not necessarily limited to) the number and the identity of the autonomous systems that have to be crossed to reach the destination networks.

The latter characteristic allows a router to enforce a routing policy that has been defined by the administrator of an autonomous system, so that, for example, this router could decide to avoid using a specific route because this route traverses autonomous systems whose degree of reliability is incompatible with the sensitive nature of the traffic that could use this route.

The forwarding of IP traffic over the Internet implies the crossing of several autonomous systems, thus yielding the activation of an EGP routing protocol. The BGP-4 (border gateway protocol version 4) protocol (RFC 4271 [9]) is currently the EGP that has been deployed over the Internet. The BGP protocol has arisen from the experience acquired during the very first stages of Internet deployment, especially through the deployment of the NSFNET (National Science Foundation NETwork), owing to the specification and the implementation of the exterior gateway protocol (EGP) (RFC 904 [10], RFC 1092 [11], RFC 1093 [12]).

1.1.4 On the BGP-4 Protocol

The principal feature of a BGP-4-enabled router consists in exchanging reachability information about IP networks (aka IP destination prefixes) with other BGP-4-enabled routers. Such information includes the list of the autonomous systems that have been crossed, and it is sufficiently specific for it to be possible to build up an AS connectivity graph from this information.

This AS connectivity graph will help BGP-4-enabled routers in avoiding routing loops (which result in the development of IP network-killing 'black holes'), and it will also help in enforcing the routing policies that have been defined by the AS administrator.

The BGP protocol relies upon transmission control protocol (TCP) port 179 (RFC 793 [13]) – a transport layer-specific protocol that supports fragmentation, retransmission, acknowledgement and sequencing capabilities.

The BGP communication between two routers can be briefly described according to the following chronology:

- The BGP routers establish a TCP connection between themselves by exchanging messages that aim to open this connection, then confirming the parameters that characterize this connection.
- Once the TCP connection has been established, the very first exchange of (reachability) information is composed of the overall contents of the BGP table maintained by each peer.
- Then, information is exchanged on a dynamic basis. This information actually represents specific advertisements every time the contents of one or the other BGP tables have changed. Since the BGP-4 protocol does not impose a periodic update of the global contents of the BGP routing table, each router must keep the current version of the global contents of all the BGP routing tables of the routers with which it has established a connection.

Specific messages are exchanged on a regular basis, so as to keep the BGP connection active, whereas notifications are sent in response to a transmission error or, more generally, under specific conditions. The receipt of a notification results in the BGP communication breakdown between the two BGP peers, but such a breakdown is smoothed by the TCP protocol, which waits for the end of the ongoing data transmission before effectively shutting down the connection.

Although the BGP-4 protocol is a routing protocol of the EGP type, routers that belong to the same autonomous system have the ability to establish BGP connections between themselves as well, which yields the following typology:

- The connections that are established between BGP routers belonging to different autonomous systems are called 'external sessions'. Such connections are often named 'external BGP' or 'eBGP' connections.
- The connections that are established between BGP routers belonging to the same autonomous system are called 'internal sessions'. Such connections are often named 'internal BGP' or 'iBGP' connections.

iBGP connections are justified by the will to provide (to the BGP routers belonging to the same autonomous system) as consistent a view of the outside world as possible. Likewise, an IGP protocol provides a homogeneous view of the internal routes within an autonomous system.

A BGP route (i.e. the reachability information that is transmitted within the context of the establishment of a BGP connection) is made up of the association of an IP prefix and the attributes of the path towards the destination identified by this prefix. Upon receipt of such information, the router will store it in the BGP routing table, which is actually made up of three distinct tables:

- The *Adj-RIB-In* table, which stores all the advertized routes received by a BGP peer. This information will be exploited by the BGP decision process.
- The *Adj-RIB-Out* table, which stores all the routes that will be advertized by a BGP peer. These are the routes that have been selected by the BGP decision process.
- The *Loc-RIB* table, which stores all the routes that will be taken into consideration by the BGP decision process. Among these routes there will be those that are stored in the *Adj-RIB-Out*.

The distinction between these three tables is motivated by the BGP route selection process. In practice, most of the BGP-4 implementations use a single BGP routing table, which will be indexed appropriately according to the above-mentioned typology.

1.1.5 The Rise of MPLS

The hop-by-hop IP routing paradigm of the old days of the Internet (as introduced in Section 1.1.1) is being questioned by the multiprotocol label switching (MPLS) technique (RFC 3031 [14]). MPLS is a switching technique that allows the enforcement of a consistent forwarding policy at the scale of a *flow*, where a flow can be defined as a set of IP datagrams that share at least one common characteristic, such as the destination address.

In this case, all the IP datagrams of a given flow [designated as a forwarding equivalence class (FEC) in the MPLS terminology] will be conveyed over the very same path, which is called a label switched path (LSP) (see Figure 1.3).

MPLS switching principles rely upon the content of a specific field of the MPLS header, which is called the *label*. Labels are the primary information used by MPLS-enabled routers

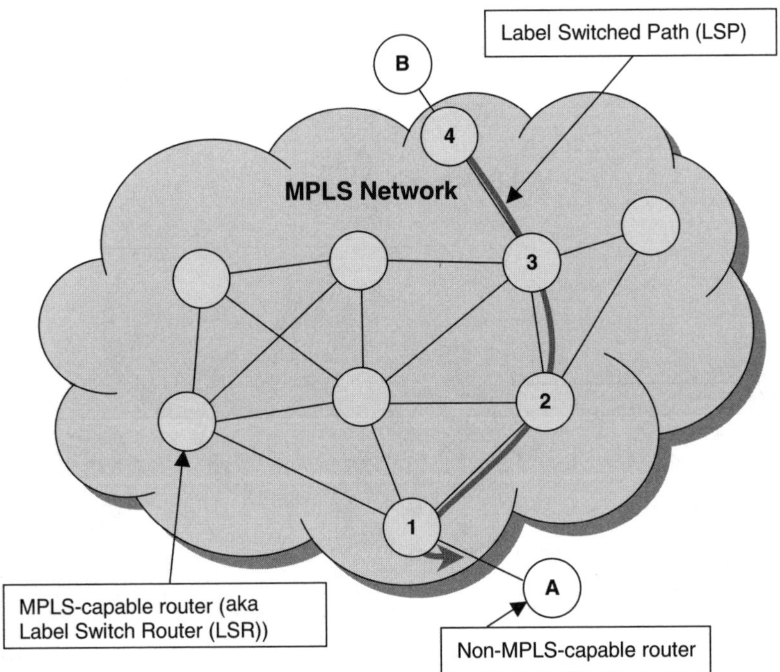

Figure 1.3 MPLS label switched paths

to forward traffic over LSP paths. MPLS has been defined so that it can be used whatever the underlying transport technology, or whatever the network layer-specific communication protocol, such as IP. The MPLS forwarding scheme is depicted in Figure 1.4.

The MPLS forwarding scheme relies upon the maintenance of label tables, called label information bases (LIBs). To forward an incoming MPLS packet, the MPLS-enabled router will check its LIB to determine the outbound interface as well as the outgoing label to use, based upon the information about the incoming interface as well as the incoming label. As per the example provided by Figure 1.4:

- Router A of the figure, which does not support MPLS forwarding capabilities, is connected to (or has the knowledge of) networks N1 and N2, which can be reached through its Ethernet 0 (E0) interface. Table 1.2 is an excerpt from its FIB, which basically lists the network prefix, the outgoing interface and the associated next hop router.
- The black arrow in Figure 1.4 suggests that an ordinary routing update (by means of a dynamic routing protocol, such as OSPF), advertizes the routes to the MPLS-enabled router [or label switch router (LSR) in the MPLS terminology], which is directly connected to router A.
- Using the label distribution protocol (LDP) (RFC 3036 [15]), router 1 selects an unused label [label 3 in the example provided by the excerpt of its label information base below (Table 1.3)] and advertizes it to the upstream neighbor. The hyphen in the 'Label' column of Table 1.3 denotes that all labels will be popped (or removed) when forwarding the

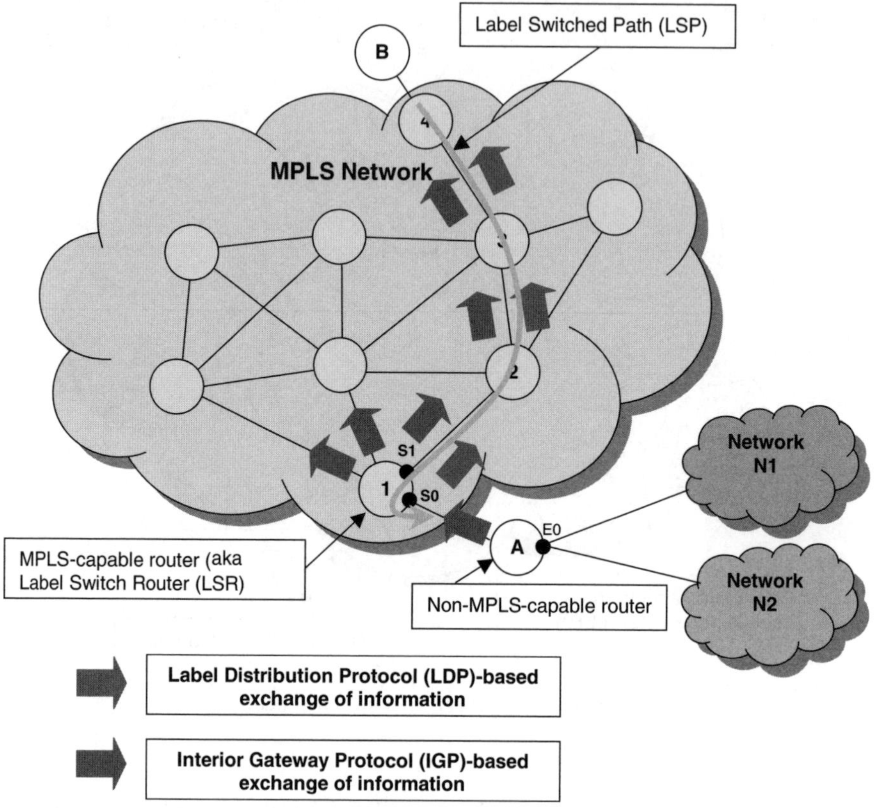

Figure 1.4 MPLS forwarding principle

packet to router A, which is not MPLS capable. Thus, an MPLS packet received on the serial 1 interface with label 3 is to be forwarded out through the serial 0 interface with no label, as far as LSR 1 which is directly connected to router A is concerned. The white arrow in Figure 1.4 (between router 1 and router 2) denotes the LDP communication that indicates the use of label 3 to the upstream LSR 2.

LSR 1 has learned routes that lead to N1 and N2 network prefixes. It advertizes such routes upstream. When LDP information is received, router 1 records the use of label 3 on the outgoing interface serial 0 for the two prefixes mentioned previously. It then allocates label 16 on the serial 1 interface for this FEC and uses LDP to communicate this information

Table 1.2 Excerpt from the forwarding information base of router A (as per Figure 1.4)

Network	Interface
N1	E0
N2	E0

Table 1.3 Excerpt from the label information base of router 1 (as per Figure 1.4)

Network	Incoming I/F	Label	Outgoing I/F	Label
N1	Serial 1	3	Serial 0	—
N2	Serial 1	3	Serial 0	—

Table 1.4 Excerpt from the label information base of router 2 (as per Figure 1.4)

Network	Incoming I/F	Label	Outgoing I/F	Label
N1	S1	3	S0	16
N2	S1	3	S0	16

to the upstream LSR. Thus, when label 16 is received on serial 1, it is replaced with label 3 and the MPLS packet is sent out through serial 0, as per Table 1.4.

Note that there will be no labels received by router B (and sent by router 4 in the figure), since the top router B is not an LSR, as illustrated by its routing table (no labels are maintained in this table). The label switched path (LSP) is now established.

Note also that MPLS labels can be encoded as the virtual path identifier/virtual channel identifier (VPI/VCI) information of an ATM cell, as the data link connection identifier (DLCI) information of a frame, in the sense of the frame relay technology, but also as 20-byte long information encoded in the 4-byte encoded MPLS header associated with each IP PDU, as depicted in Figure 1.5.

Label	EXP bits	Stack	Time To Live (TTL)
20 bits	3 bits	1 bit	8 bits

Figure 1.5 The MPLS header

MPLS capabilities are now supported by most of the router vendors of the market, and the technique is gaining more and more popularity among service providers and network operators, as the need for traffic engineering capabilities emerges. Traffic engineering is the ability to (dynamically) compute and select paths whose characteristics comply with requirements of different kinds: the need to make sure that a given traffic will be conveyed by a unique path (potentially secured), e.g. for security purposes, or the need for minimum transit delays, packet loss rates, etc.

MPLS-based traffic engineering capabilities can be seen as some of the elementary components of a global quality of service (QoS) policy.

1.2 Context and Motivation of this Book

IP service offerings (ranging from access to the Internet to more advanced services such as TV broadcasting or videoconferencing) are provisioned owing to the combined activation of different yet complex capabilities, which not only require a high level of technical expertise but also result in the organization of complex management tasks.

1.2.1 Classifying Capabilities

As stated above, IP services are provided by means of a set of elementary capabilities that are activated in different regions and devices of an IP/MPLS network infrastructure. These capabilities can be organized as follows:

- *Architectural* capabilities, which are the cornerstones for the design and enforcement of addressing, forwarding and routing policies. Such policies aim to convey service-specific traffic in an efficient manner, e.g. according to the respective requirements and constraints that may have been (dynamically) negotiated between the customer and the service provider.
- *Quality of Service* (QoS) capabilities, as briefly introduced in Section 1.6.
- *Security* capabilities, which include (but are not necessarily limited to):
  - the user and device identification and authentication means;
  - the protection capabilities that preserve any participating device from any kind of malicious attacks, including (distributed) denial of service (DDOS) attacks;
  - the means to preserve the confidentiality of (some of) the traffic that will be conveyed by the IP network infrastructure;
  - the means to protect users and sites from any kind of malicious attack that may be relayed by the IP/MPLS network infrastructure;
  - the functions that are used to check whether a peering entity is entitled to announce routing information or not, and also the features that provide some guarantees as far as the preservation of the integrity (and validity) of such (routing) information is concerned.
- *Management* capabilities, composed of fault, configuration, accounting, performance and security (FCAPS) features. Monitoring tools are also associated with such features. They are used for analysis of statistical information that aims to reflect how efficiently a given service is provided and a given policy is enforced.

1.2.2 Services and Policies

The management tasks that are performed to provision and operate an IP network or a set of IP service offerings can be grouped into several policies that define what capabilities should be activated, and how they should be used (that is, the specification of the relevant configuration parameters).

Policies can relate to a specific service [e.g. the forwarding policy to be enforced at the scale of a BGP domain to convey voice over IP (VoIP) traffic with the relevant level of quality], or can be defined whatever the nature of the service offerings (e.g. the BGP routing policy to be enforced within a domain).

The design and the enforcement of a given policy must therefore address a set of elementary questions, as follows:

- *Why*? This is what this book is about – the need for policies to facilitate the automation of sometimes tedious management tasks (configuration of routers to support different services, identification of the users entitled to access a service, etc.) that need to be checked (that is, reliability is a key characteristic of configuration tasks), as well as the dynamic allocation of (network) resources, either proactively (e.g. as part of a global network planning policy) or reactively (e.g. to address traffic growth issues).

- *What*? This is the set of capabilities that are required to enforce a policy, possibly to be inferred by the different services that may be provided. For example, a security policy may rely upon the use of filtering, encryption and firewalling capabilities.
- *How*? This is the set of techniques as well as information (in terms of valued configuration parameters) that reflects the instantiation of a given policy. This is also what this book is about – discussing and detailing the various techniques that can be used dynamically to enforce policies, as well as the provisioning of several examples of services. As an example of an instantiated policy, the QoS policy that needs to be enforced for VoIP traffic may include the explicit identification of such traffic (e.g. by means of a specific DSCP marking), as well as the whole set of configuration arguments (token bucket parameters, actions to be taken by the routers in case of in-profile and out-of-profile traffic, etc.) that define how such traffic is prioritized and forwarded. A specific chapter of this book further elaborates on this example.
- Note that timely parameters are also part of this question, like the epoch during which a policy is to be enforced (e.g. 24 hours a day, 7 days a week, etc.). 'When?' is therefore the kind of question that is addressed by these parameters.

The design, the provisioning and the operation of a wide range of IP service offerings are therefore the result of the enforcement of a complex combination of policies. Even more complex is the underlying substrate of various technologies that are solicited to provide (from the subscription phase to the actual deployment) and to manage a given service.

The foreseen development of the so-called 'triple-play' services, where data, voice and image traffics should be gracefully mixed, provided the underlying network infrastructure has the appropriate resources to convey these different traffics with the relevant level of quality, is another key driver for policy-based management and dynamic provisioning techniques.

1.2.3 The Need for Automation

Needless to say, the provisioning of a wide range of service offerings with the adequate level of quality generally takes time, because policy-based design and management are complex tasks, and also because consistency checks take time: addressing any issue that may result from the operation of conflicting configuration tasks, verifying the accessibility of the service, monitoring its availability and checking the appropriate resources are correctly provisioned on time, etc., are headaches (if not nightmares) for network engineers and operators.

In addition, several yet recent initiatives have been launched by the Internet community to investigate mechanisms and protocols that would contribute to the development of 'zero provisioning capabilities'. The objective of such initiatives is to reduce the amount of configuration tasks that require human interventions. This can be viewed as a generalization of the 'plug and play' concept.

It is therefore generally expected that the introduction of a high level of automation in the service provisioning process as well as the use of dynamic policy enforcement techniques should largely contribute to:

- a reduction in the service delivery time;
- a reduction in the overall operational expenditures (OPEX) costs associated with the delivery and the exploitation of such services: automation improves production times and is supposed to reduce the risks of false configuration which may jeopardize the quality of the impacted services.

Automation is the key notion that motivated the writing of this book.

1.3 How this Book is Organized

The organization of this book is basically twofold:

- The first part deals with the theory, where candidate protocols and architectures for the dynamic provisioning of services and the enforcement of policies within IP/MPLS infrastructures are described in detail.
- The second part of the book deals with practice, by introducing and discussing a set of examples [enforcement of QoS and traffic engineering policies, production of BGP/MPLS-based virtual private network (VPN) facilities, etc.] that aim to convince the reader about the reality of such issues and how dynamic provisioning techniques can gracefully address them.

1.4 What Is and What Should Never Be

This is not a book that aims to promote a 'one-size-fits-all' approach, where a single protocol or architecture would address any kind of concern, whatever the nature of the policy, the service and/or the environment.

This is not a book about what is going on in standardization, as far as dynamic provisioning techniques and protocols are concerned.

This is a book that aims to provide readers with a practical yet hopefully exhaustive set of technical updates and guidelines that should help service providers, network operators but also students in acquiring a global yet detailed panorama of what can be done to facilitate (if not automate) the production of services over IP/MPLS infrastructures.

And we sincerely hope you will enjoy it as much as we enjoyed writing it.

References

[1] Postel, J., 'Internet Protocol', RFC 791, September 1981.
[2] Perlman, R., 'Interconnections: Bridges and Routers', Addison-Wesley, 1992.
[3] Comer, D., 'Internetworking with TCP/IP. Volume 1. Principles, Protocols and Architecture', Prentice-Hall, 1995.
[4] Stallings, W., 'High-speed Networks, TCP/IP and ATM Design Principles', Prentice-Hall, 1998.

[5] Rekhter, Y., Li, T., 'A Border Gateway Protocol 4 (BGP-4)', RFC 4271, January 2006.
[6] Hedrick, C., 'Routing Information Protocol', RFC 1058, June 1988.
[7] Malkin, G., 'RIP Version 2', RFC 2453, November 1998.
[8] Moy, J., 'OSPF Version 2', RFC 2328, April 1998.
[9] Rekhter, Y. and Li, T., 'A Border Gateway Protocol 4 (BGP-4)', RFC 1771, March 1995.
[10] Mills, D., 'Exterior Gateway Protocol Formal Specification', RFC 904, April 1984.
[11] Rekhter, J. *et al.*, 'EGP and Policy Based Routing in the New NSFNET Backbone, RFC 1092, February 1989.
[12] Braun, H., 'The NSFNET Routing Architecture', RFC 1093, February 1989.
[13] Postel, J., 'Transmission Control Protocol', RFC 793, September 1981.
[14] Callon, R. *et al.*, 'Multiprotocol Label Switching Architecture', RFC 3031, January 2001.
[15] Andersson, L. *et al.*, 'LDP Specification', RFC 3036, January 2001.
[16] Blake, S. *et al.*, 'An Architecture for Differentiated Services', RFC 2475, December 1998.
[17] Bernet, Y. *et al.*, 'An Informal Management Model for Diffserv Routers', RFC 3290, May 2002.
[18] Heinanen, J. *et al.*, 'Assured Forwarding PHB Group', RFC 2597, June 1999.
[19] Davie, B. *et al.*, 'An Expedited Forwarding PHB (Per-Hop Behavior)', RFC 3246, March 2002.

2

Basic Concepts

2.1 What is a Policy?

Policy-based management concepts were introduced at the end of the 1990s and were standardized in the early 2000s. The notion of *policy* is generally associated with the concept of rules with various degrees of abstraction. Policies can reflect a business strategy (e.g. privilege the forwarding of corporate traffic over Internet traffic within a virtual private network), a company-wide set of rules (e.g. access to the Internet is forbidden) or a combined set of network-inferred rules that yield the specification of forwarding, routing, quality of service and/or security policies.

The aforementioned notion of abstraction refers to the definition of a scope of any given policy without explicitly describing it. According to RFC 3198 [1], a policy can be defined as 'a set of rules to administer, manage and control access to network resources', where these rules can be defined in support of business goals. The latter can also define policies as a 'definite goal, course or method of action to guide and determine present and future decisions'.

Policies defined as a set of rules follow a common information model (such as RFC 3060 [2]), where each and every rule defines a scope, a mechanism and actions. An example of such a rule could be: 'If Internet traffic exceeds 50% of the available bandwidth on the link that connects a VPN site to the network, then limit the corresponding Internet traffic-dedicated resources during certain periods'. In this example, the scope of the policy is the Internet traffic, the mechanism is the bandwidth allocation and the action consists in limiting resources used by Internet traffic during certain periods. This example also introduces the notion of the 'condition' that will trigger the application of the rule.

2.2 Deriving Policies into Rules and Configuration Tasks

Policies that are defined by network administrators need to be understood by the (network) devices that will be involved in the enforcement of the corresponding policies. This gives rise to the need for mechanisms that will process the policy-specific information so that such

Service Automation and Dynamic Provisioning Techniques in IP/MPLS Environments C. Jacquenet, G. Bourdon and M. Boucadair

devices can be configured accordingly, that is, with the configuration tasks that will have to be performed to enforce the policy. Policy-based management relies upon the following steps to derive generic policies into configuration information.

2.2.1 Instantiation

The set of rules that define a policy need to be instantiated according to the environment (e.g. the services to which the policy will be applied) where the policy will be enforced. The policy instantiation can rely upon received events or information that is descriptive of the context (Figure 2.1).

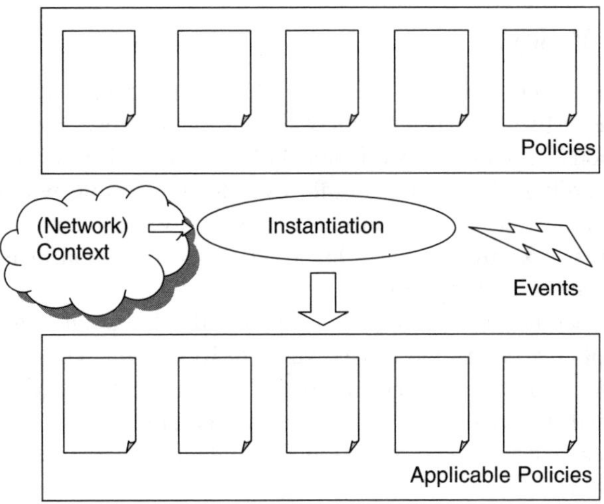

Figure 2.1 Instantiation of policies

The instantiation process requires:

- an understanding of the context-specific information, such as the importance of the mesh in the network, the operating hours, etc.;
- the processing of the incoming events (e.g. link failure) and their impact on the policies;
- knowledge of the information model.

2.2.2 Device Identification

The enforcement of policies needs not only to reflect the applicability of the policies in a given condition but also to identify (and locate) the devices that will participate in the enforcement of the policy. The relationships between the policy and the participating devices are defined in the information model. The actual location of the 'target' devices can rely upon the network topology information (as part of the information model), but also on the information depicting the forwarding paths along which traffic will be conveyed in the network.

The device identification processes require:

- knowledge of the scope of action that can be performed by a participating device [e.g. firewalls are not supposed to enforce traffic engineering policies but security policies (based upon the establishment and the activation of traffic filters, for example), while routers may not be solicited to enforce user-specific identification policies, but rather the forwarding policies that will reflect the level of quality associated with the delivery of a given service, as per user requirements];
- knowledge of the information model, as well as the (network) topology information.

2.2.3 Translation

Once the policies have been instantiated into a set of applicable policies and the target devices involved in the enforcement of such applicable policies have been defined and identified, the rules defined in the applicable policies need to be translated into device-specific configuration information. This translation process is specific to a policy and might be local to the participating device, or use a proxy capability by means of protocols such as the common open policy service (COPS) (RFC 2748 [3], RFC 3084 [4]).

2.3 Storing Policies

The information that depicts a policy needs to be stored and maintained by means of directory services. Directory services have the following characteristics:

- They provide a defined syntax for the objects they store, as well as a means to uniquely identify them (notion of distinguished names). Manipulation of the objects accessible through directory services is also defined by means of a set of allowable operations (such as ‘retrieve’ information related to a specific object, ‘modify’ the attributes of an object, etc.).
- The information model stored in directory services is hierarchical and often reflects an organizational, function-derived structure. Objects are grouped in branches, and they can have precedence defined by their position in the tree structure.
- Directory services rely upon databases that are distributed, yielding slave–master relationships. Slave databases partially or totally replicate the information stored in master databases. The master database corresponds to a central repository where policies can be managed in a centralized fashion.

2.4 Policy and Device Configuration

Figure 2.2 reflects the fact that policy-related configuration is centralized, whereas device-specific configuration information is distributed by essence.

Policies are stored in a directory and managed by a policy server. The policy server is responsible for maintaining and updating policy information as appropriate (as part of the instantiation process). Updates can be motivated by triggering events, as discussed in section 2.2.1.

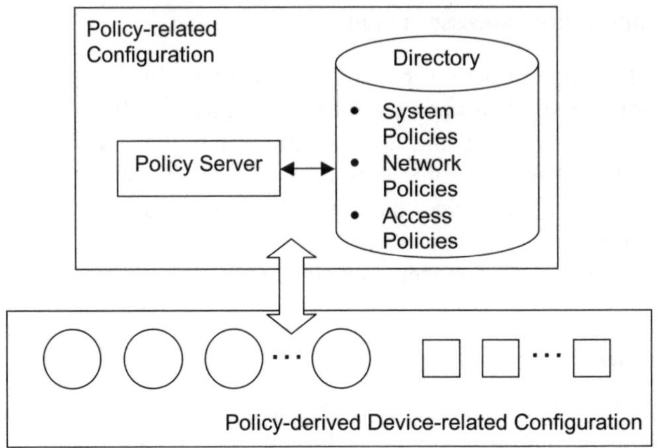

Figure 2.2 Policy configuration and policy-derived device configuration

2.5 Policy-based Management Model

Both the Internet Engineering Task Force (IETF) and the Desktop Management Task Force (DMTF) have been involved in the specification and the standardization of a policy-based management model, which now serves as a reference for the specification and the enforcement of a set of policies within networking infrastructures. Figure 2.3 outlines the different components that are introduced with this model.

Figure 2.3 depicts the relationships between the following components:

- The *policy decision point* (PDP), where policy decisions are made. PDPs use a directory service for policy repository purposes. The policy repository stores the policy information

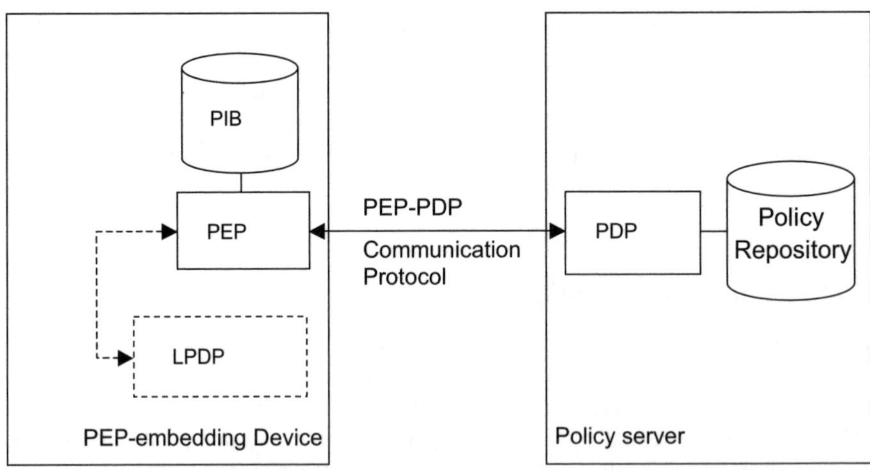

Figure 2.3 Policy-based management model

that can be retrieved and updated by the PDP. The PDP delivers policy rules to the policy enforcement point (PEP – see below) in the form of PIB elements.

- The *policy enforcement point* (PEP), where policy decisions are applied. PEPs are embedded in (network) devices, which are dynamically configured from the policy-formatted information that has been processed by the PEP. PEPs request configuration from the PDP, store the configuration information in the policy information base (PIB) and delegate any policy decision to the PDP. This is commonly referred to as the *outsourcing* mode. PEPs are responsible for deriving policy-formatted information (as forwarded by the PDP to the PEP) into (technology-specific) configuration information that will be used by the PEP-embedding device to enforce the corresponding policies accordingly. Note that PEP and PDP capabilities could be colocated.
- The *policy information base* (PIB) is a local database that stores policy information. It uses a hierarchical structure, where branches are called policy rule classes (PRCs), and where leaves are called policy rule instances (PRIs). Both PRCs and PRIs are uniquely identified by means of policy rule identifiers (PRIDs). Figure 2.4 provides a generic representation of a PIB structure, and Figure 2.5 gives an example of what a PIB can look like.
- Finally, the *local policy decision point* (LPDP) is often seen as an optional capability (from a policy-based management standpoint) that can be embedded in the device to make local policy decisions in the absence of a PDP. Examples of LPDPs include the routing processes that enable routers to dynamically compute and select paths towards a destination without soliciting the resources of a remote PDP.

The example provided in Figure 2.5 denotes a policy that basically consists in filtering out any multicast traffic sent by any source whose IP address is in the 192.134.76.0/24 range, and which is forwarded on the 239.0.0.1 and 239.0.0.2 group addresses.

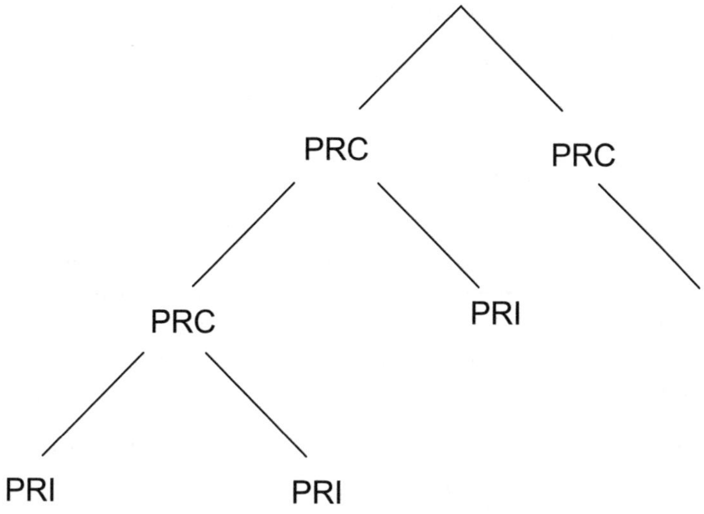

Figure 2.4 Hierarchical structure of a PIB

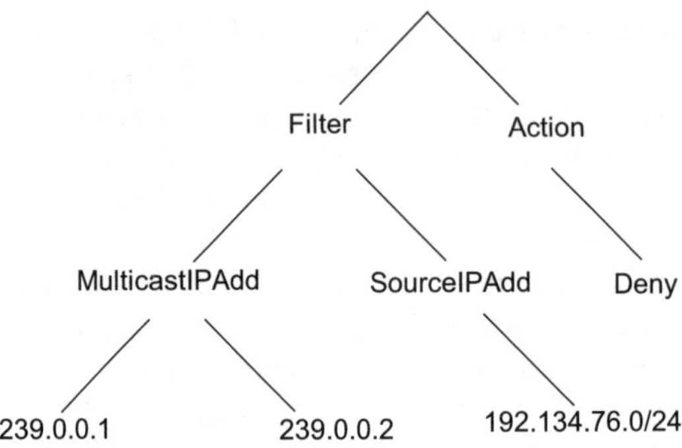

Figure 2.5 Example of an instantiated filtering policy

2.5.1 Reaching a Policy Decision

When a generic event invokes a PEP for a policy decision, the PEP generates a request that includes information related to the event. The PEP then passes the request with all the relevant policy elements to the PDP. The PDP then reaches a decision, which in turn will be forwarded to the PEP for application purposes.

Within the context of policy-based management, the PEP must contact the PDP even if no policy information is received, to retrieve the configuration information it needs upon bootstrap, for example. Both PDP and PEP should have the ability to send an unsolicited message towards each other at any time (decision change, error message, etc.).

2.5.2 Requirements for a PEP–PDP Communication Protocol

There are several candidate protocols that can be suitable for conveying policy information between PEP and PDP capabilities. This book details the machinery of some of them, such as the COPS protocol (RFC 2748 [3], RFC 3084 [4]), the remote authentication dial-in user service (RADIUS) (RFC 2865 [5]) or the Diameter protocol (RFC 3588 [6]). This section only aims to list basic requirements for such a protocol:

- The protocol needs to rely upon a reliable transport mode, to avoid undetected loss of policy queries and responses.
- The protocol should add as small an amount of delay as possible to the response time experienced by policy queries, hence stimulating fast processing capabilities.
- The protocol needs to support opaque objects to avoid protocol changes every time a new policy object has to be exchanged between a PEP and a PDP.
- The protocol needs to support a transactional way of communication, so as to stimulate the query/response formalism, including the ability to renegotiate a previous policy decision.
- The protocol should support unsolicited messaging, to allow both PEP and PDP to notify each other whenever a state change occurs.
- Communication between a PEP and a PDP should be secured, hence preserving the confidentiality of the information exchanged between both entities.

References

[1] Westerinen, A. *et al.*, 'Terminology for Policy-based Management', RFC 3198, November 2001.
[2] Moore, B. *et al.*, 'Policy Core Information Model – Version 1 Specification', RFC 3060, February 2001.
[3] Boyle, J., Cohen, R., Durham, D., Herzog, S., Raja R. and Sastry A., 'The COPS (Common Open Policy Service) Protocol', RFC 2748, Proposed Standard, January 2000.
[4] Ho Chan, K., Durham, D., Gai, S., Herzog, S., McLoghrie, K., Reichmeyer, F., Seligson, J., Smith, A. and Yavatkar, R., 'COPS Usage for Policy Provisioning (COPS-PR)', RFC 3084, March 2001.
[5] Rigney, C. *et al.*, 'Remote Authentication Dial-in User Service (RADIUS), RFC 2865, June 2000.
[6] Calhoun, P. *et al.*, 'Diameter Base Protocol', RFC 3588, December 2003.

3

The RADIUS Protocol and its Extensions

The Remote Authentication Dial-In User Service (RADIUS) protocol (RFC 2865) is one of the most popular authentication protocols used in operators' networks. Its success began with the early Livingston implementations, to offer a scalable and centralized solution to authenticate and authorize users, and possibly to report about resource usage for users connected to equipment through a log-in service (like Telnet) or to remote access servers, primarily through public switched telephone network (PSTN) or integrated services digital network (ISDN) infrastructures. Based on Livingston's early developments, the IETF has standardized its concepts and usage. The last version edited by the IETF is RFC 2865, which is based upon the same concepts than those that were described in the very beginning, but also enhances the protocol to make its implementation more suited to the evolution of remote access usages.

Nowadays, RADIUS is often seen as an obsolete protocol, which is partly true in its conception, as we will see in the next section. However, this judgement has to be moderated, since RADIUS managed to take up most technical challenges imposed by the evolution of access technologies such as large xDSL deployments and secured wireless access (IEEE 802.11i), among others. Mobile phone architectures are also using RADIUS extensively for content billing purposes, even though 3GPP enthroned the Diameter protocol as its successor.

3.1 Protocol Design

RADIUS is based upon a client/server protocol model. The client sends requests to the server, which answers back with appropriate replies depending on the initial request. Regular RADIUS architectures make the Network Access Server (NAS) support the client role, and it queries the RADIUS server as the Authentication Authorization Accounting (AAA) server, as we can see in Figure 3.1.

Service Automation and Dynamic Provisioning Techniques in IP/MPLS Environments C. Jacquenet, G. Bourdon and M. Boucadair

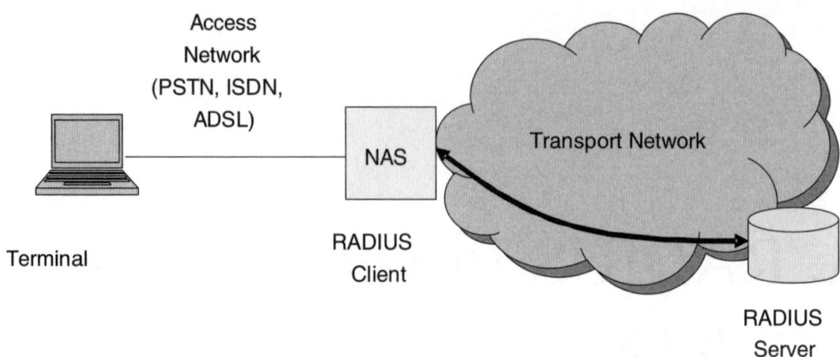

Figure 3.1 Participating devices in RADIUS architecture

RADIUS exchanges can be split into two different categories: AUTH (AUTHentication/AUTHorization) messages and accounting messages. The corresponding flows can follow different paths in the network, since purpose and constraints required by these messages are different. AUTH flows require real-time treatment to grant access as fast as possible, whereas accounting messages may be stored in a database for further exploitation, such as billing.

A RADIUS client request is triggered whenever an end-user notifies the NAS that it requires to be connected to the network. This primary notification is performed with any access protocol capable of carrying information between the end-user and the NAS, such as PPP, EAPoL with 802.1X, HTTP or even Telnet. Depending on the complexity of the state machine of the access protocol, RADIUS exchanges might vary. The very basic framework of RADIUS protocol exchanges starts with an *Access-Request* message, requesting an authentication/authorization on behalf of an end-user, and ends with an *Access-Accept* message, transmitting all required parameters for this end-user to make use of the network, assuming this end-user is entitled to access the network. Otherwise, the RADIUS exchanges will terminate with an *Access-Reject* message. We will see later that this simple request/response exchange becomes more complicated when used with EAP, for instance.

Accounting message exchanges also follow a regular client/server model, with notifications sent by the RADIUS client to the accounting server, these notifications being acknowledged by the latter.

In order to provide a way to secure RADIUS transactions, both RADIUS client and server might be configured with a shared secret, which is used in response messages (messages from server to client) to ensure message integrity and authenticity, as long as the secret is not compromised. This method does not protect the content of the message itself, but it is still an efficient way to protect a RADIUS client against man-in-the-middle replay attacks. In request messages, the shared secret is only used to encrypt the user's password.

3.1.1 Protocol Structure and Messages

RADIUS relies upon the UDP protocol, using destination ports 1812 and 1813 for AUTH and accounting messages respectively. The choice of UDP has been regularly debated, since

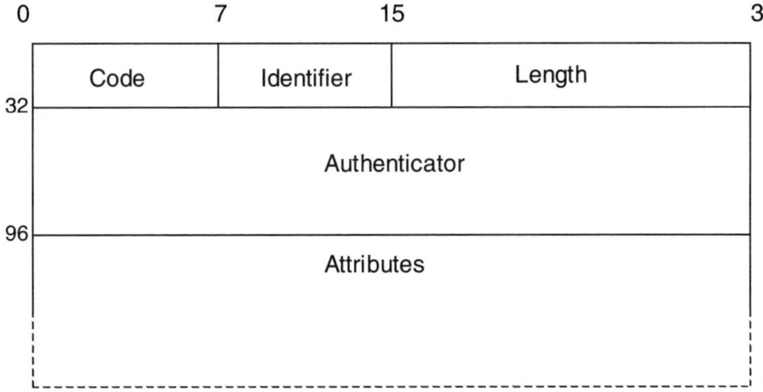

Figure 3.2 RADIUS message structure

UDP might appear to be incompatible with the required level of resilience expected by access network operators. UDP has been chosen for different reasons such as its simplicity for RADIUS implementations, and also because UDP is less resource consuming for operating systems by comparison with TCP. As a consequence, RADIUS implementers have been obliged to maintain their own timers for retransmissions, but UDP made it easier to handle backup server switching and to lower CPU consumption. Whatever one's opinion of this choice, UDP proved its suitability to RADIUS extensive usage.

RADIUS messages consist of five main fields as shown in Figure 3.2:

- *Code*: 8 bits. This field identifies the type of RADIUS message. RFC 2865 defines the different types of message, as listed below:
  - *Access-Request* (code 1);
  - *Access-Accept* (code 2);
  - *Access-Reject* (code 3);
  - *Accounting-Request* (code 4);
  - *Accounting-Response* (code 5);
  - *Access-Challenge* (code 11).
- *Identifier*: 8 bits. This field is used to identify the message exchange being performed between the client and the server. It is interesting to note that this 8-bit field length imposes a limit of 256 simultaneous requests generated by a client without answers from the AAA server. If 256 requests might appear to give a sufficient cushion to serve all requests, large-scale deployments have shown that it is not always sufficient to bear request bursts (e.g. in the case of massive disconnections).
- *Length*: 16 bits. This field indicates the total length of the RADIUS message.
- *Authenticator*: 128 bits. This field is filled up with a suite of octets generated to ensure authenticity of server-initiated messages during RADIUS exchanges. It is also used to encrypt the user's password in *Access-Request* messages.
- *Attributes*. This field completes the RADIUS message, in the limit of the path MTU between the client and the server. The attributes carry all required information for authentication, authorization and accounting messages, each attribute being encoded as a TLV (Type, Length, Value). Major attributes are described in Section 3.1.1.2.

3.1.1.1 RADIUS Message Types

Each message used during client/server exchanges has its own meaning and usage. Therefore, all attributes sent in an *Access-Request* message cannot be used within an *Access-Accept* message, for example. Each RADIUS message is defined to embed a list of mandatory attributes (to identify the end-user context, for instance), a list of optional attributes and a list of unauthorized attributes. The reader is encouraged to consult RFC 2865, and its companion documents defining new attributes, in order to obtain the detailed list. The examples provided below purposely make use of a restricted set of attributes for the sake of simplicity.

Access-Request (Code 1)

The *Access-Request* RADIUS message is the starting point for any further exchange occurring between the client and the server considering AUTH messages. This message carries a sufficient set of attributes necessary properly to identify the end-user requesting access to the network, as well as one password attribute required by the server to authenticate the end-user. Also, the *Access-Request* message has to embed an NAS identifier to determine from which device the end-user is requesting network access. Additional information might be provided, such as network coordinates (*Calling-station-ID*, *NAS-Port*), type of connection (*NAS-Port-Type*) or type of service requested (*Service-Type*).

Receipt of *Access-Request* messages by the AUTH server is acknowledged by sending back an *Access-Accept*, *Access-Challenge* or *Access-Reject* message, depending on the success, incompleteness or failure of the AUTH process respectively.

The authenticator field is filled out with the *Request Authenticator* randomly generated by the access server each time a new RADIUS exchange starts over. A *Request Authenticator* value must be used only once during the lifetime of the shared secret in order to avoid attacks by sending response messages making use of the same authenticator. The *Request Authenticator* is used to constitute reply messages afterwards, to ensure their integrity.

Access-Accept (Code 2)

The *Access-Accept* message is usually the expected answer coming back from the AUTH server (except when challenges are required). This message conveys the authentication successfulness information (inherently symbolized by the *Access-Accept* message itself) and simultaneously the authorization parameters, represented by attributes piggybacked in the packet. These authorization parameters are used to configure the remote end-user (e.g. the user's IP address) to make proper use of the network, but also to apply a predefined set of policies enforced by the NAS (e.g. a specific filter to apply).

Access-Accept messages terminate the client/server exchange between the NAS and the AUTH server. As a matter of fact, the network operator cannot rely upon *Access-Accept* messages to be assured that the user's connection actually started. *Access-Accept* messages can be interpreted as *Access-Reject* if authorization attributes are not applicable by the NAS: if the RADIUS server injects bogus attributes (invalid, malformed, not supported) in *Access-Accept* messages, the end-user might get an unexpected service from the NAS. It can also be lost (UDP is not a reliable transport mode), or silently discarded if the message is improperly built. However, acceptance of the end-user connection often means the beginning of the accounting process. An *Accounting-Request* message is, most of the time, sent after the connection setup.

The Authenticator field is used with the *Response Authenticator*, which is a one-way MD5 hash function applied to the concatenation of Code, Identifier and *Request Authenticator* fields from the *Access-Request* and response attributes, followed by the secret shared between the access server and the authentication server. Therefore, the access server can verify the integrity of the AUTH server's reply message.

Access-Reject (Code 3)
The *Access-Reject* message is used when the connection authorization cannot be granted by the AUTH server. This can occur whenever the authentication fails, or if the requested service cannot be provided (e.g. IP address exhaustion, connection type not compatible with requested service, etc.). The Authenticator field is used in the same way as *Access-Accept* messages.

Accounting-Request (Code 4) and Accounting-Response (Code 5)
The type of *Accounting-Request* or *Accounting-Response* message is carried within a mandatory attribute, *Acct-Status-Type*. The most common values for this attribute are: *Start*, *Stop*, *Interim-Update*, *Accounting-On*, *Accounting-Off*. Accounting messages are used to report actual usage of an end-user's connections, at its beginning (*Start* message), during the connection lifetime (*Interim-Update* message) or at connection termination (*Stop* message). These messages carry particular attributes with usage information, such as connection lifetime, volume transferred in both directions and termination cause (*Stop* message). Other accounting messages are also used to report the accounting activity of the RADIUS client. *Accounting-On* and *Accounting-Off* messages are used to notify the activation/deactivation of the accounting process. These messages are often used whenever the NAS is shut down and rebooted, to refresh connection status in the server connection database and ensure its consistency.

Access-Challenge (Code 11)
The *Access-Challenge* message is used whenever the authentication server requires a multiphase authentication, such as EAP. *Access-Challenge* is sent by the server upon receipt of an *Access-Request* from the client. An *Access-Challenge* message usually embeds a query for which the NAS will have to get the answer through the end-user or not. The answer is carried over a new *Access-Request* message, for which the AUTH server may reply with an *Access-Accept* message if authentication is successful, and an *Access-Reject* message if not. The exchange might continue with other *Access-Challenge* messages, depending on what is required by the AUTH server.

The Authenticator field is used in the same way as *Access-Accept* and *Access-Reject* messages.

3.1.1.2 RADIUS Attributes

Each RADIUS attribute is encoded using a TLV format, with 8-bit encoded type and length fields (see Figure 3.3). The attribute value can reach a maximum of 254 bytes length. In this encoding we see one of the main drawbacks of the RADIUS protocol. The type can only have a maximum of 256 values, which is the optimistic approach since only values from 1 to 191 are really usable, with 1–102 code types already assigned by IANA. A huge number of attributes are not standardized, and are considered to be 'vendor specific'. Multiple sets of

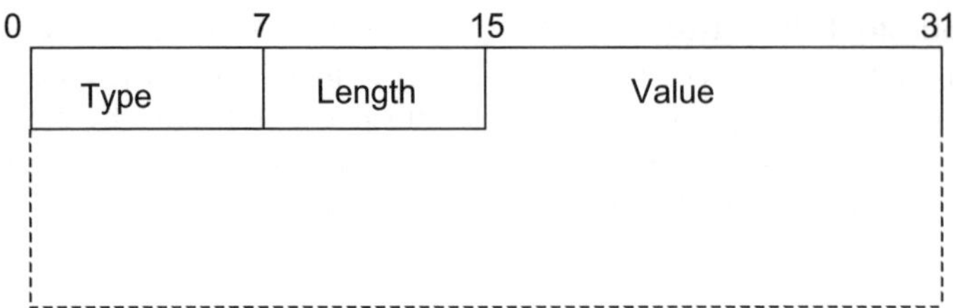

Figure 3.3 Attribute field format

these attributes are often listed in so-called 'dictionaries'. Most of the time, RADIUS servers implement several dictionaries to be compatible with the largest number of RADIUS client implementations.

RADIUS attributes are used to carry all information that the NAS and the AAA server have to exchange for identification, authentication and accounting purposes, but also to configure the end-user terminal or network access (authorization). A description of some widely used attributes follows.

User-Name (Code 1)

This attribute is used to identify the user's identity, as configured in the user's terminal. This identity is often presented following the rules defined in RFC 4282 (previously RFC 2486), the network access identifier. As this is not reliable information with which to obtain the user identity, it is often accompanied with one of the password attributes to perform authentication.

User-Password (Code 2) and CHAP-Password (Code 3)

These attributes convey the password used to authenticate the end-user. Both attributes propose password secrecy either by ciphering *User-Password* with the Authenticator field or by replying in *CHAP-Password* to a NAS-generated challenge with a ciphered declination of this challenge calculated with the user's password. The *CHAP-Password* attribute can be used in conjunction with the *CHAP-Challenge* attribute which embeds the challenge value (if not present, the *Request Authenticator* field is used instead).

NAS-IP-Address (Code 4)

This attribute indicates the NAS IP address that has generated the initial RADIUS message. This attribute is particularly useful when RADIUS proxies are used, since the authentication server cannot rely upon the source address field of the IP packet. With this information, the authentication server (or RADIUS proxy) may activate proper behavior such as using a predefined RADIUS dictionary for vendor-specific attributes, for instance (see the *Vendor-Specific* attribute section). This attribute can be replaced with *NAS-Identifier* (code 32), which has the same usage, the NAS identity being encoded as a string instead of an IP address.

NAS-Port (Code 5)

This attribute gives an indication of the physical connection from which the end-user is connected. This information is particularly useful for locating the end-user with reliable network information. Services that require the end-user to be connected from a certain location (for security reasons, or to force a service to be available at a specific location) require this kind of data, which can be combined by the authentication server with the identity claimed by the end-user to get its service.

Framed-IP-Address (Code 8) and Framed-IP-Netmask (Code 9)
These attributes carry the basic IP network configuration of the end-user by providing a valid IP address (or IP subnet) and the appropriate network mask. Using this IP address, the end-user will be able to send and receive routed packets in the IP network. It might be required for the NAS to keep trace of IP addresses allocated to end-users to avoid IP address spoofing. In this case, the NAS will have to drop packets sourced with an IP address different from that allocated. Since RADIUS cannot send these parameters directly to the end-user, this is realized by means of the access protocol (IPCP in the case of PPP, for instance).

Filter-Id (Code 11)
This attribute, sent by the AUTH server in *Access-Accept* messages, is an indication for the NAS to apply a locally defined policy to the end-user. The attribute designates a policy number that has been previously provisioned in the NAS. The format of this attribute is not detailed in standard documents, and therefore implementers might use textually encoded policy rules that can be locally interpreted by the NAS to enforce the required policy. This alternative is, however, barely used.

Vendor-Specific (Code 26)
The very widespread usage of RADIUS required the creation of new attributes, but the code space scarcity forced IETF and implementers to use a particular attribute reserved for specific vendor-inferred usage: the *Vendor-Specific* attribute (VSA). This attribute gives the opportunity to extend the code type by starting the value field with a 16-bit encoded *Vendor-Id* (as defined in the SMI network management private enterprise code), then followed by a sub-TLV. Even though the *Vendor-Specific* attribute payload length can reach a maximum of 192 bytes, this makes it possible to create enough attributes to address all needs.

On the other hand, the massive usage of VSA contributed to creating a huge list of incompatible attributes between different RADIUS device vendors: it quickly became a nightmare for network operators as well as authentication server vendors, as it makes it necessary to maintain multiple dictionaries that have to be properly employed depending on the RADIUS client vendor. One typical example can be found with the configuration of DNS addresses to the end-user device, which have to be conveyed in *Vendor-Specific* attributes, whereas DNS addresses are common configuration parameters for Internet access nowadays. Figure 3.4 depicts an example of *Vendor-Specific* attribute encoding, as proposed in RFC 2865.

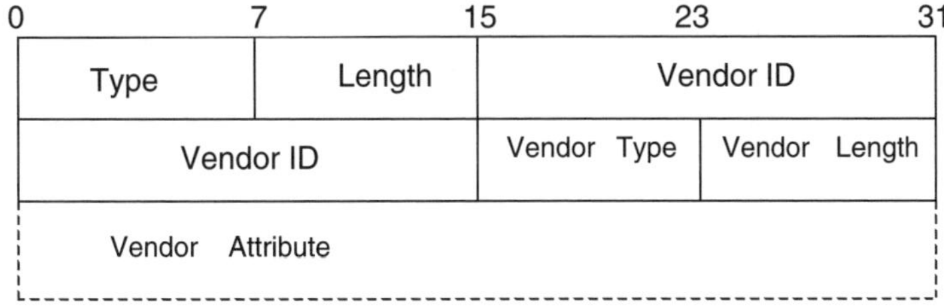

Figure 3.4 *Vendor-Specific* attribute format

3.1.1.3 Availability and Reliability

RADIUS Message Exchanges

Because of its very basic structure, RADIUS is the example of the protocol family where simplicity serves robustness. UDP is heavily used in lots of signaling protocols. However, it is known to be highly unreliable for multiple reasons:

- UDP does not provide any mechanism to detect packet loss;
- UDP is unable to ensure packet sequencing;
- UDP does not provide any traffic control mechanisms to avoid overflowing between two peers.

Usually, protocols relying upon UDP implement specific mechanisms to detect message loss by introducing sequencing fields. That is the case, for instance, with L2TP (RFC 2661) which implements this for control messages, lost data messages being supposed to be handled by the upper layers. When this mechanism is not implemented, the usual work-around is to acknowledge each message with a dedicated reply packet. None of these mechanisms has been systematically implemented for RADIUS. Messages are just partially sequenced in order for peers to match query and answers, but there is no general sequencing field pertaining to the whole RADIUS exchange.

Moreover, only a partial workaround is provided by the RADIUS state machine, since only some RADIUS messages require a reply from the client or from the server:

- The RADIUS server has to answer to an *Access-Request* message with an *Access-Accept* message or with an *Access-Reject* message in the case of failure (see Figure 3.5). The server may answer with an *Access-Challenge* message if the AAA server requires another authentication round (see Figure 3.6). Of course, the request is silently ignored if the request message is considered invalid (code undefined, invalid length, etc.).

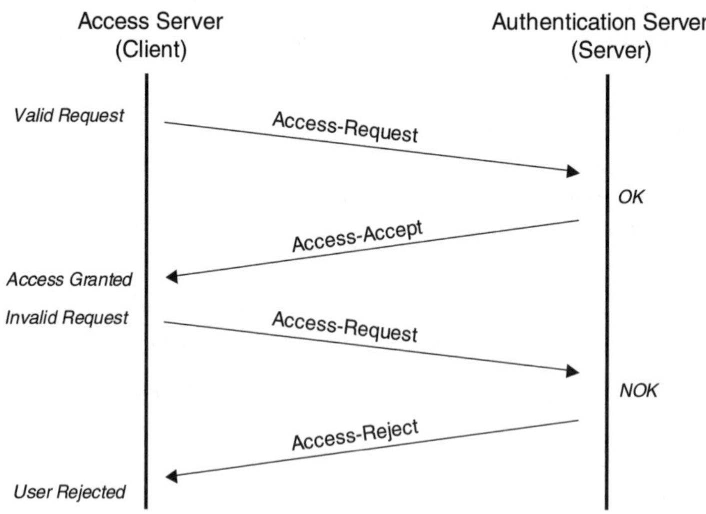

Figure 3.5 Simple RADIUS exchange to access the network

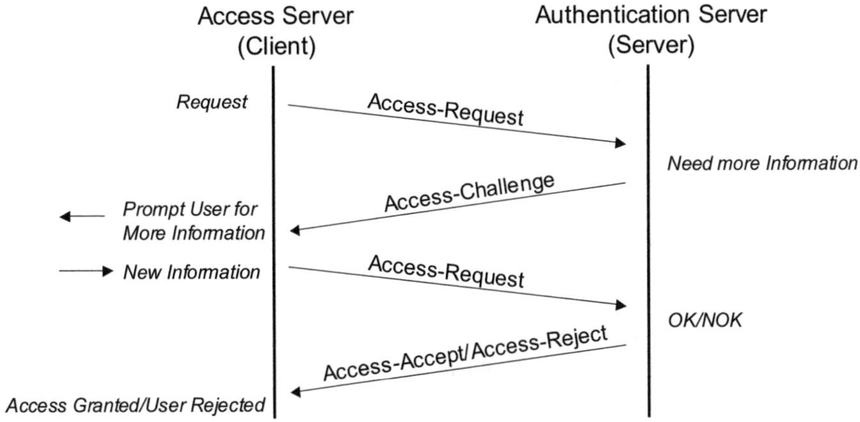

Figure 3.6 RADIUS exchange to access the network with *Access-Challenge* messages

- The RADIUS client has to answer to an *Access-Challenge* message with an *Access-Request message.*
- *Access-Accept* messages are never acknowledged in any way, even though they convey important authorization information.

Access-Accept message loss is difficult to detect: the RADIUS server usually internally reserves network resources, with no way of knowing if these resources are actually used. One can argue that accounting messages can provide this kind of information, which is only partially true, as accounting messages are not always sent to the same equipment as access messages. Moreover, accounting messages can follow a batch process, opposed to the real-time process required for session and authorization management. Retrieving allocated information requires the accounting database to be synchronized with the authorization history.

Accounting message exchanges do not suffer the same drawback. As the piece of accounting information is sent from the client to the server, the latter must only acknowledge it. A specific *Accounting-Response* message has been defined with the sole purpose of acknowledging the actual receipt of accounting data (see Figure 3.7). However, RFC 2866 opens up the possibility of conveying *Vendor-Specific* attributes along *Accounting-Response* messages, but usually this message does not carry any additional attribute.

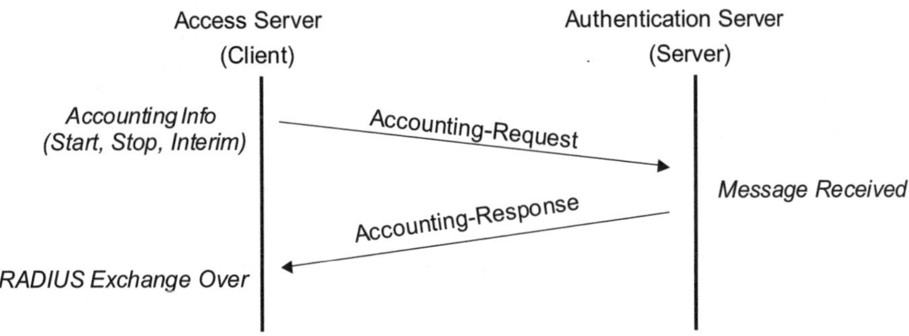

Figure 3.7 RADIUS accounting messages exchange

Retransmission Rules

RADIUS requires the maintenance of retransmission timers. Using the Identifier field, each peer maintains an internal state of ongoing exchanges, and more specifically of pending messages. The two basic parameters usually configurable in RADIUS-capable equipment are the retransmission timer and the number of successive retries. When a message is not answered once the retransmission timer has expired, the same message is issued again, and so on, until the configured number of retries is reached.

Usually, sending multiple consecutive messages denotes different issues: the network is down, the RADIUS server is down or the RADIUS server is overloaded. In this case, the client can try to reach another configured server. Some RADIUS server implementations reply with *Access-Reject* messages when receiving an *Access-Request* when they are almost saturated. This is something to avoid for two reasons:

- rejected end-users will retry to connect immediately, sustaining the server load as a side effect (as opposed to a no-response behavior where the end-user waits for its authorization, thus not creating additional burden on the server side);
- sending a message (accept or reject) always consumes server resources.

Other parameters are also available to the network administrator, providing a wide panel of curative tweaks. For instance, some vendors implement a 'server deadtime' temporization, which is a general timer that might be used by the RADIUS client, during which it considers a non-responding server as dead. After such a timer expires, the client tries to reach the nominal server again.

The definition and usage of timers are not specified in RADIUS. That is a really efficient way to offer a wide range of possibilities and combinations between multiple timers. The network administrator is free to set up the appropriate timer values depending on the acceptable level of resilience and local constraints.

Another characteristic of the RADIUS protocol is the lack of a keep-alive mechanism, clearly designated in RFC 2865 as harmful for the whole protocol, as keep-alive message processing is not scalable, at a time when CPU resources were scarce. This is something that can be argued, since other protocols such as L2TP, defined at the same period of time, make use of keep-alive messages without any history of scalability issues. In its structure, RADIUS has been designed to be entirely connectionless, running in a purely transactional fashion. RADIUS clients' requests are then used as probes to determine the aliveness of peers: if this is an efficient way to save one specific message every minute or so, this also has the drawback of lowering the service quality as real end-users will fail to connect, since their connection request is intended to help the (network) AAA chain to fix itself. Nowadays, this protocol conception is deprecated, as network operators are in competition for the best service and availability.

3.1.2 Forces and Weaknesses

Message Loss Barely Impacts upon the Service Offered for End-users

As a matter of fact, the RADIUS protocol is a success story: even though protocol reliability is not fully satisfying, operational problems are rare. These problems have never been seen to be fundamentally crucial by network operators so far. Generally, when a RADIUS message is lost, only one end-user is affected, with no compromise of the AAA infrastructure as a whole. When the number of requests becomes significant, losing one message does not have any impact on the service: the end-user usually may not even see any service disruption if s/he runs an automatic connection client. During the connection lifetime, RADIUS message losses do not compromise the service: the end-user, again, will not notice any problem if a RADIUS server goes down. To sum up, all these ingredients explain why RADIUS is accepted with its handicaps:

- One message loss does not compromise the service provided by the ISP.
- One message loss will only have a (limited) impact on a single end-user.
- One message loss might not even be seen by the end-user if s/he is connected through a router or with an automatic connection client.
- When a RADIUS server is down, no sessions already existing are affected, implying no service degradation for end-users.

RADIUS Failures Have to be Handled with Care by Network Operators

Here is a non-exhaustive list of possible situations that need particular care from network administrators:

- *Access-Accept message loss from the server to the client.* In this case there is an impact for the end-user, who will not be serviced. Moreover, successive connection attempts will have to be handled specifically by the server, since it internally maintains used resources for each end-user (such as an IP address, for instance). The only way rapidly to free resources is to keep trace, in real time, of pertaining accounting messages. In this case, the server will be able to release users' resources after a reasonable delay during which it is intended to receive accounting information. Having a RADIUS server being able to handle accounting messages may require a specific engineering design, as these messages might also be required for the billing chain.
- *Accounting message lost from the client to the accounting chain.* This situation might have an impact similar to the previous case, where the resource for a particular end-user is not freed, blocking further connection to the network. To avoid resource usage state inconsistencies, operators are obliged to purge their databases to make sure that disconnected end-users have no more resources allocated by the network. This issue arises where lost messages are not retransmitted (to spare RADIUS client resources, for instance). Losing accounting messages also affects the billing chain, since it becomes impossible for the operator to bill the end-user for the related connection.
- *The RADIUS server goes down.* Even with a backup server being able to treat RADIUS queries, all problems are not solved. The backup server is not aware of ongoing connections, creating a problematic situation where it is difficult (or impossible) consistently to continue the provisioning of specific and unique resources such as IP addresses. The problem gets even worse when the nominal server goes up again. Some operational methods exist to manage this situation: using accounting *Interim-Update* messages is a way for the server regularly to have a refreshed status of ongoing sessions, even though this adds a consequent burden for both the client and the server. Another method consists in integrating sophisticated proxies which are able to store messages

during server failures, or to install a mirroring accounting server that transparently listens and stores accounting messages.

- *The RADIUS client goes down.* In this case, end-users will usually have a service failure since the RADIUS client is also the access server. But as the client loses all connection contexts, the server does not have any trace of terminated connections, and it becomes impossible to bill clients for the service provided until the moment the access server failed. Again, a workaround can be found in sending regular *Interim-Update* messages. However, this will not provide an efficient way for the server to free resources reserved for disconnected end-users. Therefore, the RADIUS client must send an *Accounting-Off/On* message to the RADIUS server to indicate that resources previously allocated to end-users connected to the client have to be freed.

It is highly recommendable for the network administrator to enforce prioritization rules between RADIUS messages in the case of congestion: some messages deserve more attention for operators than others. For instance, it might be reasonable to drop new *Access-Request* messages in order to privilege accounting messages (especially *Start* messages) of sessions already established. However, *Access-Request* messages issued after an *Access-Challenge* have to be prioritized, or the complete sequence of previous messages will be lost.

An Incomplete Set of Messages

As a counterpart to being simple, RADIUS offers an incomplete set of messages, forcing RADIUS implementers and network administrators to elaborate and enforce various tricks to reach their goals. One of the most striking examples is the ambivalent meaning of the *Accounting-Request* message, which is simultaneously used to give session and usage information to the accounting chain, and used by the RADIUS server to manage the end-user session. One can consider the AUTH chain to have real-time constraints, as accounting tickets may deserve a batch treatment. For this reason, most RADIUS client implementations propose to set up different RADIUS servers for AUTH or accounting messages. Using RADIUS dynamically to detect session aliveness requires a very responsive accounting chain that is permanently connected to resource management servers used for authorization purposes.

Diameter offers more opportunities to create and extend the message vocabulary of the AAA chain. In spite of the definition of extensions to the RADIUS protocol at the IETF, it has been admitted as a specific rule not to create any new message, because of Diameter. As network administrators would like to have a complete RADIUS toolkit to handle particular features (such as credit control, for instance), no new message definition is allowed for RADIUS, even though it might ease the implementation. As a consequence, a significant amount of vendor-specific extensions are proposed, making use of existing, and not always adapted, RADIUS messages, worsening protocol weaknesses.

Compatibility Issues

As mentioned before, RADIUS offers a high degree of liberty for implementers to define new message attributes that might be necessary to run networks. As it is an undeniable advantage for vendors and network administrators, who do not have to wait for standardized features, this causes a real compatibility issue between all RADIUS-capable equipment.

For a network operator using RADIUS equipment built by different vendors, it is a challenge to synchronize the equipment to offer the same services, with identical attribute format, for both clients and servers. The solution can often be found in adding a RADIUS proxy in the chain, to behave as a translator between various vendor-specific attribute dictionaries.

3.1.3 Authorization and Provisioning with RADIUS

One of the primary goals of RADIUS is to carry authorization parameters after successful authentication, usually in the *Access-Accept* message (we will see later that other types of message can also be used). Authorization parameters can be divided into two major groups: parameters that will be transmitted by the access protocol (PPP/IPCP, DHCP, ...) to the end-user terminal in order to use the authorized network properly, and parameters that have to be taken into account by the access server to enforce user-specific policy rules. It is interesting to note that parameters transmitted to the end-user terminal can also be used by the access server to ensure that the remote terminal actually uses the right parameters. This situation arises for IP addresses that are assigned to the end-user terminal: since this address parameter (*Framed-IP-Address*) is transmitted from the authentication server to the access server with RADIUS and from the access server to the end-user with the access protocol, the access server uses this parameter to enforce antispoofing policies, to make sure that the end-user terminal actually uses a valid source address in IP packets.

3.1.3.1 Authorization Parameters

Among the parameters that have been defined in the RADIUS base protocol (RFC 2865), a classification of NAS or terminal parameters can be done. These parameters are those that may be used in *Access-Accept* messages, and more specifically focused towards plain IP network access provided by ISPs. Parameters defined for administrative usage or to provide a command line interface are not described.

(i) NAS Authorization Parameters

- *Service-Type*. As this parameter may be used by the client in *Access-Request* messages as a hint for the access server to provide a certain type of service, it is also often used in *Access-Accept* messages in order to define the operation currently performed, or the operation that needs to be done next. Here is a list of values carried by a *Service-Type* attribute used in *Access-Accept* messages, and usually used in the ISP context:
  - *Framed*: the end-user has to be connected with a framed layer-2 protocol (such as PPP);
  - *Call-back Framed*: the same as *Framed*, but the user is first disconnected and then called back by the access server;
  - *Authorize-Only*: defined in RFC 3576, its usage is detailed in Section 3.2.3.2. The *Authorize-Only* service type is sent by the RADIUS server to trigger a new authorization phase that has to be initiated by the RADIUS client with an *Access-Request* message. This service type is particularly useful to ease Diameter/RADIUS translations.
- *Framed-Protocol*. This indicates the protocol to use between the terminal and the access server (PPP, SLIP, GPRS PDP Context, etc.).

- *Framed-Routing.* This is used when the authorized terminal happens to be a router. The value indicates the routing role of the terminal (route listener, route sender or both).
- *Filter-Id.* This parameter is very important, probably the most powerful to be used with RADIUS to communicate authorization policies and rules. The *Filter-Id* parameter is indicated by the authentication server, conveying a locally defined reference that can be interpreted by the NAS. Section 3.1.3.2 talks more extensively about the usage of this parameter.
- *Framed-MTU.* When the access protocol does not negotiate this parameter, the *Framed-MTU* is used to set the maximum transmission unit to be applied by the NAS for the remote terminal. *Framed-MTU* might be different from the MRU value negotiated between PPP peers, for instance, since PPP-LCP is run before RADIUS exchanges occur.
- *Framed-Compression.* This is the type of compression to be used between the access server and the terminal (TCP/IP VJ header compression, IPX header compression, Stac-LZS compression).
- *Callback-Number.* The number to be used to call the end-user. This attribute is only relevant when the terminal is reachable through a PSTN/ISDN network.
- *Framed-Route.* This parameter is used to provision IP routes for the end-user to be configured on the NAS. The content of the value field is implementation dependent, but routes should be noted in a human-readable manner including prefix, subnet (Ascend/xx notation, where xx denotes the number of high-order bits positioned), the IP address of the gateway and a set of metrics.
- *Vendor-Specific.* With *Filter-Id*, this is a very interesting attribute to convey complex authorization parameters that can be applied by the NAS (see Section 3.1.3.2).
- *Session-Timeout.* The value of this attribute specifies the authorized connection duration in seconds. However, some NAS vendors round off this value to the nearest minute in order to lower the surveillance burden when thousands of connections are managed by the same equipment.
- *Idle-Timeout.* The same as *Session-Timeout*, but the timer is reinitialized every time traffic is flowing for the end-user.
- *Termination-Action.* This attribute is used by the authentication server to inform the NAS about the next operation to realize. This attribute is used when authentication/authorization operations are incomplete. Since *Access-Accept* messages terminate normal RADIUS exchange, this attribute is used to continue or to restart the process.
- *Port-Limit.* This provides the maximum number of ports to be assigned for a specific end-user. This might be used for multilink PPP, when an end-user makes use of multiple ISDN channels to increase connection bandwidth.

(ii) Terminal Authorization Parameters

- *Framed-IP-Address.* This attribute is sent by the authentication server to assign an IP address to the end-user terminal. The actual IP address assignment of the terminal has to be performed with the access protocol. For instance, this can be done with IPCP in the PPP case, or by DHCP. As already mentioned, this parameter can also be used by the access server to enforce antispoofing filtering policies.
- *Framed-IP-Netmask.* The same usage as the *Framed-IP-Address*, this is the network mask that has to be associated with the IP address.

- *Reply-Message.* This attribute is used to display a message to the end-user. It is not a pure authorization parameter, but it can be used to continue the authentication/authorization stage when used in *Access-Challenge* messages, or simply be used as an authorization notification in *Access-Accept* messages. This notification has to be transferred to the end-user thereafter, owing to the access protocol.
- *Vendor-Specific.* This can also be used to convey information to provision the end-user terminal with some specific configuration parameters. For example, DNS addresses to be configured to the end-user terminal are conveyed by RADIUS *Vendor-Specific* attributes.

3.1.3.2 Conveying Specifically Elaborated Policy Rules

The current set of RADIUS messages makes it possible to provision terminals as well as access servers with policy rules that can by transmitted in *Access-Accept* or *Change-of-Authorization* messages. The latter, defined in RFC 3576, will be explained more thoroughly in Section 3.2.3. For RADIUS implementations that do not support RFC 3576, *Access-Accept* messages are the only provisioning means that can be used. Since *Access-Accept* messages are not acknowledged, it is not always possible to be 100 % sure that authorization and provisioning parameters have been correctly transmitted and applied. As this is not really a problem for the initial user connection/authorization request, this may become problematic if the *Access-Accept* has been sent within a complementary authorization round, which occurs when the end-user session is already established. Fortunately, solutions exist to solve this issue:

- The RADIUS client could be configured to send successive requests if it does not receive any response from the RADIUS server.
- Accounting messages could be sent as soon as the authorization changed, simultaneously acknowledging the *Access-Accept* message and providing an explicit notification detailing the policy rules applied. This solution is implementation specific.

It is important to note that the IETF RADIUS extensions working group is currently defining new attributes that are specifically designed to provision filtering rules. This is discussed in Section 3.2.6.1. Even though these attributes are not yet finalized, and are yet to be adopted by the IETF community, they are very likely to be used in the future to convey precise provisioning and authorization rules with RADIUS.

This section will focus on the two standardized attributes briefly mentioned in Section 3.1.3.1, and which can be used to build specifically elaborated policy rules: *Vendor-Specific* and *Filter-Id* attributes. The transport of a complete set of provisioning parameters might not be possible in a single attribute, since the TLV format (Figure 3.3) imposes a limit of 253 bytes. However, it is possible to convey multiple *Vendor-Specific* or *Filter-Id* attributes in one single RADIUS message. Therefore, proprietary syntax rules have to be implemented when a large policy rule has to be split into multiple attributes.

(i) Using Vendor-Specific Attributes

The *Vendor-Specific* attribute has a specific format described in Figure 3.4 and is particularly suited to configure very different authorization parameters. To define a *Vendor-Specific* attribute, the implementer needs a *Vendor-Id* number, registered in the SMI network management private enterprise code (a list of registered companies can be found at the following URL: http://www.iana.org/assignments/enterprise-numbers). The *Vendor-Id* attribute is a 32-bit field. Considering that there are less than 25 000 registered companies at this time, it leaves some room for newcomers. The *Vendor-Type* field can be freely used within the *Vendor-Id* space to dissociate different parameter types such as filtering rules, bandwidth limitation boundaries, NAT/PAT parameters, etc. The *Attribute-Specific* value is defined to make sense with a particular *Vendor-Type*, and must follow specific syntax and semantic rules in order to be properly interpreted by the NAS. The RADIUS server also has to implement this syntax and semantics, especially if it has dynamically to transfer policy rules, following specific end-user session characteristics. For instance, monitoring and controlling a specific end-user may require specific IP parameters to be described in the policy rule that depend on the subscribed service.

(ii) Using the Filter-Id Attribute

The *Filter-Id* format is very liberal, as the value field is text encoded and is initially used to designate a particular policy rule already listed in the NAS. There are two different ways to use this attribute:

- *NAS locally defined policy.* Both RADIUS server and NAS have the knowledge of the same set of filtering rules, listed by name or by number (or both) and transported in text format in a human-readable manner.
- *Dynamically interpreted policy.* The RADIUS server is able to elaborate dynamic policy rules in a way that the NAS is able to interpret and to apply the rules in conformance with the granted service. As this behavior is not explicitly mentioned in RFC 2865, there is no technical reason to forbid this usage. However, this is not a very practical way to provision configuration parameters since it requires the implementation of a specific rule parser on the NAS.

Whatever the *Filter-Id* usage, the RADIUS protocol operations must not be affected by this attribute. It is therefore not possible to trigger another round of RADIUS authorization messages depending on the value of this attribute, for instance.

The *Filter-Id* attribute is very sensitive to compatibility issues, as the value field usually has a local meaning and is interpreted depending on the equipment used. This has to be taken into consideration, especially if services have to be extended to roaming users, and/or if new incompatible equipment is deployed in the network. For this specific case, the *Vendor-Specific* attribute has two advantages: the attribute cannot be misinterpreted (correctly interpreted or simply rejected), and the attribute can be translated by a proxy if different equipment brands are involved.

(iii) Example

The following example describes a method to use *Filter-Id* and *Vendor-Specific* attributes to provide the same service. The case presents a registered user *Omer* who is entitled to connect to the Internet with a limited speed of 2048 kbit/s for downloads and 512 kbit/s for uploads, applied to the IP traffic. The provisioning of this bandwidth limitation is done during connection establishment, by parameters transmitted in the *Access-Accept* RADIUS message.

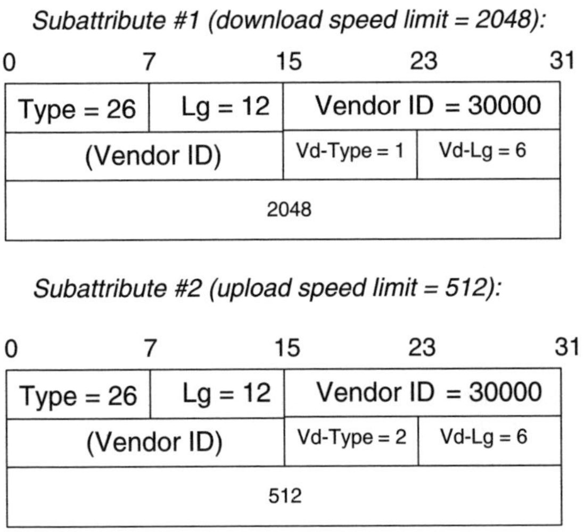

Subattribute #1 (download speed limit = 2048):

0	7	15	23	31
Type = 26	Lg = 12	Vendor ID = 30000		
(Vendor ID)		Vd-Type = 1	Vd-Lg = 6	
2048				

Subattribute #2 (upload speed limit = 512):

0	7	15	23	31
Type = 26	Lg = 12	Vendor ID = 30000		
(Vendor ID)		Vd-Type = 2	Vd-Lg = 6	
512				

Figure 3.8 Example of two *Vendor-Specific* attributes defined for bandwidth limitation

Vendor-Specific Attributes

For the purpose of this example we have defined two *Vendor-Specific* attributes, especially created for the virtual 'Foobar' company, which was given the enterprise code 30 000 (*Vendor-Id*). The first *Vendor-Specific* attribute created by the Foobar company has been allocated the type number 1 to define the download bandwidth boundary, as the type number 2 defines the upload bandwidth boundary. The choice of the company is to encode the speed limit in 32 bits, using the integer format. The two newly created attributes are shown in Figure 3.8.

Filter-Id Attribute with Predefined Local Policy

In every NAS on which a user has the possibility of being connected, a common set of services is configured, spanning the whole possible download/upload offers the ISP is able to provide (Table 3.1).

When the user *Omer* is connecting to the network, the authentication server will have to indicate the '2048/512' offer that corresponds to the *Filter-Id* number 3. In this example, the *Filter-Id* value is UTF-8 encoded in a 2-byte length field. The attribute will be as shown in Figure 3.9.

Table 3.1 Table of service offers indexed with the *Filter-Id* attribute

Filter-Id	Download limit	Upload limit
1	512	128
2	1024	256
3	2048	512
4	8192	512

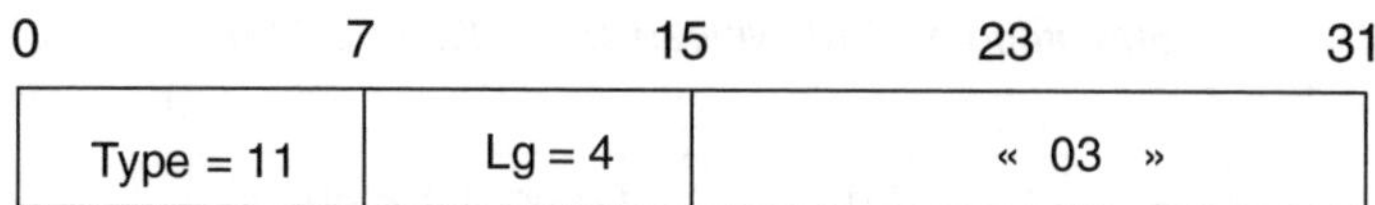

Figure 3.9 *Filter-Id* attribute with predefined local policy

***Filter-Id* with Dynamic Interpretation**

In this example, the *Filter-Id* value field is parsed by the NAS with the following grammatical rule:

```
Download Bandwidth Limit = string;
Upload Bandwidth Limit = string;
Value = "Service:" Download Bandwidth Limit "/"Upload
Bandwidth Limit;
```

If the grammatical rules are correctly parsed during analysis of the attribute, the NAS converts strings into real values to be applied as the bandwidth limit for the user. The attribute will be as shown in Figure 3.10.

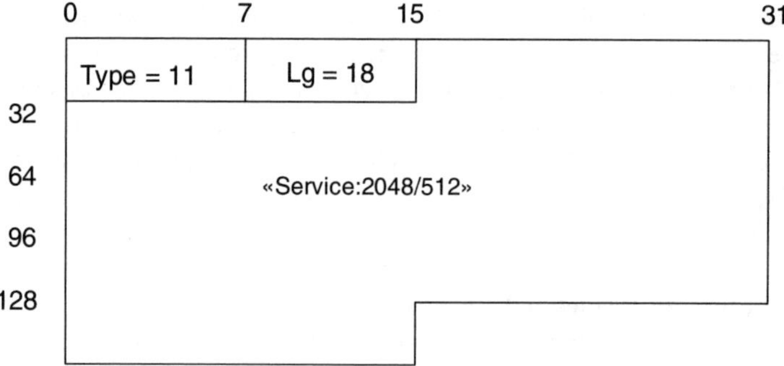

Figure 3.10 *Filter-Id* attribute with dynamic interpretation

3.2 RADIUS Extensions

3.2.1 EAP Support with RADIUS

The Extended Authentication Protocol (EAP) was primarily designed for PPP (RFC 2284) but quickly became the standard used for layer-2 links other than PPP. EAP is an access protocol that makes use of dedicated authentication methods and therefore has to interact with authentication protocols. RFC 2869 was the first effort to standardize a smooth interworking between RADIUS and EAP, and RFC 3579 came up afterwards as a complete solution set, updating the EAP part of RFC 2869.

In order to carry EAP messages from the authenticator to the authentication server, two new RADIUS attributes have been specifically defined: *EAP-Message*, which conveys the actual EAP information, and *Message-Authenticator*, which is used to check the message integrity of request messages, because the original RADIUS Authenticator field is only relevant for replies. Checking the integrity of requests is a way to prevent spoofing attacks targeted against the RADIUS server.

The other challenge for a successful interaction was to adapt the RADIUS protocol to the multipass EAP exchanges. As mentioned earlier, the RADIUS protocol is usually restricted to a request/response authentication, since EAP methods usually require multiple yet successive exchanges. To achieve this, *Access-Challenge* messages are sent by the authentication server, as long as message exchanges have to be pursued. As in CHAP or PAP authentication cases, successful or invalid authentication is terminated either by an *Access-Accept* or by an *Access-Reject* message respectively.

The integration of EAP support within RADIUS is a key to use the latter for layer-2 and layer-3 authorization: EAP is now widely deployed with 802.1X (which is a layer-2 authentication/authorization standard), and RADIUS can be used to integrate additional layer-3 authorization attributes if the 802.1X authenticator has the corresponding layer-3 capabilities. This case is an example of how RADIUS can easily extend usages.

The complete description of EAP can be found in RFC 3748, but this introduction will describe in a quick and simple manner what typical EAP exchanges might occur between a peer and its authenticator (the NAS in our case):

(a) The authenticator (NAS) is the initiator of EAP exchanges, by first sending an EAP *Request* message claiming for peer identity.
(b) The peer replies to the authenticator with the required information, by sending an EAP *Response* message.
(c) The authenticator sends a new request, requiring a response from the peer.
(d) The (b) + (c) phases are repeated until all EAP method phases are completed.
(e) If authentication is successful, the authenticator sends an EAP *Success* message to the peer; if it fails, it sends an EAP *Failure* message.

As can be seen above, there are only four messages defined in EAP: *Request, Response, Success, Failure*. EAP defines multiple types for *Request* and *Response* messages, in order to specify the nature of the request or response.

An authentication server is not required for EAP to work, as authentication can be completely handled by the authenticator. However, it is most of the time desirable to make use of a centralized entity designated to perform the authentication on behalf of the authenticator. The authenticator may be used in this case as a RADIUS client, and will have to transmit EAP messages between the peer and the RADIUS server. Note that the RADIUS server may itself delegate the EAP method processing to an external server, since authentication computation procedures can be managed independently of RADIUS protocol exchanges.

3.2.1.1 *EAP-Message* and *Message-Authenticator* Description

The *EAP-Message* (attribute 79) is an attribute used by the authenticator to encapsulate EAP messages coming from the peer to the authentication server, and is used to convey replies from the authentication server to be sent to the peer. The *EAP-Message* attribute can be inserted in

Access-Request, *Access-Challenge*, *Access-Accept* and *Access-Reject* RADIUS messages. It is possible to have multiple *EAP-Message* attributes in a single RADIUS message, as the maximum size of an attribute may not be sufficient to carry the original EAP message. In this case, *EAP-Message* attributes are concatenated to form one EAP message. However, it is not possible to send one EAP message within multiple successive RADIUS messages, as this would break the state machine of the RADIUS protocol.

On the other hand, it is not possible for a RADIUS message to carry more than one EAP message either. The EAP/RADIUS interaction has been built around the idea that every message of one protocol matches to the other. There is no possibility of message aggregation, nor reduction of exchanges between the peers.

The *Message-Authenticator* message (attribute 80) has been defined to ensure integrity and authentication of RADIUS messages used to convey EAP information within *Access-Request*, *Access-Challenge*, *Access-Accept* and *Access-Reject* messages, but can also be used to reinforce protection of these RADIUS messages, even when not used within the context of EAP. The value field of the *Message-Authenticator* attribute is the calculation of a MD5 hash of the whole packet, including the Authenticator field and the shared secret used as the key for integrity check. The essential innovation of this attribute is to protect the server from spoofed *Access-Request* messages, whereas the Authenticator field could just be used by the RADIUS client to ensure the integrity of the RADIUS server's replies.

3.2.1.2 EAP and RADIUS Exchange Overview

There is a infinite combination of exchanges that might occur between the peer, the authenticator and the authentication server. Most of the cases that can be encountered are described in RFC 3579, but we will present here the typical framework of an EAP/RADIUS message interaction (Figure 3.11):

1. The NAS (authenticator) sends an EAP *Request/Identity* to the peer.
2. The peer answers with an EAP *Response/Identity* message.
3. The NAS relays this information to the authentication server within an *Access-Request* message.
4. The RADIUS server sends back an EAP *Request*, hence initiating the actual EAP authentication method process, within an *Access-Challenge* message.
5. The NAS relays the EAP *Request* to the peer and waits for the peer's answer (which will be sent in another *Access-Request* message, by which an *Access-Challenge* has to be followed).
6. The peer sends back its answers using an EAP *Response* message, with a subtype adapted to the chosen method, and containing the relevant information.
7. The NAS relays the EAP message in an *Access-Request* message.

Steps 4 to 7 are repeated until the EAP method phases are completed or fail:

- If the authentication server considers that authentication is successful, it encapsulates the EAP *Success* message in an *Access-Accept* message and sends it to the NAS which will enforce the RADIUS authorization parameters that have to be sent along, and will simultaneously send the EAP *Success* message to the peer (step 8).

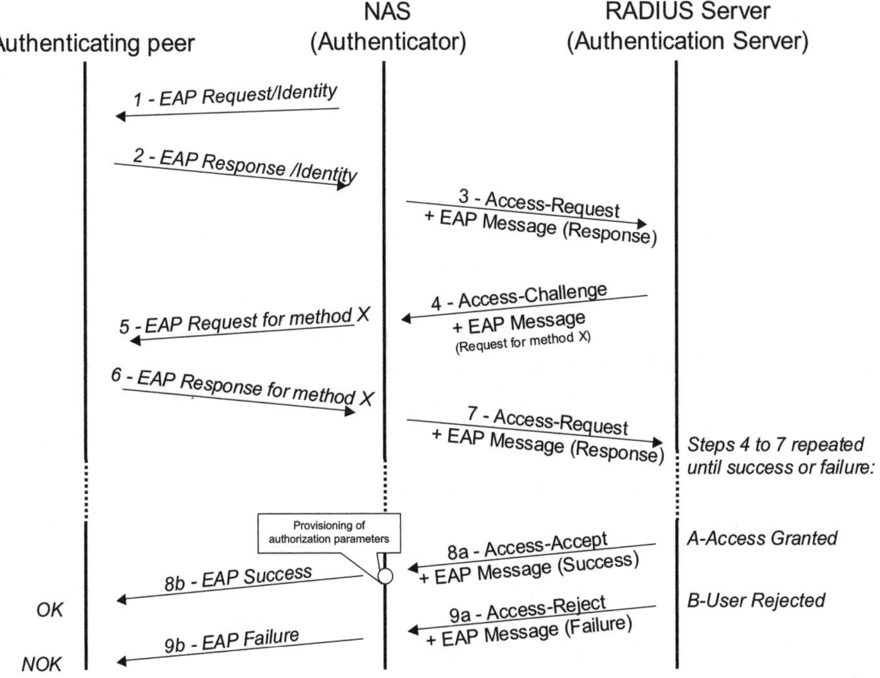

Figure 3.11 Generic EAP and RADIUS exchanges

- If the authentication fails, the authentication server encapsulates an EAP *Failure* message within an *Access-Reject* message. The NAS will relay the EAP *Failure* message to the peer and will take appropriate actions to keep the peer in the appropriate unauthorized state (step 9).

3.2.2 Interim Accounting

The notion of 'interim accounting update' messages appeared in RFC 2869, along with numerous additional notions, as already seen. In any standard authentication/authorization/accounting process with RADIUS, the client sends an *Accounting-Request* message with an *Acct-Status-Type* = *Start* once the session has been authorized by an *Access-Accept* message and the NAS has enforced appropriate authorization parameters.

The RADIUS server is intended to send back an *Accounting-Response* message to acknowledge receipt and treatment of the message. By the end of the session, the RADIUS client sends an *Accounting-Request* message with an *Acct-Status-Type* = *Stop*, acknowledged again by the server with an *Accounting-Response* message. As RADIUS does not implement a connected mode as Diameter does, the server gets the status of the end-user at the end of the session, hopefully if no network perturbations have impeached the transmission of the *Stop* message.

The benefits of and need for an intermediate status update are then obvious: by sending periodic accounting messages, the server is able to keep track of active session states more precisely, with the advantage of giving a hint of actual resource consumption. This is helpful for credit control applications, and also to reduce gain loss whenever end-user consumption cannot be retrieved because of network failure, for instance. As RFC 2866 prepared the possibility of *Accounting-Request* messages with *Acct-Status-Type* = *Interim-Update*, its usage is defined in RFC 2869 with the appearance of a new *Acct-Interim-Interval* attribute sent in *Access-Accept* messages.

This attribute is valued with the interval, expressed in seconds, that separates two interim updates. Beyond simple indication of the frequency of update messages, it is also a hint given by the server to the client to activate the interim update process. However, this mechanism can also be statically configured on the client, and is defined to override the value provided with the *Access-Accept.* This last point has been argued lately at the IETF, since it is impossible remotely to provision a specific behavior for a restricted number of end-users, as long as a general parameter is set up in the NAS. The last consensus tends to change this behavior by allowing the *Acct-Interim-Interval* to override the general parameter, if the new value is bigger than the predefined one.

RFC 2869 requires that intervals between two interim updates must not be smaller than 60 seconds, and should be smaller than 600 seconds. These lower boundaries have been defined to protect the server from bursts of accounting messages, but, in return, they prevent the accounting server from getting a frequent session status. Changes of authorization might also trigger the emission of an interim update towards the accounting server. However, this behavior has not been fully adopted yet, and new standard definitions seem to privilege the consecutive emission of *Stop* and *Start* messages.

Indicating session status changes is possible with consecutive *Stop* and *Start* messages by using the *Acct-Multi-Session-Id* attribute as a unique identifier for all authorizations granted during a session lifetime. It is important to note that this method requires each authorization change to be seen as a session change. In addition to this, a new *Acct-Termination-Cause* value would have to be defined in order to specify that there is no session disruption. In the future, it will be acceptable to foresee that both *Interim* and *Stop/Start* behaviors will coexist: even though the *Stop/Start* method is a clean way to achieve accounting for a single session experiencing different authorization phases through time, it is also predictable that it will have a strong impact on deployed servers that might improperly process *Stop* messages, always interpreting it as an indication of session termination. From this point of view, using interim update messages is less aggressive.

One future extension of the interim update message can also come from the need for instantaneous session status updates. ISPs may require the precise status of an end-user session to be obtained for multiple reasons:

- 'Ghostbusters' processes, which are activated when a session is trying to connect as another one using the same credentials is already online. This situation can occur whenever an account has been robbed, or when the accounting *Stop* message of the first session has been lost for some reason. Instantaneous querying of the session status is a way to figure out exactly what is going on.
- Real-time consumption checking performed by the end-user, who might pay on a volume basis, and who is encouraged to check his/her credit regularly in order to avoid overconsumption penalties.

- Credit control mechanisms, which can have a dedicated behavior with end-users close to their credit limit. By requesting more frequent updates from the RADIUS client, usage measurement is accurately performed without requiring an important update frequency from the NAS. This is also a way to go below the 60 seconds boundary required by RFC 2869.

3.2.3 Dynamic Authorization

The standardization effort has made RADIUS a particularly attractive protocol to dynamically change authorization parameters for individual sessions. These possibilities are described in RFC 3576 and have different forms: *Disconnect-Request* and *Change-of-Authorization* (CoA) messages. The former is used to cease authorizations previously granted to the end-user by invoking the NAS to disconnect the end-user immediately, and the latter requests the NAS to change authorizations for the indicated session, without disconnection. The usage of the *Disconnect-Request* message is inherently limited, and therefore this section will be more focused on the CoA message.

The first characteristic of these two messages is their place in the protocol architecture. The traditional client/server RADIUS model makes the NAS the client, whereas *Disconnect-Request* and CoA are server-initiated, unsolicited messages. RADIUS clients have to listen to the specific UDP port 3799 in order to receive these messages. Upon receipt of a *Disconnect-Request* message, the client will send back a *Disconnect-ACK* or *Disconnect-NAK* message with an indication of the possible error cause. In the same manner, the *CoA-Request* message requires a *CoA-ACK* or *CoA-NAK* reply.

CoA messages can be used in two different modes in order to achieve dynamic authorization changes:

- *Push mode*: the *CoA-Request* message directly embeds authorization parameters to be enforced by the NAS.
- *Pull mode*: the *CoA-Request* is a hint sent to the client for the sole purpose of requesting a new authorization from the RADIUS client. This mode has been defined to ease Diameter translation.

3.2.3.1 Changing Authorization in Push Mode

This mode is the native mode of the Change-of-Authorization message, as described in Figure 3.12. It is similar to the *Access-Accept* message, but used in an unsolicited manner with some additional restrictions relative to embedded attributes that might not have the same meaning. The authenticator field is valued with the same rules employed for accounting messages, keeping in mind that the *Authenticator Request* does not come from the client but from the server, and the *Authenticator Response* comes from the client.

CoA-Request messages must embed at least one or more attribute precisely and uniquely to identify the session on which change of authorization parameters will apply. These attributes can be, among others, the *User-Name*, *Framed-IP-Address* or *Acct-Session-Id*, for instance.

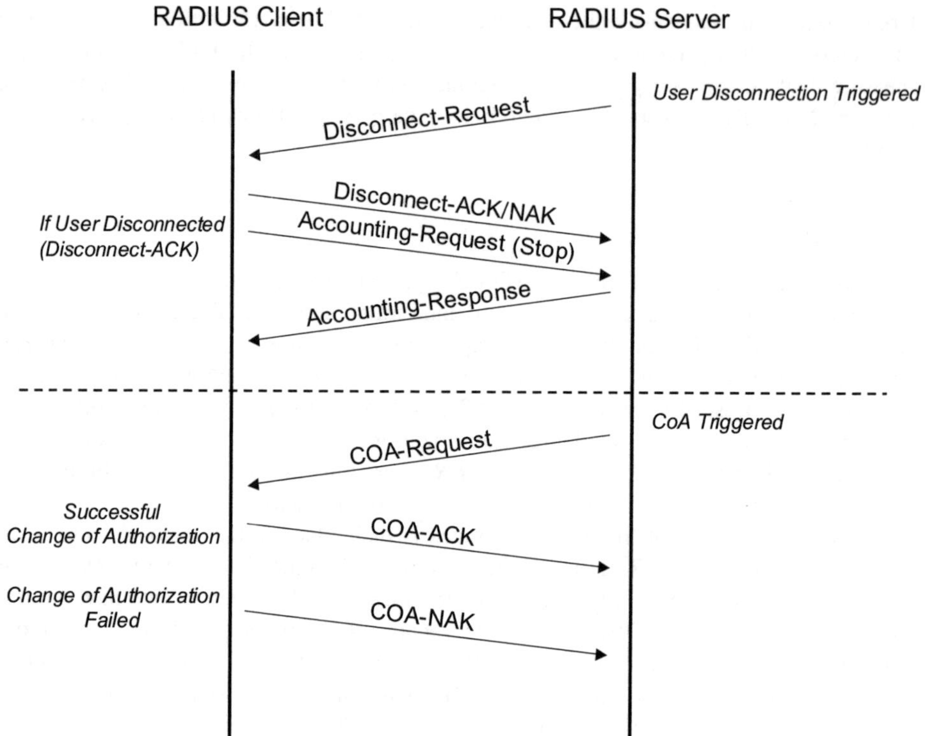

Figure 3.12 Disconnect and CoA message exchanges

All authorization-related attributes described in Section 3.1.3 are usable within *CoA-Request* messages, and can be dynamically applied without requiring successive disconnection/connection to get authorizations for desired services. Upon successful application of new authorization parameters, the NAS sends a *CoA-ACK* message to the server. The server may be forced to rely upon this message to perform the corresponding billing operations, as the standard does not specify any particular behavior for accounting. The RADIUS client is not required to send any accounting information to the server. As mentioned in the previous section, there are, however, two possibilities to alert the accounting server of the change of authorization: sending an interim update accounting message, or sending successive accounting *Stop* and *Start* messages.

If the NAS is not capable of changing the authorization parameters as requested, a *CoA-NAK* message is sent back to the RADIUS server, with an *Error-Cause* attribute indicating the cause of failure.

3.2.3.2 Changing Authorization in Pull Mode

This mode is used when the RADIUS server wants to change one session authorization by means of a regular *Access-Request/Access-Accept* exchange. To do so, the server sends a *CoA-Request* with a *Service-Type* attribute specifically set to *Authorize-Only*. Upon receipt

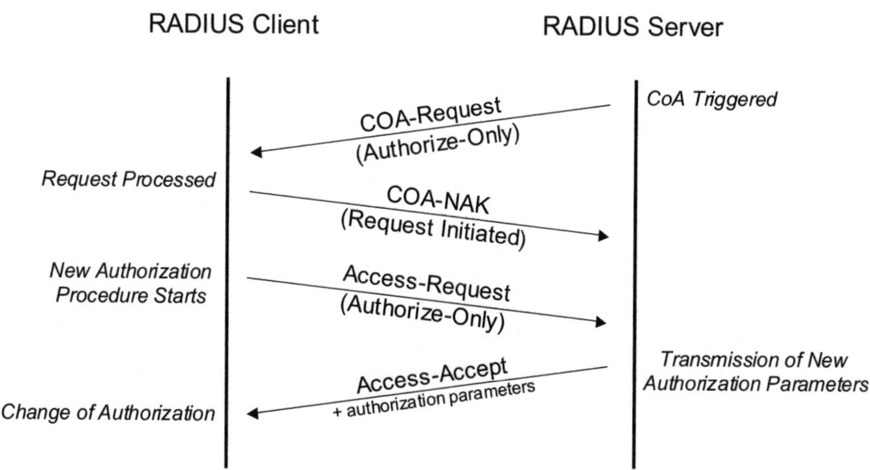

Figure 3.13 Changing authorization in pull mode

of this message, the client will acknowledge this request by sending back a *CoA-NAK* with an *Error-Cause* attribute set to *Request Initiated.* Eventually, the NAS will send an *Access-Request* message with a *Service-Type = Authorize-Only* to the RADIUS server, which will send an *Access-Accept* with the new authorization parameters in return. This is shown in Figure 3.13.

If the new authorization parameters are not applicable by the NAS, the behavior is not clearly described. One might expect the *Access-Accept* to be treated as an *Access-Reject* message, which will terminate the ongoing session. We can also expect the *Access-Accept* to be treated as an invalid message, and therefore silently discarded without any session disruption (however, experts do not encourage this). The two opposite behaviors are documented in standardization notes, and this may create potential problems as each implementer will have its own interpretation. This is another source of interoperability issues between implementations. An *Accounting-Request* message could provide an indication of whether or not authorization parameters have been applied, but the way to use it is not precisely defined yet (see Section 3.2.2).

3.2.4 Using RADIUS for Assignment, Prioritization and Filtering with VLANs

Issued in 2006, RFC 4675 defines a set of attributes defined to support VLAN, priority and filtering attributes. This RADIUS extension is intended to be used in a context where each user has an IEEE-802 port assigned (either physical or 'virtual', as IEEE-802.11 allows). The bridge to which the port belongs is then the RADIUS client, and may request from the AAA server an instruction specifically to provision VLAN filtering and prioritization for each user. It is therefore possible for network administrators to control end-user access to a restricted set of VLANs, and possible to manage traffic prioritization between VLAN IDs.

RFC 3580 already allows provision of individual VLAN access control for an end-user attached to an IEEE-802 port. In this case, the tunneling attributes are used, but they provide a coarse precision of access control tuning. With RFC 4675, the method proposed in RFC

3580 is not deprecated at all, but the new attributes described below are defined to provide better accuracy in dynamic provisioning of subscribers when attached to a IEEE-802 network.

Four attributes are defined in RFC 4675:

- *Egress-VLANID (code 56).* The RADIUS server indicates in this attribute an egress VLAN that is allowed to be accessed by the end-user. It is possible to enumerate a complete set of authorized VLANs by sending consecutive *Egress-VLANID* attributes in the same authorization packet. The attribute format proposes a tag field with which it is possible to specify whether VLAN frames have to be tagged or not.
- *Ingress-Filters (code 57).* This attribute activates filtering for VLAN ID(s) specified in the *Egress-VLANID* attribute(s), using a field indicating whether filtering is enabled or not. When filtering is disabled, the end-user associated with the pertaining port has full access to the desired VLAN.
- *VLAN-Name (code 58).* Similar to the *Egress-VLANID* attribute, it indicates authorized VLAN by its name instead of its number.
- *User-Priority-Table (code 59).* This attribute is composed of an 8-byte encoded table that determines a correspondence between the priority of frames received by a port with the priority to be enforced for the end-user. Therefore, without modifying the priority information conveyed within 802 frames, it is possible to assign a local and personalized traffic prioritization. The format of the table is defined in IEEE-802.1d documents.

Among these four attributes, only *Ingress-Filters* and *User-Priority-Table* are considered to be 'pure authorization attributes'. Therefore, these two attributes are exclusively transported within *Access-Accept* and *Change-of-Authorization* messages. *Egress-VLANID* and *VLAN-Name* are also authorization attributes, but can also be used as a hint sent by the NAS to the AAA server to provision access with maximum efficiency, when possible.

Using the VLAN filtering and prioritization mechanism in a network can be reported by using accounting messages. *Egress-VLANID*, *Ingress-Filters* and *VLAN-Name* can be inserted within *Accounting-Request* messages to log their usage and application on the NAS.

It is worth noting that these attributes can be used 'as is' by Diameter without any other translation than those required by the protocol format.

3.2.5 Filtering IP Traffic

As seen in Section 3.1.3, it is possible dynamically to provision filter rules into a NAS, even though it is not really convenient or reliable to be used across a large network, or to be used across multiple networks. For instance, the *Filter-Id* attribute requires either filter rules to be preprovisioned into the NAS or a specific filter rule interpreter to be implemented within each NAS. With RFC 4005, Diameter does not suffer this problem: the *NAS-Filter-Rule* attribute has been defined to convey complex filtering rules in a standardized format owing to IPFilterRule syntax defined in RFC 3588. With RFC 4849 issued in April 2007, it is now possible to use the same attribute with RADIUS.

One of the core features proposed is to provide a standardized way to provision specific per-user filters on the NAS. Previously, this could be done by other means: using the *Filter-Id* attribute, or by defining on-purpose *Vendor-Specific* attributes. Note that the *Filter-Id* attribute may have an ambiguous meaning, since its value field has not been precisely defined and could therefore embed a sequence of parameters to provide a complex semantic. The RADIUS *NAS-Filter-Rule* (code 92) is defined to sweep away good and bad habits by providing a unified attribute with a comprehensive and elaborate grammar to define filter rules to be enforced by the NAS. The *NAS-Filter-Rule* is able to define rules to permit or deny various traffic flows restricted to layer-3 and layer-4 IP protocols (see Section 4.4.2 for details). The rules can be complex and may not fit the 253-byte string field limit imposed by the RADIUS attribute format. In this case, multiple *NAS-Filter-Rule* attributes can be inserted in the same message, the concatenation of all string fields giving the complete set of rules to be interpreted (the string field is ASCII encoded, and the separator between filter rules is the NULL byte). RFC 2865 mentions that multiple attributes of the same type must not be reordered, even by proxies in the chain. This rule ensures that rules will be conveyed without degradation along the AAA chain.

No accounting specific attribute has been defined to report the effective enforcement of a filter on the NAS. It is the same *NAS-Filter-Rule* attribute that is used within *Accounting-Request* messages. Whenever a filter is applied, either at session startup or during a session, an accounting message must be sent with this attribute, indicating that it has been taken into account. If the filter rule is not applicable, the session fails if it is still in its startup phase. If the session is already up, the session's authorizations remain unchanged and no accounting message is sent.

With several methods to enforce filtering policies within an NAS, a rigorous consistency in attribute usage is required. Firstly, it is not recommended to use the *Filter-Id* attribute and the *NAS-Filter-Rule* attribute at the same time. Even though this is not prohibited, precedence has not been defined to give priority to one attribute over the other. Therefore, it is likely that behavior across a large network would not be consistent, especially if different NAS types were used. The ongoing work at the IETF is preparing a new attribute for filtering traffic, more powerful and versatile. This will be discussed in Section 3.2.6.1. Even though this new attribute is not standardized yet, RFC 4849 explicitly mentions that the *NAS-Filter-Rule* attribute will have to be ignored if both attributes are present in the same message.

As compatibility with Diameter is required for latest RADIUS attributes, translation between the RADIUS *NAS-Filter-Rule* (code 92) and the Diameter *NAS-Filter-Rule* (code 400) will occur on concatenation and split procedures, to fit the limits of each protocol. When the Diameter attribute is translated into RADIUS, the Diameter *NAS-Filter-Rule* string field is split into 253-byte portions, which makes the same number of RADIUS *NAS-Filter-Rule* attributes to be inserted, in the same order, within the equivalent RADIUS message. A specific translation procedure applies from RADIUS to Diameter when multiple *NAS-Filter-Rule* attributes are present. In this case, RADIUS attributes are concatenated into a single string field which generates one Diameter *NAS-Filter-Rule*. It might be that translation is impossible, especially if the 4096-byte RADIUS boundary is reached. In this case, translation is considered to be unfeasible and an error message is generated.

3.2.6 Future Extensions

The IETF is still engaged in the standardization of new RADIUS extensions. Some proposals are intended to clarify the usage of RADIUS, or to update some RFC documents to take into account emerging operators' needs. We will describe in this section some extensions that have a direct impact on end-user service provisioning, especially concerning features such as traffic filtering, prepaid extensions and bandwidth management. Not all extensions presented herein are integrated as official standardization working documents yet, and some of them may not succeed in becoming a standard in the future. However, having a look at these works will give the reader a better knowledge of the directions to be taken by RADIUS in the coming years.

3.2.6.1 Extended Filtering Attributes

Throughout the different sections of this chapter we have already identified three different ways to enforce specific filtering policies onto network access servers: *Vendor-Specific* Attributes, *Filter-Id* and *NAS-Filter-Rule*. The latter combines the advantages of being standardized and offering a decent level of sophistication. However, it is still limited to layers 3 and 4 of the IP protocols, which can be a constraint when provisioning complex policy rules for individual user access.

Specific work has been conducted to enrich the *NAS-Filter-Rule*, basically by extending the RFC 3588 IPFilterRule syntax, using augmented BNF (ABNF) for syntax specifications (RFC 2234). A new *NAS-Traffic-Rule* attribute is proposed, which spans from layer-2 to layer-5 (HTTP) traffic. Moreover, this attribute also embeds actions in order to redirect identified flows. The same concatenation rules used with *NAS-Filter-Rule* apply, and it is likely that Diameter translation would work the same way. At this stage of the work, a specific attribute is proposed for accounting (*Acct-NAS-Traffic-Rule*), but it will likely be dropped following the same solution documented in RFC 4849 and RFC 4675, which consists in using the same attribute number for authorization and accounting, since no confusion can exist in the context of their usage.

If the definition of the *NAS-Traffic-Rule* succeeds in the future, its usage will probably deprecate all other attributes defined for filtering. It is even possible that *NAS-Filter-Rule* may never be used if *NAS-Traffic-Rule* shows up soon, since NAS and AAA server implementation supporting the former would not yet be widely deployed. This situation is even more likely since RFC 4849 explicitly specifies that *NAS-Filter-Rule* will have to yield to *NAS-Traffic-Rule* if both attributes are present in the same message. Considering *Filter-Id* and VSA attributes, they will still be used for a long time anyway, because they are already used in specific situations, combined with a lack of will for additional investments in a chain that already works and makes money.

3.2.6.2 Bandwidth Parameter Attributes

Besides attributes created to provide additional filtering capabilities such as those described above, the ability to provision bandwidth limitations to the end-user session is presented in Ref. [1]. As usually seen with RADIUS, this possibility is already available with most of the devices available on the market, making use of *Vendor-Specific* attributes. The purpose of this proposition is to define three new attributes to define the user's bandwidth profile:

- The *Ingress-Bandwidth* attribute defines in its 32-bit value field the bandwidth available to the user for ingress traffic, i.e. traffic going to the end-user. The bandwidth rate is represented in the value field in kilobits per second.

- The *Egress-Bandwidth* attribute has the same usage as *Ingress-Bandwidth* for traffic going in the opposite direction (from the end-user).
- The *Bandwidth-Profile-Id* attribute provides an opened text field in which the administrator can designate a bandwidth profile reference to be applied to the corresponding end-user. This attribute is similar to the *Filter-Id* attribute but is expected to be used from a bandwidth management standpoint. It is not intended to be used in roaming environments, where references are usually locally defined.

Ingress-Bandwidth and *Egress-Bandwidth* attributes can be used in *Access-Request*, *Access-Accept*, CoA and *Accounting-Request* messages. In *Access-Request* messages, these attributes are used as a hint to the authorization server to recommend a particular bandwidth, considering the limitations of the NAS or possibly the user's link. In *Access-Accept* and CoA, these attributes are used by the NAS to provision available bandwidth for the end-user; in *Accounting-Request* messages, these attributes are used to indicate to the accounting chain the actual bandwidth applied for the session.

This proposition is not stabilized yet, and it is still unclear whether this proposal will be standardized some day, even though there are obvious needs coming from network operators who are continuously dealing with proprietary attributes and subsequent incompatibilities.

3.2.6.3 Prepaid Extensions for RADIUS

This proposition [2] is not an official standardization item yet, but it is in the scope and the objectives of the RADIUS extension working group to provide a solution that can be usable for credit control applications, as it has been defined within the Diameter credit control application (RFC 4006). The purpose is to define additional attributes and usage guidelines to support new charging models such as duration-based, volume-based and one-time-based services. The prepaid solution defines a new method for provisioning various credits for the end-user sessions, the NAS being in charge of managing and measuring the usage of these credits in real time, reporting this usage to the prepaid server and requiring, if needed, additional credits.

The ongoing work is far from being completed, and it is not even sure that this proposition will become a plain standardization document; for this reason, we will not describe this solution further. The lack of alternative propositions within the IETF makes it almost impossible to rule on this proposition: the architectural consequences are definitely important for the future of RADIUS, and it seems reasonable to have multiple solutions proposed and reviewed by a college of experts.

3.2.7 RADIUS and its Future

Besides protocol mechanisms specified in RFC 2865, multiple RFCs have been written as companion documents to describe precisely how RADIUS has to be used for particular purposes. Table 3.2 shows what RFCs define the RADIUS protocol, its extensions and its usage.

Beyond weaknesses inherited from the very first stages of its birth, RADIUS is still very flexible. This property makes its use very adaptable for multiple purposes. In addition,

Table 3.2 RADIUS standardization documents

Title	RFC	Status	Comments
Remote Authentication Dial-In User Service (RADIUS)	2058	Standards Track	Made obsolete by RFC 2138. Initial protocol definition
RADIUS Accounting	2059	Informational	Made obsolete by RFC 2139. Definition of accounting information carried with RADIUS
Remote Authentication Dial-In User Service (RADIUS)	2138	Standards Track	Made RFC 2058 obsolete, but itself made obsolete by RFC 2865. Protocol definition
RADIUS Accounting	2139	Informational	Made RFC 2059 obsolete, but itself made obsolete by RFC 2866. Definition of accounting information carried with RADIUS
The Network Access Identifier	2486	Standards Track	Made obsolete by RFC 4282. Standardized method for user identification, defines the syntax for the network access identifier (user identity presented by the client for network authentication)
Microsoft Vendor-specific RADIUS Attributes	2548	Informational	This RFC gives an overview of vendor-specific attributes defined by Microsoft for RADIUS
RADIUS Authentication Client MIB	2618	Standards Track	Made obsolete by RFC 4668. SNMP MIB definitions for RADIUS client authentication functions
RADIUS Authentication Server MIB	2619	Standards Track	Made obsolete by RFC 4669. SNMP MIB definitions for RADIUS server authentication functions
RADIUS Accounting Client MIB	2620	Standards Track	Made obsolete by RFC 4670. SNMP MIB definitions for RADIUS client accounting functions
RADIUS Accounting Server MIB	2621	Standards Track	Made obsolete by RFC 4671. SNMP MIB definitions for RADIUS server accounting functions
Implementation of L2TP Compulsory Tunneling via RADIUS	2809	Informational	Discussion about implementation issues when performing L2TP compulsory tunneling using the RADIUS protocol
Remote Authentication Dial-In User Service (RADIUS)	2865	Standards Track	Makes RFC 2138 obsolete. Base protocol definition
RADIUS Accounting	2866	Informational	Makes RFC 2139 obsolete. Definition of accounting information carried with RADIUS

Table 3.2 (*Continued*)

Title	RFC	Status	Comments
RADIUS Accounting Modifications for Tunnel Protocol Support	2867	Informational	Updates RFC 2866 by defining new accounting attributes, and by adding new values to the *Acct-Status-Type* attribute
RADIUS Attributes for Tunnel Protocol Support	2868	Informational	Definition of new attributes to support compulsory tunneling with RADIUS. Does not update RFC 2809 which is more specifically focused for L2TP
RADIUS Extensions	2869	Informational	Definition of new attributes for various purposes such as EAP, ARAP, *Accounting-Interim*, *Service-Type* and accounting attributes
Network Access Servers Requirements: Extended RADIUS Practices	2882	Informational	Description of practices and interpretations of implemented RADIUS features
RADIUS and IPv6	3162	Standards Track	Definition of new attributes to support IPv6 characteristics
IANA Considerations for RADIUS (Remote Authentication Dial-In User Service)	3575	Standards Track	Updates RFC 2865 about guidance to the IANA regarding registration of values related to the RADIUS protocol
Dynamic Authorization Extensions to Remote Authentication Dial-In User Service (RADIUS)	3576	Informational	Definition of new messages to the RADIUS protocol, in order for a RADIUS server to or change of authorization notification
RADIUS (Remote Authentication Dial-In User Service) Support For Extensible Authentication Protocol (EAP)	3579	Informational	Updates RFC 2869 for the rules to follow in interacting with EAP
IEEE 802.1X Remote Authentication Dial-In User Service (RADIUS) Usage Guidelines	3580	Informational	Suggestions about the RADIUS protocol usage when used by 802.1X authenticators
Remote Authentication Dial-In User Service (RADIUS) Attributes Suboption for the Dynamic Host Configuration Protocol (DHCP) Relay Agent Information Option	4014	Standards Track	Definition of a DHCP RADIUS suboption that could be transmitted from the RADIUS server to the DHCP server through the DHCP relay agent
The Network Access Identifier	4282	Standards Track	Makes RFC 2486 obsolete. Standardized method for user identification, defines the syntax for the network access identifier (user identity presented by the client for network authentication)

(*continued*)

Table 3.2 (*Continued*)

Title	RFC	Status	Comments
Chargeable User Identity	4372	Standards Track	Defines new attributes to convey a specific identity for roaming situations
RADIUS Extension for Digest Authentication	4590	Standards Track	Extension to support digest authentication for HTTP-style protocols such as SIP
RADIUS Authentication Client MIB for IPv6	4668	Informational	Makes RFC 2618 obsolete. IPv6 support
RADIUS Authentication Server MIB for IPv6	4669	Informational	Makes RFC 2619 obsolete. IPv6 support
RADIUS Accounting Client MIB for IPv6	4670	Informational	Makes RFC 2620 obsolete. IPv6 support
RADIUS Accounting Server MIB for IPv6	4671	Informational	Makes RFC 2621 obsolete. IPv6 support.
RADIUS Dynamic Authorization Client MIB	4672	Informational	SNMP MIB definitions for RADIUS client authorization functions, when RFC 3576 is used
RADIUS Dynamic Authorization Server MIB	4673	Informational	SNMP MIB definitions for RADIUS server authorization functions, when RFC 3576 is used
RADIUS Attributes for Virtual LAN and Priority Support	4675	Standards Track	New attributes for filtering and prioritization to VLANs for IEEE-802 access
RADIUS Delegated-IPv6-Prefix Attribute	4818	Standards Track	Defines a new attribute to convey an IPv6 prefix to be delegated by the NAS
RADIUS Filter Rule Attribute	4849	Standards Track	Definition of a new RADIUS attribute inherited from Diameter for IP traffic filtering.

RADIUS is evolving within the RADEXT working group at the IETF to support new features and to extend its usage. The protocol has the ability to carry new accounting information, and has also evolved significantly to offer sophisticated authorization functions. With its ability to designate every user context maintained in an NAS, RADIUS is a privileged medium to enforce individualized policies during the authorization process.

In spite of its lack of reliability, it seems very unlikely that RADIUS will disappear from the networks within the next 5 years. There are still a lot of functionalities to add (credit control, extensible attribute definition, etc.), operators are still providing new requirements and it is a good sign that the standardization process started again in 2004 to define extensions to the protocol (even though IETF imposed a restricted margin of evolutions). Even a second-generation evolution such as RFC 3576 for dynamic provisioning is being reworked again for new improvement.

However, one can easily foresee the end of its evolution, because 3GPP and ETSI have chosen Diameter for the IMS model, to achieve fixed–mobile convergence. Although RADIUS was left without any real competitor for years, the growing popularity of Diameter will likely push RADIUS aside in the next 3 years for new usages and new network AAA chain deployments, of course depending on a wide adoption of IMS evolutions by telcos.

References

[1] Lior, A. *et al.*, 'Network Bandwidth Parameters for Remote Authentication Dial-In User Service (RADIUS)', draft-lior-radius-bandwidth-capability-01, July 2005.

[2] Lior, A. *et al.*, 'Prepaid Extensions to Remote Authentication Dial-In User Service (RADIUS)', draft-lior-radius-prepaid-extensions-08.txt, July 2005.

4

The Diameter Protocol

As the IP protocol became the reference technology to build new networks, the need for a very reliable AAA protocol was crucial. From the simple requirement of providing an access on a remote access server in a secured manner, or a connection to the Internet or to the company's VPN, needs have strongly shifted to support a more sensitive and demanding exploitation of networks. New phone access technologies have evolved to support (and rely on) the IP technology, and therefore this IP technology had to provide an AAA protocol with new capabilities that RADIUS was unable to offer. Firstly, 3GPP defined the IP Multimedia Subsystem (IMS) as the architecture design guidelines for the mobile phone network control plane. 3GPP decided to rely on the Diameter protocol to carry AAA functions, and to have a role in provisioning appropriate resources in network equipment. In the beginning, however, 3GPP needs were more oriented to RADIUS evolution of NASREQ aspects. Next, ETSI/TISPAN decided to adopt the IMS architecture design for fixed networks built upon IP, meaning that Diameter is now a central piece of technology to be used in the core of new voice architectures, for both mobile and fixed networks, thus leading to a real Next-Generation Network (NGN). Diameter is under the spotlight more than ever, as historic network operators plan to replace the PSTN service with VoIP technologies, owing to the IMS.

The next sections of this chapter aim to present the requirements that have defined the Diameter protocol, followed by a presentation of the protocol itself, as well as some use cases in different network environments, unfortunately not relying upon real-life experience since operational usage of Diameter is really poor at the time of writing.

4.1 Learning from RADIUS Deficiencies

As seen in Chapter 3, RADIUS is a very powerful protocol, very flexible and easy to deploy. But RADIUS also has a set of irretrievable drawbacks when the transport of AAA messages has to be reliable, or when the whole architecture must span to different networks

Service Automation and Dynamic Provisioning Techniques in IP/MPLS Environments C. Jacquenet, G. Bourdon and M. Boucadair

in order to provide roaming users an identical connection service wherever they are. The RADIUS standard is also very delicate to extend since its field size drastically restricts the numbering space, although this is possible with non-standard proprietary extensions.

From this standpoint, IETF realized that RADIUS would not be suitable for sophisticated AAA operations to come, and decided to initiate a process to specify a new protocol, with backward RADIUS compatibility (a restricted compatibility, of course, considering the inherent drawbacks mentioned above). Requirements were edited from various standardization working groups:

- NASREQ: specific requirements related to the network access server, in order to provide at least the same level of service as provided by RADIUS;
- ROAMOPS: specific requirements related to roaming operations, as the new AAA protocol is supposed to service roaming users;
- MOBILEIP: specific requirements related to the mobile IP technology, in order to provide appropriate messages and features that could be usable for users connected to mobile IP.

The resulting set of requirements (RFC 2989 [1]) were eventually used to evaluate candidate protocols proposed to fulfill evaluated AAA operations. These requirements were classified in five distinct categories: general requirements, authentication requirements, authorization requirements, accounting requirements and specific mobile IP requirements.

4.1.1 General Requirements

A set of 14 general protocol capabilities was listed, taking back what RADIUS was already able to propose, but also including new features such as:

- *Scalability.* The new AAA protocol must be designed in order to support millions of simultaneous users, thousands of requests and connected devices pertaining to the AAA chain. Therefore, all fields must provide enough numbering space to make this possible.
- *Fail-over.* The protocol must include a mechanism to detect server failures and take appropriate actions to connect to a backup node.
- *Mutual authentication* between client and server.
- *Transmission level security.* The protocol must provide hop-by-hop security at the transmission layer, i.e. providing authentication, integrity and confidentiality of messages exchanged between peers.
- *Data object confidentiality.* This requirement is the first part of the end-to-end security the AAA protocol has to provide, in order to prevent intermediate nodes of the AAA chain from analyzing the content of AAA exchanges.
- *Data object integrity.* This requirement is the second part of the end-to-end security the AAA protocol has to provide, in order to prevent intermediate nodes of the AAA chain from modifying data objects conveyed between a client and a server.
- *Certificate transport.* The protocol must provide the capability to convey certificates between peers.

- *Reliable AAA transport mechanism.* This is a set of requirements representing a major step forward compared with what RADIUS provides. Basically, each message has to be explicitly acknowledged without any consideration of the meaning of this message; the protocol application ensures retransmissions.
- *Run over IPv4.*
- *Run over IPv6.*
- *Support of proxy and routing brokers.* The AAA protocol must embed in its structure all mechanisms to include third-party actors that may operate proxies or routing brokers not located on the forwarding path but having the ability to redirect messages to the proper entity.
- *Auditability.* The protocol must provide traces of the major actions done on AAA packets over the traversed nodes.
- *Ability to carry service-specific attributes.* This requirement is also an important improvement compared with the RADIUS capabilities. The AAA protocol has to provide hooks for extensions that may be defined for any application that requires an AAA function.

Besides classical AAA requirements, the AAA protocol has to propose specific features for mobile IP support, such as an appropriate encoding of mobile IP registration messages, an ability to cross proxy-firewalls without altering the quality of AAA exchanges and the ability to permit the allocation of local home agents.

4.1.2 Authentication Requirements

Beyond general requirements, specific needs have to be taken into account, relative to specific usage of the AAA protocol for authentication purposes. Firstly, the protocol must provide at least the same level of functionalities as RADIUS natively embeds, as well as some additional features that may ease the management of user sessions connected to a NAS:

- *NAI support.* The protocol must be able to manage network access identifiers that can be used by users.
- *CHAP support.* CHAP is commonly used for PPP access and must therefore be supported to allow a smooth transition, not impacting upon end-user terminal requirements.
- *EAP support.* EAP is becoming the standard authentication framework for all kinds of access. The AAA protocol must support the EAP exchange framework.
- *PAP support.* Even though it is dangerous to use PAP as an authentication protocol, there are still a large number of users employing it. The AAA protocol must be able to support it.
- *Re-authentication on demand.* This requirement describes the ability of an AAA server to trigger an authentication procedure the AAA client will have to initiate towards a designated user.
- *Authorization only.* The protocol must not require user credentials to be transmitted during the authorization phase.

4.1.3 Authorization Requirements

Besides authentication, there is also a complete set of authorization requirements:

- *Static and dynamic IP address assignment.* The authorization process of the AAA protocol must be able to transport a static IPv4/v6 address (the server sends this address explicitly), or to transport a hint for the AAA client dynamically to assign an IPv4/v6 address to the end-user.
- *RADIUS gateway compatibility.* The AAA protocol must be compatible with RADIUS attributes.
- *Reject capability.* This requirement makes it possible for an intermediate node such as a proxy to take non-authorization decisions based on its own set of criteria, either by blocking access requests or by blocking access authorization messages.
- *Re-authorization on demand.* This requirement refers to the ability of a server to update authorization information on the AAA client for a specific end-user. This update can be triggered either by the server itself or upon a specific request sent by the AAA client.
- *Support for access rules, filters and restrictions.* This is an important requirement the AAA protocol has to fulfill. The AAA protocol must provide means to convey authorization parameters to enforce specific rules on the access device in order to control the end-user activity or the quality of service provided to the end-user.
- *State reconciliation.* This imposes the requirement on the AAA protocol to provide means to help in state registration of resource allocations for end-user specific authorizations.
- *Unsolicited disconnect.* This requirement makes the server able dynamically to trigger an end-user disconnection.

4.1.4 Accounting Requirements

A set of specific accounting requirements has been edited as well:

- *Real-time accounting.* The AAA protocol must be able to report events simultaneously with their appearance.
- *Accounting record extensibility.*
- *Batch accounting.* The ability to store an ensemble of accounting records to send them simultaneously to the server;
- *Guaranteed delivery.* This requirement mandates the server explicitly to acknowledge accounting messages, to notify the ability to take it into account.
- *Accounting timestamps.* Every accounting event should have the possibility to be sent with a time hint.
- *Dynamic accounting.* This requirement makes accounting behavior compatible with dynamic authorization. With dynamic accounting, any authorization state that dynamically changes during an end-user session lifetime can be reported to the accounting server.

4.1.5 Diameter is Born

From the whole set of requirements listed above, multiple protocols were potential candidates, expecting appropriately to answer to AAA challenges. Four protocols were mentioned as potential candidates:

- SNMP (RFC 3411 [2]). This protocol was rejected because of the huge engineering work that would be required to make it a serious AAA protocol, even though it was still acceptable for accounting.
- COPS (RFC 2748 [3]). This very powerful protocol was seriously considered, taking into consideration that only few changes would be required to make it a serious AAA protocol. COPS was particularly appreciated for its data model.
- RADIUS++. An extension of the RADIUS protocol to make it compatible with AAA protocol requirements could have been possible, but it would have required significant engineering work. Moreover, the final work would likely result in a protocol comparable with Diameter.
- Diameter (RFC 3588 [4]). However, some changes to the initial protocol specification were needed to make it compatible with AAA protocol requirements.

Diameter has been considered as the best candidate to support AAA relationships. A complete evaluation of each of these protocols is made in RFC 3127 [5].

4.2 Diameter: Main Characteristics

The Diameter base protocol (RFC 3588 [4]) has been conceived as a peer-to-peer protocol that provides support for reliable exchange of information between Diameter nodes, whereas RADIUS was built as a transactional protocol following a client–server model. The Diameter base protocol provides the minimum set of functionalities required for any AAA protocol, including support for accounting, while specific service information is handled at the application level. Diameter is based upon TCP or SCTP running over IPv4 or IPv6 (whereas RADIUS runs over UDP), using destination port 3868, thus ensuring a reliable delivery of the message at the transport layer, even though acceptance of specific messages requires an explicit applicative acknowledgment, especially for accounting messages. A Diameter client is not required to support both transport protocols, but intermediate nodes or servers are mandated to allow a connection with both protocols. Using a reliable transport protocol also eases failover mechanisms, as a network failure can be detected without having to wait for an applicative acknowledgment. Explicit applicative acknowledgment is needed, however, since a network reliable delivery does not mean that the Diameter process was able to treat the message. This is particularly crucial for accounting messages that are directly linked to the billing system, and can lead to a loss of revenue.

The use of TLS or IPSec is also required, in order to ensure security between nodes. Even though TLS or IPSec is sufficient for hop-by-hop security, it is not a satisfactory method for protecting the payload all over the AAA signaling path: a proxy-broker, connected to the Diameter client of network A and to the Diameter server of provider B, will be able to read sensible payload and correlate identity with resource usage. The Diameter specification encourages the usage of an end-to-end security mechanism.

4.2.1 Diameter Network Entities

The Diameter protocol describes three different categories of nodes:

- *Diameter client.* Designates the Diameter node in charge of providing access to a resource for which Diameter messages have to be generated. The access device is likely to be a Diameter client, but it can also be a service platform commanding a policy enforcer that can control the transport plan.
- *Diameter agent.* Designates a large category of nodes implied in routing and treating Diameter messages.
- *Diameter server.* Designates the terminal node in charge of final AAA operations.

Diameter agents can have various roles, depending on their location in the AAA chain (see Figure 4.1):

- *Relay agent.* The relay agent is able to route Diameter messages towards the appropriate node, depending on the targeted realm to which the message is destined. This routing operation is realized using the *Destination-Realm* information embedded in Diameter messages and the agent action notified inside the message. Therefore, the relay agent alters the Diameter message by inserting or removing routing information.
- *Proxy agent.* The proxy agent performs an operation comparable with that of the relay agent, except that the proxy Diameter is able to modify Diameter messages in order to enforce a specific (and local) policy. This policy can be related to admission control, provisioning or resource usage.

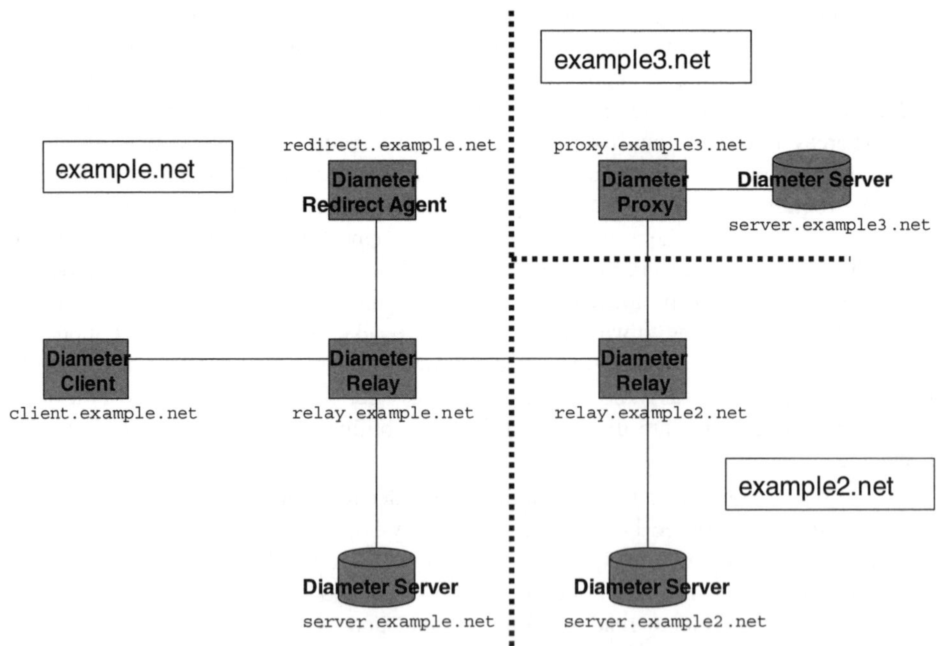

Figure 4.1 Diameter chain example

- *Redirect agent.* The redirect agent is solicited by other agents to find the path towards the targeted server. It is envisioned that active Diameter nodes in a network will not be able to store all routing information for Diameter messages. Interrogating a redirect agent is a way to dedicate a specific resource to keep route information, and to communicate it to other Diameter agents.
- *Translation agent.* The translation agent role is to convert any AAA protocol already existing (or to come) to Diameter. This is particularly true for RADIUS or TACACS+ protocols, which are required to follow a specific translation to be compatible with Diameter, even though this translation is eased by attribute compatibility.

4.2.2 Diameter Applications

One important feature of Diameter is to provide the support of different *applications*. A Diameter application can be described as a set of messages, attributes and message treatment rules that are defined to fulfill requirements for a specific usage. So far, five different applications have been defined to extend Diameter: mobile IP (RFC 4004 [6]), NASREQ (RFC 4005 [7]), credit control (RFC 4006 [8]), EAP (RFC 4072 [9]) and SIP (RFC 4740 [10]). Diameter also defines a new capability exchange possibility; nodes can advertise what applications are supported during connection setup, thus ensuring that all nodes of a AAA chain will support a common set of functionalities. A specific 'wildcard' application code has been defined for relay agents, since relay agents are intended transparently to support all current and future applications. The Diameter base protocol natively provides the accounting application. However, it is possible that particular accounting messages do not exist, and therefore additional accounting applications can be defined. Defining a new Diameter application should not be abusively employed, and existing applications have to be reused as often as possible. The Diameter base protocol also defines a set of generic commands that are related to session management and connection management.

4.2.3 Sessions and Connections

Diameter introduces two important notions that could usually be employed indistinctively in other circumstances, when talking about end-user connectivity management with RADIUS, for instance. Diameter connections and sessions have been precisely defined and must be carefully employed:

- *Connection.* A Diameter connection is the peering relationship that two Diameter nodes maintain in order to exchange Diameter commands. A Diameter connection is built over IPSec and TCP or SCTP as transport protocol. A Diameter connection is not required to span the whole AAA chain from the client to the server; a connection has to be conceived as the primary communication relationship between two nodes. It is foreseen that AAA exchanges cross multiple Diameter connections.
- *Session.* A session can be understood in the same sense as end-user sessions when connecting to a network access server. In the RADIUS case, the session is handled by the RADIUS client for which it has to query a server to ensure authenticity, apply authorization parameters and provide accounting. With Diameter, the session notion is

quite similar and its reference is unique along the AAA chain. Diameter provides more reliable session management messages than RADIUS does, especially by separating session aliveness monitoring from session accounting. Diameter defines specific messages to manage user sessions that were also added to the RADIUS protocol within RFC 3576, to change authorization parameters or dynamically disconnect end-users. But Diameter is still better able to manipulate sessions, as session termination can be notified to the AAA server without sending accounting information: RADIUS-based AAA chains still rely upon *Accounting-Request/Stop* messages to determine whether a session is terminated or not.

4.2.4 Diameter Routing

The elaborated routing capability is one of the most interesting features provided by Diameter. It is possible to build a complex AAA chain topology over different networks, and rely on an efficient forwarding intelligence to enable a Diameter client to reach the appropriate server, as long as a Diameter path exists between them. As routers provide a forwarding path over networks to exchange packets between them across multiple autonomous systems, Diameter provides the same idea for AAA message forwarding.

To route request messages, Diameter nodes make use of two specific AVPs: *Destination-Host* and *Destination-Realm*. The *Destination-Host* AVP designates the Diameter server that must be reached to answer the request. The *Destination-Host* address is encoded as a Fully Qualified Domain Name (FQDN), such as 'diameterserver.example.net'. Each Diameter peer maintains a peer table in order to determine to which peer a Diameter message has to be sent depending on the *Destination-Host* AVP content. If the *Destination-Host* is the machine on which the message has been received, the message is processed locally. If the *Destination-Host* is known, the peer table is consulted and the message is forwarded to the corresponding peer. The peer table comprises all nodes with which the Diameter node has established a connection. For each node entry, the peer table maintains useful indications such as peer status, connection expiration time and TLS activation, as shown in Figure 4.2.

When the *Destination-Host* AVP is not present in the message, or if the host is not present in the peer table, the Diameter node has to use the *Destination-Realm* AVP to route the message towards a node that will be able to take charge of the Diameter message towards its final destination. This is possible through the realm-based routing table which defines how messages have to be processed. If no realm seems able to route the message, the Diameter message cannot be routed and an error is returned to the message initiator.

The realm-based routing table (Figure 4.3) provides some kind of superset indication of the direction in which the Diameter message has to be sent. The realm-based routing table

Host Identity	Status	Static/Dynamic	Exp. Time	TLS
server.example.net	open	dynamic	1h20m01	disabled
relay1.example.net	open	static	12h40m05	disabled

Figure 4.2 Peer table example

Realm Name	App. ID	Local Action	Server Identifier	Static/Dynamic	Exp. Time
example2.net	RELAY	RELAY	relay1.example.net	static	12h10m51
example.net	NASREQ	RELAY	server.example.net	static	5h47m33

Figure 4.3 Realm-based routing table

makes a correspondence between a realm name that is expected to treat the message with a set of Diameter nodes that are able to route the message in order to reach this *Destination-Realm*. The realm information is initially extracted from the network address identifier (NAI) that was initially used to identify the session requester. The NAI format is defined in RFC 4282 [11] and usually follows a recognizable format such as 'username@example.net'. The realm part of the NAI ('example.net') is used by the Diameter client to fill the *Destination-Realm* field. The realm-base routing table also provides an indication of the application supported by the Diameter node to which the message will be sent. If this node is a relay and not a server, the wildcard application ID will be used, signaling that any kind of request can be routed through this node. It is important to note that the Diameter peer (server identifier) indicated in the realm-based routing table has to be present in the peer table to be valid. When it is time to forward the message to another peer, the peer table has to be consulted to make sure that forwarding is possible.

The first request for a session is expected to be targeted towards a *Destination-Realm* (using the *Destination-Realm* AVP), and not towards a specific destination (explicit server address), except if there are specific security constraints. Using the realm to route messages is a way for Diameter relays and proxies to perform load-balancing between nodes, and also to determine the best server to use depending on the supported application. If the *Destination-Host* is explicitly specified, it is possible to reach a Diameter server that is unable to run the application requested to process the message. However, as soon as a first exchange between a client and a server for a session is done, the *Destination-Host* can be used in order to reach immediately the node initially contacted during the first round-trip.

Figure 4.4 gives an example of Diameter message routing for two simple cases where:

- user@example.net wants to connect to his/her home network as s/he is already directly connected to it;
- user@example2.net wants to connect to his/her home network as s/he is connected to a foreign network.

In case 1, when user@example.net wants to connect to the network (1a), the B-RAS acts as a Diameter client and determines how to treat the access request. Firstly, the realm name is extracted from the user identity: example.net is the targeted realm. The Diameter client checks its realm-based routing table, and a lookup in the realm-based routing table (Figure 4.3) indicates that the message has to be directly forwarded to server.example.net, as long as the request applies to the NASREQ case (which is the case in this example). Afterwards, the Diameter client checks its peer table to get information on server.example.net (Figure 4.2). Using the corresponding entry, the Diameter client is able to send the appropriate message (1b) to the Diameter server.

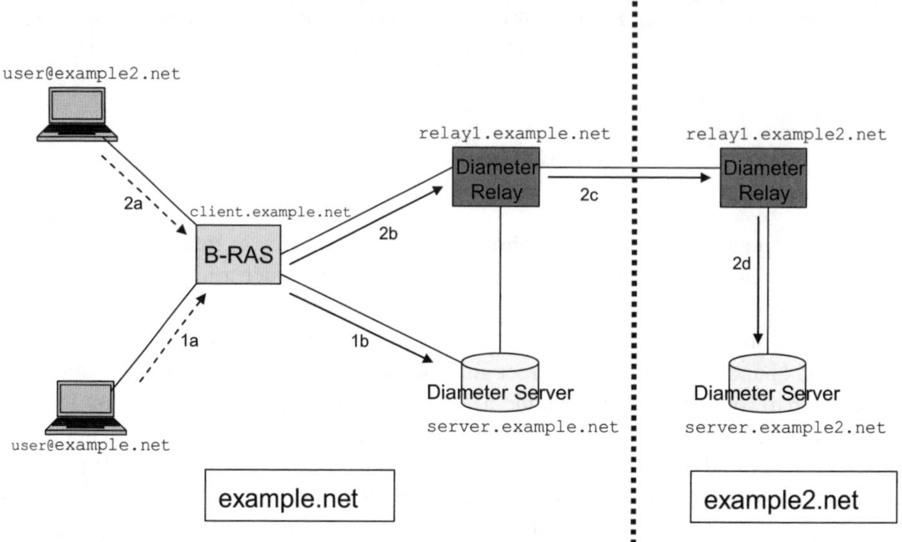

Figure 4.4 Complete example of message routing

Case 2 is quite similar to case 1, but the realm name extracted from the user identity (2a) is different from the local network. The Diameter client checks its realm-based routing table and determines that requests targeted to example2.net have to be forwarded to relay1. example.net, whatever the Diameter application. The peer table provides necessary information to reach relay1.example.net, and the Diameter message is sent towards it (2b). When relay1.example.net receives the message, it checks how to treat this message. The *Destination-Realm* AVP mentions that example2.net is the targeted domain, and a lookup in its own realm-based routing table indicates that the message has to be forwarded (2c) towards relay1.example2.net (following parameters indicated in its own peer table). When relay1. example2.net receives the Diameter message, the *Destination-Realm* AVP has not changed and is used again to find a corresponding entry in its own realm-based routing table. The Diameter server 'server.example2.net' has to be reached to treat the message and is forwarded accordingly (2d). Once the message arrives on server.example2.net, it is locally processed. The response message has to follow the same way back to the diameter client (B-RAS in this example), since every Diameter peer in the path keeps the state of routed messages in order to ensure reliable delivery between the client and the server. Keeping message states is particularly efficient for resending messages from intermediate nodes without requiring specific action from termination nodes.

4.2.5 Peer Discovery

Static configuration of peer and realm tables is a rather fastidious task that network administrators should get rid of by using DNS. Dynamic peer discovery is optional in Diameter implementations, but interesting when the Diameter client needs to identify the peer to which it has to send its requests, and also when an agent needs to discover another agent to handle a message that needs to be routed to a server. Dynamic discovery of Diameter peers is not a preliminary action that is performed when a peer starts up: it has to be performed when required, i.e. when a message has to be routed to another node and no routing entries in the peer table or realm-based routing table can be used to forward the message. In this case, a specific request can be sent to a DNS, for instance, to get the FQDN of the Diameter peer to reach in order to forward the pending message.

4.2.6 Peer Connection Maintenance for Reliable Transmissions

One of the main distinctive approach characteristics of Diameter compared with RADIUS is the appearance of a connected relationship between two peers that have directly to communicate between each other. The RADIUS approach was to consider that consecutive non-acknowledged messages might be considered as a peer failure. This led to an ensemble of negative side effects: failure detection delay is bad since a peer has to realize several tentatives to discover and react accordingly; there is an unclear meaning of protocol messages since a protocol acknowledgement also has an applicative usage; actual useful messages are used as hints to give a slight flavor of reliability to the protocol, accepting that operational AAA messages are used for that purpose.

The Diameter approach is totally different: firstly it makes use of a connected IP transport layer, and provides protocol-specific messages to acknowledge each request and dedicated messages to test and maintain active connections between peers. This is achieved owing to *Device-Watchdog* messages that are sent over the connection at regular intervals, when no traffic is seen during the corresponding period. *Device-Watchdog* messages are useless when AAA traffic is going through the connection, since other messages avail themselves of transport failure procedures: requests are stored at each node as long as the corresponding answer is not received. Thus, if the peer to which requests have been sent is not responding, pending requests can be sent to another node, avoiding message loss along the AAA path.

4.3 Protocol Details

This section will describe more precisely the message format used by Diameter, as well as command codes used for the base application.

4.3.1 Diameter Header

The Diameter header format is built as shown in Figure 4.5.

The fields are as follows:

- *Version.* Indicates the version number of the Diameter protocol used (0x1).
- *Message length.* Total length of the message including all header fields.
- *Command flags.* Four flags have been defined:
  - the (R)equest bit is used to indicate that the embedded command has to be interpreted as a request (when this bit is not set, it is an answer);

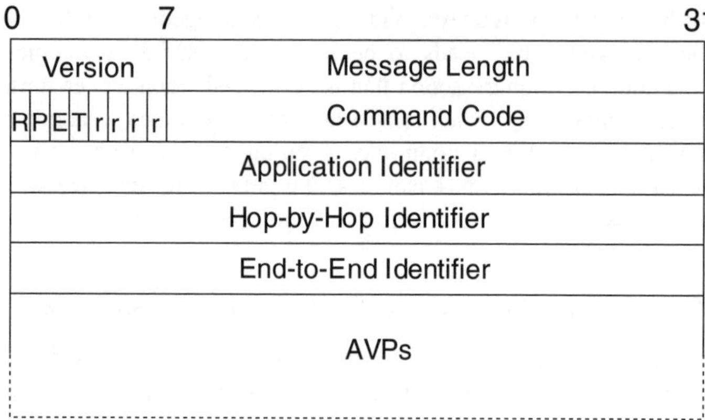

Figure 4.5 Diameter message header

  - the (P)roxiable bit is an indication that the message can be relayed by a proxy or a Diameter relay, without needing a local processing;
  - the (E)rror bit is positioned on answer messages to indicate to the sender that it is malformed (the corresponding message also contains an error code);
  - the Re-(T)ransmission bit is positioned when a message is potentially a duplicate that is sent when an error occurred during transmission of the original message.
- *Command code.* Indicates the command code giving the meaning of the present message. Command code numbering is managed by the IANA.
- *Application identifier.* The application identifier is used by peers to determine the Diameter application context into which the message requires to be processed.
- *Hop-by-hop identifier.* This identifier represents a unique numbering of the present request, between two directly connected nodes. This identifier is kept between requests and answers between two nodes: this eases the work of each node in mapping answers to requests and identifying requests that have not been acknowledged.
- *End-to-end identifier.* This identifier is used to detect duplicates, using the *Origin-Host* AVP to ensure uniqueness of this identifier between a Diameter client and a server. The number is forged using time information in order to make the number unique when device reboot occurs.
- *AVP.* Attribute value pairs used as parameters to complete the action determined with the command code.

The construction of the Diameter header follows the same philosophy used for RADIUS, i.e. a command code that determines the nature of the action to be performed, augmented by pertinent attributes (AVP). This way, Diameter can claim some kind of compatibility with other AAA protocols such as RADIUS, as it is easy to implement a translation agent that just has to translate the command code and the Diameter header into RADIUS protocol, and to move the attributes from one message to the other without (or with minimal) adaptation. For that reason, the AVP numbers ranging from 1 to 255 are reserved for backward compatibility with RADIUS, whereas AVP numbers starting from 256 are specific to Diameter.

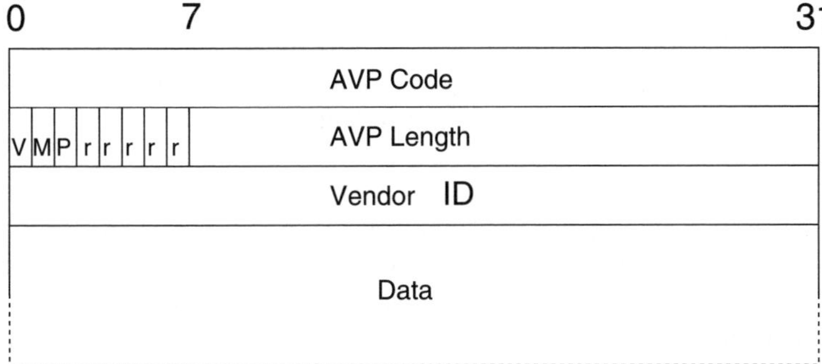

Figure 4.6 Diameter AVP header

4.3.2 AVP Format

AVPs are similar to attributes defined for RADIUS. Diameter AVPs are not *stricto sensu* attribute value pairs formed with typical type/value fields, because the whole attribute is augmented with an 8-bit bank, for which 3 bits are already defined in version 1 of Diameter, and other bits can be defined by forthcoming Diameter applications (Figure 4.6). A Diameter AVP size must align on a 32-bit block, even though this might require additional padding, especially when the value transported is a string.

- *AVP code.* With a 32-bit encoded value, the AVP code identifies the attribute. As mentioned above, AVP codes below 256 are reserved for RADIUS compatibility, and other values are managed by the IANA. Having a large range of possible attribute codes is a real advantage for further development, as standardization bodies will be less reluctant or shy to assign forever a part of an almost unlimited resource.
- *AVP length.* Gives the total length of the AVP, including all fields, but excluding the optional padding that could be added for the 32-bit alignment.
- *Data.* This is the field encoding the useful value that corresponds to the AVP code.

The AVP bits are used to help Diameter peers to handle AVPs properly:

- The (P)rotection bit indicates that the content is protected for end-to-end security. However, considering the ongoing work at the IETF to update the Diameter base protocol, this bit will likely disappear in the future (becoming a 'reserved' bit), since work on solution for end-to-end security has never ended.
- The (M)andatory bit is used to indicate that the AVP must be supported by the node that received it, with the exception of relay agents which are not supposed to interpret such fields. Diameter nodes such as proxy, client or server must support the AVP transmitted with the M-bit, or the message has to be rejected. Using the M-bit for an AVP is defined by the application rule that makes use of the AVP. For interoperability reasons, an AVP sent with the mandatory bit may not be accepted by the corresponding peer if the current application does not require it. AVPs sent with the M-bit cleared are considered as informational, and can be silently discarded if not supported.

- The (V)endor-specific bit indicates whether the AVP is standardized (bit cleared) or defined by a vendor for its specific usage (bit set). When the V-bit is set, the *Vendor-Specific* field is present in the AVP format to indicate the official SMI company number used by the vendor, in the same way as with RADIUS.

Diameter allows usage of grouped AVPs. A grouped AVP is an ensemble of multiple AVPs that are concentrated into one AVP. For instance, this is particularly useful when it is required to send multiple values corresponding to the same session: all AVPs included in the group are therefore implicitly linked to each other. A grouped AVP is defined as such in an application or in the base protocol. As an example, the *Proxy-Info* AVP is defined as a grouped AVP of *Proxy-Host* AVP and *Proxy-State* AVP in the base protocol. AVPs included in a grouped AVP are not defined as 'sub-AVPs': these are plain AVPs that can be used alone. On the other hand, it is also possible to imagine several cascaded group layers within the same AVP in a single message.

4.3.3 Command Codes

As we have seen above, Diameter messages are distinguished by two parameters: the command code and the (R)equest bit. Formally, there are different messages for requests and answers, but the Diameter protocol makes them share the same command code; the R-bit is then used to differentiate a request from an answer.

The Diameter base protocol only defines seven different command codes, corresponding to seven different kinds of request with their corresponding answer, for a total of 14 different messages. Only a limited number of command codes have been defined in order to restrict the base protocol to the very minimum features, i.e. Diameter connection management, generic session operations and accounting.

(i) Diameter Connection Management

- *Capabilities-Exchange-Request/Answer* (CER/CEA, command code 257). These two messages must be used as soon as the transport layer is brought up, in order for two directly connected peers to exchange their capabilities, such as authentication and accounting application identifiers, supported vendor attributes or security mechanisms, as well as some specific parameters such as IP address to be used by each node to identify the source of each message (which can be different from the IP interface address, especially when a loopback address is used, for instance). If capabilities exchanged show that there is an incompatibility between the peers, the connection cannot be effective and the transport layer is torn down. Because of a scope limited to the connection, *Capability-Exchange* messages cannot be forwarded in any manner through a Diameter agent to a remote peer.
- *Device-Watchdog-Request/Answer* (DWR/DWA, command code 280). These messages have been defined to improve detection of transport connection failure. The more quickly peers can react to a failure in the AAA chain, the more efficiently and transparently failover procedures can be activated. *Device-Watchdog* messages act as periodic probes sent over the connection to determine if the remote Diameter instance is still up (which would not be detected by a traditional ICMP message) when no traffic has been monitored over the connection for a while. These messages are used to monitor connections and therefore cannot be routed through a Diameter agent.

- *Disconnect-Peer-Request/Answer* (DPR/DPA, command code 282). To improve peer behavior if one of the nodes needs to shut down a connection, *Disconnect-Peer* messages have been defined to give the opportunity to anticipate a foreseen Diameter service disruption. A *Disconnect-Peer-Request* message is sent whenever a peer wants to inform the other peer that the connection needs to be terminated. By doing so, the other peer can anticipate this by setting up an alternative route to forward Diameter messages. A *Disconnect-Peer-Answer* is therefore sent back to the requester to inform that the message has been taken into account. This message is also useful for preventing disconnected peers periodically to request a new connection if the service has been terminated for administrative or technical reasons (maintenance, overload, etc.). These messages are limited to the scope of the local connection, and cannot be routed through a Diameter agent.

(ii) Generic Session Operations

- *Abort-Session-Request/Answer* (ASR/ASA, command code 274). Similar to the *Disconnect-Request* message used in RADIUS (see Section 3.2.3), the *Abort-Session-Request* is used by a Diameter server to trigger a session disconnection on the access device. This kind of action can be performed, for instance, where authorization previously granted is no longer valid. Upon receipt of this request, the access device responds with an *Abort-Session-Answer*, including the status of the requested operation in the *Result-Code* AVP, mentioning whether the action was a success or not. It is interesting to note that, once the session is terminated, the access device has to issue a *Session-Termination-Request* to the Diameter server that can also be followed by an *Accounting-Request* message sent to the Diameter accounting server.
- *Session-Termination-Request/Answer* (STR/STA, command code 275). The *Session-Termination-Request* message is used by an access device to notify the Diameter server that the indicated session is no longer active. The Diameter server acknowledges this message with a *Session-Termination-Answer*. This way to proceed is a real improvement compared with how RADIUS works: each time a session is terminated, the authorization server is notified of the event, not needing to rely upon a hypothetical accounting message that could be received and treated by a dedicated non-real-time accounting server. Moreover, it is possible that some sessions are not required to be accounted (even though accounting is mandatory most of the time). By separating session management from accounting messages, Diameter helps in resource management reliability.
- *Re-Auth-Request/Answer* (RAR/RAA, command code 258). The *Re-Auth-Request* message is used by the service when it is required to ask for a targeted session to be re-authenticated or re-authorized. However, not all access devices support dynamic user re-auth, and these messages have to be specifically supported by applications that may require their usage. By initiating a re-auth of the user, it is possible to refresh the user authorizations and credits for service usage, from a service management initiative.

(iii) Accounting Messages

- *Accounting-Request/Answer* (ACR/ACA, command code 271). Accounting messages have exactly the same meaning when used with RADIUS. The *Accounting-Request* message is used by an access device to report accounting information towards a server, and is acknowledged by an *Accounting-Answer* message.

To improve reliability in connection management, the Diameter protocol defines the *Origin-State-Id* AVP that can be used as a hint sent by access devices to Diameter servers, in order to determine whether an access device has restarted. The *Origin-State-Id* is set during Diameter device startup with an incremental value that can be based upon the current time, for instance, or upon another parameter as long as the value increases between reboots. When this AVP is sent within a CER message, this can be a hint for the connected node to determine whether the access device has lost its sessions, and if corresponding states maintained have to be cleaned up. Since CERs are only sent one hop away, this can only be used by proxies or a server directly attached to the access device. A Diameter server has to wait until the first session request to infer that the access device has rebooted, and that corresponding states can be deleted.

4.3.4 Accounting

Diameter accounting differs from RADIUS because of an additional record type EVENT (using the *Accounting-Record-Type* AVP), besides usual START, STOP and INTERIM records. An EVENT record is used to notify the accounting server of the completion of a service that does not have a beginning or an end, but has to be billed as a whole. As a point of comparison, an EVENT accounting record can be seen as the concatenation of START and STOP records sent in an atomic message. The EVENT record can also be used to notify a session establishment failure to the accounting server.

The server also has means to change the access device accounting behavior in the case of failure, using the *Accounting-Realtime-Required* AVP. This AVP can be sent by the authorization server or by the accounting server in an *Accounting-Answer* message, and has to be interpreted by the access device to define its behavior whenever a problem occurs in sending accounting reports. Here is a list of the possible behaviors:

- *Deliver and grant.* If the accounting record cannot be properly sent immediately to an accounting server, the access device has to stop the corresponding session.
- *Grant and store.* As long as the service accounting information can be stored in the access device, the service can be granted to the session. Once the accounting service starts again, accounting records are sent. If access device storage resources are exhausted, the service cannot be granted anymore and corresponding sessions are terminated. This is the default behavior recommended to be used with Diameter if the *Accounting-Realtime-Required* AVP is not present in server messages.
- *Grant and lose.* In this case, accounting information is used by the server as a hint, and it is not required to offer the service. The service is not affected by accounting server connectivity problems.

4.4 Diameter Network Access Application (NASREQ)

The NASREQ application was one of the first reasons for the existence of Diameter, in order to improve RADIUS for remote access management. As seen in Section 4.3.3, command codes defined within the Diameter base protocol specification are not sufficient for an access device to require a server to perform authentication and grant related authorizations for a new session. The Diameter network access application (NASREQ) (RFC 4005 [7]) defines only one new command code for this purpose: *Authentication-Authorization-Request/Answer* (AAR/AAA, command code 265). This new message is an equivalent of *Access-Request/Access-Accept/Access-Reject* messages used in RADIUS, but also replaces *Access-Challenge* messages which are used when multi-round authentication/authorization is required, with EAP for instance.

Apart from the new AAR/AAA message, the NASREQ application relies upon the following Diameter messages defined in the base protocol:

- *Re-Auth-Request/Re-Auth-Answer*;
- *Session-Termination-Request/Session-Termination-Answer*;
- *Abort-Session-Request/Abort-Session-Answer*;
- *Accounting-Request/Accounting-Answer.*

It is possible that some authentication protocols require a multi-round authentication: EAP is a well-known example of this particularity, the number of round-trips between the end-user (therefore the access device) and the authentication server being highly dependent on the EAP method used (see Section 3.2.1). Doing multi-round authentication with RADIUS requires a workaround making use of the *Access-Challenge* message that could be sent right after an *Access-Accept* has been received from the server. Using the *Access-Challenge* message was motivated by the interdiction of sending an *Access-Request* after an *Access-Accept* message.

Diameter does not make use of *Access-Challenge* messages anymore, and instead makes it possible to send an *AA-Request* message after an *AA-Answer* coming from the server. As defined in the Diameter base protocol, the *Result-Code* AVP included in the latter message thus indicates this property with a value code set to DIAMETER_MULTI_R-OUND_AUTH.

4.4.1 AVP Usage for NASREQ

The NASREQ application is essentially focused on RADIUS compatibility and makes use of the majority of RADIUS attributes (with an attribute value below 256), forbidding the use of some of them, replacing others. Additional AVPs have been defined for the specific use of Diameter:

- *CHAP AVPs. CHAP-Auth* (code 402), *CHAP-Algorithm* (code 403), *CHAP-Ident* (Code 404), *CHAP-Response* (code 405) are used when the end-user wants to authenticate using the challenge authentication protocol (CHAP). The *CHAP-Auth* AVP is a grouped AVP embedding the three aforementioned, and must replace the use of the RADIUS *CHAP-Password attribute.*
- *NAS authorization AVPs. NAS-Filter-Rule* (code 400) and *QoS-Filter-Rule* (code 407) are two very important AVPs that enrich authorization possibilities. These attributes are described further in Section 4.4.2. The *NAS-Filter-Rule* attribute has been recently standardized for RADIUS, and a similar effort is engaged for QoS management to extend what already exists for Diameter and to make it available for RADIUS. These efforts are carried on within the DIME working group (IETF).

- *Tunneling AVP.* The *Tunneling* AVP (code 401) is a grouped AVP that embeds all other AVPs defined for tunnelling with RADIUS (defined in RFC 2868, see Chapter 3). This grouped AVP can be seen as a replacement of the 'tag' field used in tunnel attributes: the basic purpose of this field is to give the tag value to tunnel attributes that refer to the same tunnel configuration. Therefore, a single message could convey multiple tunnel configuration attributes. Using multiple *Tunneling* AVP is a more consistent way to achieve the same thing with Diameter.
- *Accounting AVPs.* Besides AVPs already defined in the Diameter base protocol and for RADIUS, NASREQ provides a new *Accounting-Auth-Method* AVP (code 406) to indicate in accounting messages the authentication method used by the end-user (PAP, CHAP, EAP, etc.). Four new attributes are also defined to replace those used with RADIUS to report data traffic usage: *Accounting-Input-Octets* (code 363), *Accounting-Output-Octets* (code 364), *Accounting-Input-Packets* (code 365) and *Accounting-Output-Packets* (code 366). These replacements are required because of the field format change which is now unsigned 64 bits, instead of the 32-bit value previously used. It is important to note that the *Accounting-Record-Type* (code 480) defined in the base protocol to identify the record type of the accounting message (*Start*, *Stop*, *Interim*, *Event*) replaces the *Acct-Status-Type* used with RADIUS.

4.4.2 Enhanced Authorization Parameters

The two new AVPs, *NAS-Filter-Rule* and *QoS-Filter-Rule*, have been defined to enforce enhanced policies from the authorization server to the access device for a particular session. It is recommended that usage of the *Filter-Id* RADIUS attribute be avoided in the NASREQ context, because of its loose definition which makes it almost impossible to have a deterministic behavior: each vendor is free to implement its own syntax. Other RADIUS authorization parameters such as *Session-Timeout* are still present and usable with Diameter, and will therefore not be described in this section.

Both *NAS-Filter-Rule* and *QoS-Filter-Rule* AVPs are defined in NASREQ, but the attribute format is defined in the base protocol. The *NAS-Filter-Rule* AVP is interpreted by the access device in order to apply traffic filters relative to the protocols transmitted on a session, whereas the *QoS-Filter-Rule* is used to treat the user traffic by marking or metering it, as defined in the DiffServ Architecture (RFC 2475 [12]). The format of these AVPs is described below.

The value field of these AVPs must comply with the following syntax, but is interpreted differently:

action dir proto ***from*** *src* ***to*** *dst* [***options***]

(i) NAS-Filter-Rule AVP

action: permit or deny.
dir: from the terminal (in) or to the terminal (out).
proto: IP protocol specified by its number, the 'ip' keyword can be used as a wildcard.
src and ***dst***: specifies the address and netmask, as well as the port numbers, using the syntax address[/bits_ masked]<address[/bits_masked]> [ports].

options:

- frag: the criteria is matched if the IP packet is not the first fragment of a datagram.
- ipoption <spec>: spec is used to identify the part of the IP packet header that may match the filter. Spec can be 'ssrr' for strict source routing, 'lsrr' for loose source routing, 'rr' for record route and 'ts' for timestamp. Multiple header options can be mentioned, separated with commas.
- tcpoption <spec>: similar to ipoption, tcpoption is used to apply filters on specific TCP fields of the IP datagram. Spec values can be 'mss' for maximum segment size, 'window' for TCP window advertisement, 'sack' for selective ACK, 'ts' for timestamp and 'cc' for T/TCP connection count.
- established: filter matches TCP packets that have the RST or ACK bits positioned.
- setup: filter matches TCP packets that have the SYN bit positioned, but not the ACK bit.
- tcpflags <spec>: filter matches TCP packets that have positioned bits specified in spec: fin, syn, rst, psh, ack, urg. The filter can also specify bits that must not be positioned by prepending a '!' before mentioning the bit reference.
- icmptypes <types>: filter matches ICMP packets that have the type(s) specified in types. Multiple type values can be specified in a single filter, separated by commas. The type values are: 0 (echo reply), 3 (destination unreachable), 4 (source quench), 5 (redirect), 8 (echo request), 9 (router advertisement), 10 (router solicitation), 11 (TTL exceeded), 12 (bad IP header), 13 (timestamp request), 14 (timestamp required), 15 (information request), 16 (information reply), 17 (address mask request), 18 (address mask reply).

(ii) QoS-Filter-Rule AVP

action: tag or meter.
dir, **proto**, **src** and **dst** similar to the *NAS-Filter-Rule*.
options:

- The 'DSCP <value>' option has to be used with the tag action to change the DSCP field of the header of IP packets that match the rule.
- The 'metering <rate> <value_under> <value_over>' option has to be used with the meter action to change the DSCP field of the IP packet header to 'value_under' if the traffic throughput is below the defined 'rate', or to 'value_over' if it goes beyond. It is interesting to note that 'value_over' can explicitly mention a 'drop' action to destroy IP packets without marking it.

Access devices have to behave the same way for filters provisioned through *NAS-Filter-Rule* or *QoS-Filter-Rule*. The first filter entry that matches the inspected packet stops the evaluation over the filter list. Comparison of packets with filters is done following the place of each filter in the list. The place of a filter is determined by its order of installation within the access device. If the last evaluated rule is a permit and no rules are matched, the packet is dropped. On the other hand, if the last evaluated rule is a deny and no rules match, the packet is forwarded. This behaviour is very similar to *Access-List* evaluation rules usually configured in IP routers. A single authorization message can include multiple filter rules, and this is foreseen to be the usual case. The place of each filter in the packet will determine the place of the corresponding filter in the access device evaluation list.

4.4.3 Enhanced Authorization Examples

1. The example provided in Section 3.1.3.2 to offer different bandwidth profiles for end-users can be provisioned using two *QoS-Filter-Rule* AVPs sent within an *AA-Answer*, for instance. The purpose of this example is to set a bandwidth limit of 512 kbits/s for download and 128 kbits/s for upload traffic. The way to achieve this is to use the meter action of the *QoS-Filter-Rule*, and to drop packets that go beyond the specified limit. Other packets can be marked with the default 000000 DSCP value, corresponding to the best effort.

   The rule for download traffic could be

   Meter out ip from 0.0.0.0 to 0.0.0.0 metering 512 000000 drop

   The rule for upload traffic could be

   Meter in ip from 0.0.0.0 to 0.0.0.0 metering 128 000000 drop

   However, using rules of this kind has some limits: it becomes impossible to enforce an additional policy to the traffic because the traffic successfully evaluated over this rule will stop the evaluation if there are remaining filters after. On the other hand, the rate limit will not work if other rules are previously matched before evaluating this one. Finally, these rules also override the DSCP field that could have been positioned by other equipment in the forwarding chain. This might be a drawback if the NAS session is used to convey traffic between two company premises that require prioritization. For this reason, bandwidth limitation is still tricky to achieve even with Diameter, and using the *Filter-Id* AVP may remain the most reliable way to achieve it, even though its usage is not recommended. It is also possible to define *Vendor-Specific* attributes for that purpose, but it is no more roaming friendly than the *Filter-Id* AVP.
2. The example below shows how to authorize only HTTP exchanges towards a single service platform, to pay additional services that could further change the filtering options and all kind of traffic for the end-user. The service platform is represented by the address 10.0.10.10, and the DNS address that must be reached by the end-user to get HTTP redirection is 10.0.11.11. Once the end-user is connected and opens its web browser, the first DNS request leads to the service platform whatever the URL. The user is constrained to exchange traffic to these two addresses as long as other authorization parameters are provisioned onto the access device. The HTTP traffic to the service platform is identified by use of TCP (protocol number = 6) and port 80.

   The rules to open outgoing traffic to the DNS server and the service platform are:

   Permit in 6 from 0.0.0.0 to 10.0.10.10 port 80
   Permit in ip from 0.0.0.0 to 10.0.11.11
   Permit out 6 from 10.0.10.10 to 0.0.0.0
   Permit out ip from 10.0.11.11 to 0.0.0.0

   Each packet that does not match any of these four rules will be dropped. To allow the end-user to reach any address among the Internet space, and to prevent him/her from sending packets to administrative IP addresses of network equipment using the 192.168.0.0/16 subnet, two solutions are possible:

- A RAR message with a *Re-Auth-Request-Type* AVP set to 'AUTHORIZE_ONLY' is sent by the authorization server (upon request of the service platform, most likely), including new filters in the payload. The new rules will have to override already existing filters. This way to proceed is, however, not specified in the Diameter base protocol or NASREQ.
- A RAR message is sent to the access device to trigger the creation of a new subsession that will require a new AAR/AAA message for which new filters will apply.

It is important to note that temporary access authorization to network resources is far better managed by using the Diameter credit control application (RFC 4006 [8]) that has been defined for this purpose. *NAS-Filter-Rule* and *QoS-Filter-Rule* are better suited to be applied when permanent traffic filtering is required. Bandwidth management, as in the example shown above, is usually performed using application extensions. ETSI/TISPAN and 3GPP use new AVPs that have been defined for that purpose. Even though it is encouraged to use existing messages and AVPs to realize a specific service, it is not always the most suitable way to achieve it. Diameter extensibility is a real strength that operators must use wisely: creating a specific application, or defining *Vendor-Specific* AVPs can be the best and simplest solution, and should be even easier to do with Diameter than with RADIUS.

4.5 Diameter Credit Control Application

The credit control application (RFC 4006 [8]) is a very interesting application for Diameter. However, its complexity and the study of all possibilities it offers would warrant a book dedicated solely to this topic. Implementing a credit control feature in access devices and AAA servers is the ultimate way to give a business face to provisioning in IP/MPLS networks: the purpose of this application is not another way to provide fancy user traffic management. Credit control 'philosophy' is based upon real-time service usage monitoring, performed by the access device, for every end-user that has opened a session related to a service that can be consumed with a control provided by the service provider through its AAA server. On the service provider side, the AAA server has to provide, for each session requiring it, a granted quota of credits that will be consumed at the access device. The access device is in charge of monitoring this resource consumption, to request additional credits for the end-user for service continuation, and to report unused credits that will be credited to the end-user account. This can be seen as an alternative accounting method that has been designed for real time, whereas usual accounting procedures as defined in the base protocol are more likely to be used for batch processing.

Credit control is defined to work in two different modes: credit allocation and direct debiting. When used in credit allocation mode, each credit control request will cause a reservation on end-user credit resources. The actual debiting of the reserved credit is realized when the service has been provided. When used in the direct debiting mode, the required credit is directly calculated and debited from the user's account. This mode is particularly suitable for one-time services, which are paid as a whole and are immediately due without refunding policy.

The credit control application is agnostic to the type of service for which the end-user is ready to pay, but both access device and credit server have to agree on which service the network has to provide. The server must have the service knowledge to be able to calculate the appropriate credit required by the access device to provide a decent service, and to make its calculations based upon a rating that has been defined previously by the service provider. The access device cannot perform this task, especially when the end-user is not connected to its home network: the credit calculation must be realized by a centralized unit with a consistent behavior over traversed networks. However, access device and server have to share a similar vision of the service to provide, otherwise granted resources can exceed the units corresponding to the paid service, or, even worse, the service may not be satisfactory because of credit burning faster than expected.

Nor is the AAA server aware of the complexity of service to be provided: the access device is in charge of providing the respective service, monitoring its usage and requesting credit for end-user access. The credit control application may rely upon other Diameter exchanges that might previously occur to make the service possible. For instance, credit control can make use of AA-R and AA-A messages defined in NASREQ to credit specific resources for the established session. But credit control can also be invoked after exchange of SIP messages. Credit control Diameter exchanges are materialized by two new messages (regrouped in one command code): *Credit-Control-Request* (CCR) and *Credit-Control-Answer* (CCA) with command code 272.

Multiple services can be provided simultaneously for a single end-user by means of subsessions. Each subsession will have its own credit control context (as well as its own subsession context): sharing the available credit is the challenge of the credit server, which has to split the resource with the appropriate granularity such that a service does not reserve too many resources to the detriment of other services. On the other hand, splitting the credits into too many small units might create an important credit management overhead because of frequent credit updates.

The resources provided by the credit server to the access device is not only money, but more generally one can talk about granted 'units'. Units can be of different nature: they can be money, but also remaining time, amount of data to be exchanged or any units that could be defined for a particular purpose. Credit control is not only session based – it can also be event driven, for instance upon receipt of an SIP message to signal emission of SMS: this service can be achieved in a one-shot direct operation, not requiring a session context to be maintained.

It is important to note that credit control messages and AVPs as defined in RFC 4006 [8] have been defined upon request of 3GPP for IMS. 3GPP as well as ETSI/TISPAN have themselves added multiple AVPs to be used in the context of credit control to provide conversational services within a fixed–mobile convergence framework.

4.6 Diameter in NGN/IMS Architecture for QoS Control

This section does not pretend to provide an exhaustive view of when and how the Diameter protocol is used within NGN architectures – and accurate description of what next-generation networks (NGNs) actually are would alone require more than one book. We will start here with a simplified and overall view of what a NGN is, and we will focus on the QoS control architecture as defined within ETSI/TISPAN.

4.6.1 What is an NGN?

Considering the extreme complexity of NGN architectures, one might question the purpose of all this. The goal that NGN intends to achieve is to make networks, wherever in the world, interoperable, for both data and voice (and actually all services), whatever the access method (mobile or fixed lines), relying upon IP networks. Therefore, deploying NGN architectures is the way to achieve fixed/mobile convergence.

Nowadays, NGN is often associated with IMS: this is true, but IMS is only a part of the NGN architecture. IMS was first defined within the 3GPP (Third-Generation Partnership Project), the standardization body in charge of defining the third generation of mobile architectures. The purpose of the IMS is to provide a multiservice, multiaccess architecture that is secure and reliable. IMS has been designed to be an enabler for providers to propose real-time and non-real-time services, where the user is mobile and wants to use multiple services simultaneously. The IMS blocks were firstly defined in Release 5 of the 3GPP architecture. The next release (R6) issued in 2004 enhanced the previous release of IMS by including dynamic policy control enhancements for end-to-end QoS. One of the key characteristics of the IMS architecture is to be adaptable whatever the transport network, as long as it works over IP. For this reason, IMS has been adopted by ETSI to standardize NGN architecture for fixed lines (whereas 3GPP focuses on mobile access), to define the TISPAN (Telecommunications and Internet converged Services and Protocols for Advanced Networking) architecture. 3GPP IMS blocks and interfaces are reused by ETSI/TISPAN, with adaptations to the context and thus some changes in the naming of interfaces.

NGN functional architecture as defined within ETSI/TISPAN (see ETSI ES 282001 [13]) defines two fundamental distinct layers: the transport layer and the service layer. The service layer consists of several components, IMS being one of them, designed to be in charge of multimedia conversational services. Other components have already been defined, such as the PSTN/ISDN emulation subsystem. The service layer model is open, and new components can be defined as required (such as the content broadcasting subsystem, for instance). Splitting NGN architecture into different layers helps in separating functions and mutualizing infrastructure to optimize investments. The goal is also to provide end-users with a seamless view of subscribed services whatever the access mode. An overview of NGN architecture elements is shown in Figure 4.7.

The service layer interacts with other layers and entities:

- user equipment;
- transport layer;
- other networks.

The transport layer itself comprises the three following components:

- *Transfer functions.* These take on all actions that can be applied to traffic coming from or to users' equipment and other networks. Several functions have been identified and described within ETSI/TISPAN, with specific roles. For instance, the layer-2 termination function (L2TF) is in charge of the layer-2 termination procedures at the edge of the access network. Another important function is the border gateway function (BGF) which is in charge, among other functionalities, of NAT/NAPT actions, packet marking, resource allocation and bandwidth reservation.
- *Network attachment subsystem (NASS).* This component takes care of functionalities to attach the terminal to the network. Among its functionalities we have: dynamic

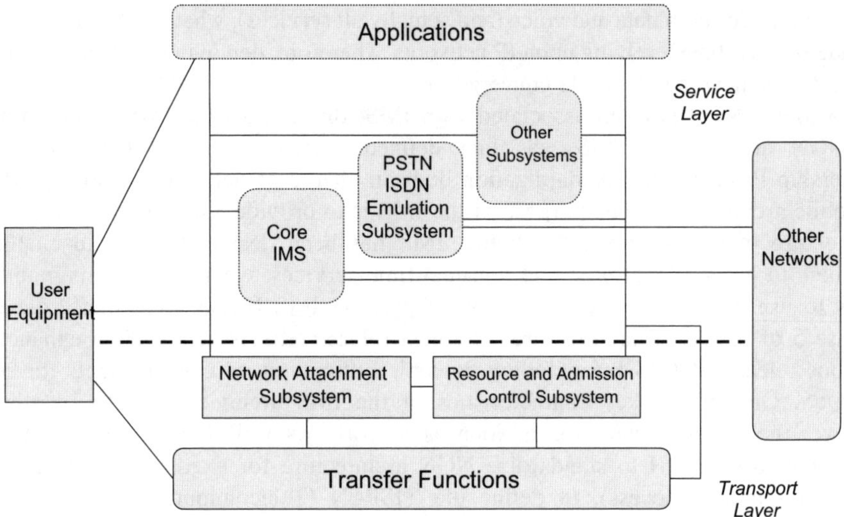

Figure 4.7 Overview of NGN Architecture in ETSI/TISPAN

provisioning of IP addresses (e.g. DHCP), authentication to access to the IP layer (distinct from the service attachment/authentication), location management.

- *Resource and admission control subsystem (RACS).* This component provides functionalities for admission control and gate control (such as NAT) by interworking with other components of the network to check network resource availability and capabilities. This entity will also check if the user's profile is compatible with the requested resources. The RACS is in charge of collecting resource requests from the service layer, and of pushing the corresponding policies to the transfer equipment.

All these components are themselves composed of smaller parts, each assuming a subset of standardized functionalities. The goal of standardization bodies, apart from describing these blocks and their usage, is to define interfaces to make them interact appropriately across entities, components and layers.

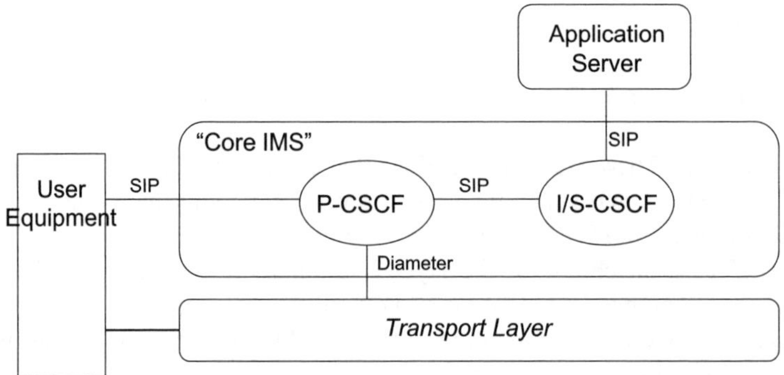

Figure 4.8 A simplified view of core IMS interactions with the user's equipment and the transport layer

Talking about protocols, the session initiation protocol (SIP) (RFC 3261 [14]) is the privileged protocol to be used within the 'core IMS' and by user equipment (see Figure 4.8). SIP is preferred for the service layer, to control interactions between the user equipment (UE) and the application server (AS). SIP messages cross several entities within the service layer, such as the call session control function (CSCF). These entities are themselves able to interact with other layers in the NGN architecture, but then SIP is not the protocol chosen to interface entities, and Diameter is often privileged.

We will now focus on the QoS control architecture by describing interactions between entities, implying Diameter.

4.6.2 QoS Control in ETSI/TISPAN Architecture

QoS control follows a very straightforward logic in NGN architecture (Figure 4.9).

The reader will note the appearance of a new component, the application function (AF). This is the generic terminology used to designate the entity located in the service layer that is in charge of interacting with the transport layer. In an IMS architecture, the AF would be the P-CSCF (Proxy-Call Session Control Function), the role of which (among others) is to extract from SIP messages the network capabilities required to provide the service, and to ensure that these capabilities are present.

Here is an overall description of the procedure that applies to enforce QoS policies in the network, in order to provide the service requested according to the network policy rules and capabilities:

1. *Network attachment.* During this phase, the user equipment attaches itself to the network. This includes authentication, authorization based on the user's profile (including provisioning of the IP address) and start of accounting related to network attachment. It is important to understand that this phase is not related to any service whatsoever. Attaching to the network and having an IP address to communicate is not considered as a service per se, as it does not require any interaction with the service layer. During the network attachment phase, the NASS can communicate to the RACS resource information that is specific to the entity that has just connected.

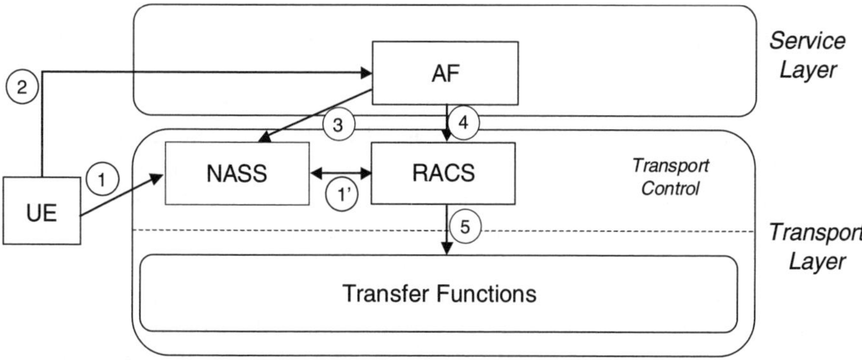

Figure 4.9 QoS control mechanisms in NGN architecture

2. *Service request.* This phase occurs whenever the user equipment sends requests to get a service from the network. The most common service might be thought to be placing a phone call, using VoIP with IMS. This step consists in exchanges of SIP messages with the service layer. In the case of IMS, the AF is the P-CSCF which is in charge of receiving SIP messages from the UE, forwarding it to other entities of the service layer and, depending on the result of the service request, interworking with the transport layer to reserve resources.
3. *RACS location.* This phase is optional and consists in the AF getting information to connect with the right equipment to connect to the RACS. For this, the AF just requires an IP address or a FQDN, and this information can be provided to the AF by the NASS.
4. *Resource request.* Once the service application has performed the first verifications to check if the end-user is entitled to use the requested service, the application function requests from the RACS the appropriate network resources in order to get the corresponding quality of service. The RACS will then check if the requested resource is available. At this stage it is interesting to note that the RACS can ask the NASS for specific information on the connected entity requesting the resource if this information is not available during phase 1.
5. *Policy provisioning.* Once the RACS has verified that the network resource is available to provide the service, it communicates with entities of the transfer plan within the transport layer to push the policies. Once these steps have been successfully completed, the UE is notified that the requested service is available, and it can start to use it.

The big picture provided above requires a little more description, in order to see the protocols and interfaces that are involved in these phases. We will now look more thoroughly into the RACS and the transfer functions.

Here is a brief description of the entities that appear in Figure 4.10:

- *A- RACF (Access-Resource and Admission Control Function).* This entity has two main roles in the RACS:
  - Firstly, it provides the admission control function by checking if the QoS resources are available when a request is received by the SPDF for an access line. For instance, one might consider that having simultaneous video calls depends on the characteristics of the end-user's access line (or subscription). In this case, the A-RACF will perform admission control by granting or rejecting the service, because of resource scarcity. In this role, the A-RACF is capable of installing some policies onto the RCEF and on the access node.
  - Secondly, the A-RACF has the role of network policy assembly (NPA). Multiple SPDF can query a single A-RACF for admission control, and the role of the latter is to ensure the consistency of all resource requests received, taking into consideration access network policies that have to be applied on each access line. By combining all requests received for an access line, the A-RACF ensures that a consistent quality of service will be maintained.
- *SPDF (Service-based Policy Decision Function).* This entity is the entry point for application functions, for all the RACS functions. The AF queries the RACS through the SPDF, and the SPDF applies a logical decision policy based on the network policies that are stored within. Depending on the request, on who sends the request and on the network policies, the SPDF may request admission control from the A-RACF. The SPDF will then build the policies to be enforced on the BGF, and will push these policies. Being

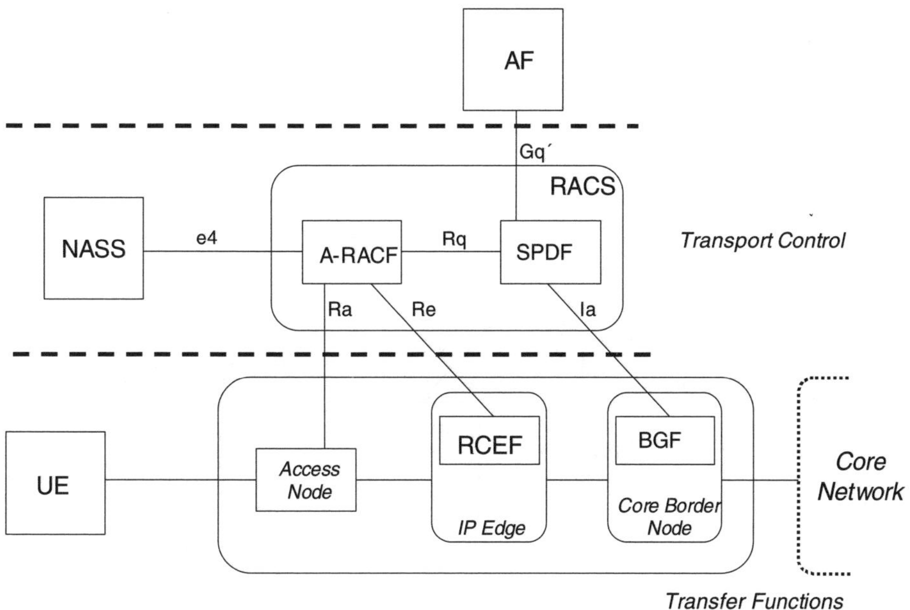

Figure 4.10 A view of the RACS and surrounding entities

responsible for the logic of the RACS, the SPDF is in charge of orchestrating event sequences in order to give a consistent report to the AF, and to manage appropriately the network resources.

- *RCEF (Resource Control Enforcement Function).* This entity placed in the transfer plan is in charge of enforcing policies that have been pushed by the A-RACF. The possible policies that the RCEF is able to apply are: opening or closing of gates in order to allow user flows to be forwarded or not; marking of IP packets according to the rules defined by the A-RACF; application of traffic bandwidth restriction for upstream and downstream traffic, to remain within the authorized boundaries.
- *BGF (Border Gateway Function).* This entity is part of the transfer plan through which the end-user traffic goes. The BGF is placed at the edge of the core network (it can be combined with the IP edge where the RCEF is placed), and is in charge of forwarding the user's IP packets and to apply flow policies on it. The BGF is capable of working on 'microflows', i.e. flows generated by applications (following the 5-tuple IP source address, IP destination address, IP source port, IP destination port, Protocol). The BGF is also the functional entity capable of handling NAT.

Going back to the example provided in Figure 4.9 and applying it with the details of Figure 4.10, we can now have a more thorough look at the event sequence that occurs within the RACS between all these elements and their corresponding entities in the service layer and those in charge of transfer functions (see Figure 4.11):

- *Step 1.* The AF sends a resource request to the RACS, and more specially to the SPDF since it is its sole point of contact. The SPDF computes the AF request, and decides whether or not to solicit the A-RACF for admission control.

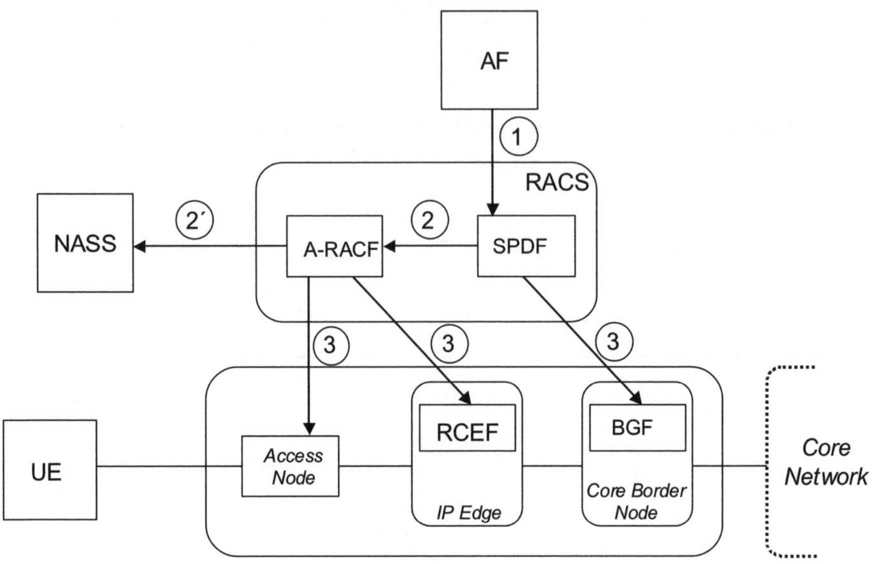

Figure 4.11 QoS control mechanisms with RACS internal/external exchanges

- *Step 2.* If required, the SPDF sends an admission control request to the A-RACF.
- *Step 2′.* If the A-RACF does not have access line information, it queries the NASS in order to get it.
- *Step 3.* A-RACF pushes appropriate policies to the RCEF and access node, as the SPDF pushes its policies to the BGF.

The decision of 'what policy to push where' basically depends on the type of policy to apply, and on the location where the resources have to be granted in order to provide the appropriate QoS for the service. This makes it possible to have multiple operators providing services at their own level: as seen above, A-RACF and SPDF can be part of different domains. Therefore, the network access operator will push appropriate policies to the equipment it is in charge of (IP edge nodes, access nodes), whereas the core network operator will push its own policies on its equipment (onto the BGF). Here is a summary of what policies can be pushed to RCEF and BGF:

- RCEF: opening/closing gates, packet marking, policing for downlink and uplink traffic;
- BGF: opening/closing gates, packet marking, policing for downlink and uplink traffic, per-flow resource allocation, NAT, hosted NAT traversal, usage metering.

To communicate between each other, interfaces have been defined between these entities, with a specific role, and a protocol is associated with each of them. Here is a summary of the RACS interface and the associated protocol:

- *e4 (between NASS and A-RACF).* This interface based on Diameter has been defined to communicate connectivity information that is collected by the NASS [in the connectivity location register function (CLF)], and to transmit it to the A-RACF for admission control.

- *Gq′ (between AF and SPDF).* This interface based on Diameter is used by the application function to send resource requests to the RACS. Diameter is particularly well adapted for this purpose, as it offers all the features required by entities that may not be in the same domain.
- *Rq (between SPDF and A-RACF).* This interface based on Diameter is used by the SPDF to send resource queries to the A-RACF. As for Gq', Diameter is especially adapted to the situation where SPDF and A-RACF are located in different domains.
- *Re (between A-RACF and RCEF).* This interface is used for the A-RACF to push policies onto the IP edge node at the RCEF point. The protocol to be used for this interface has not been chosen in Release 1 of ETSI/TISPAN.
- *Ra (between A-RACF and access node).* This interface is used for the A-RACF to push policies into the access node. The protocol to be used for this interface has not been chosen in Release 1 of ETSI/TISPAN.
- *Ia (between SPDF and BGF).* This interface is based on H.248.1 version 3 [15], also known as Megaco (RFC 3525 [16]), and is used by the SPDF to push QoS policies onto the BGF.

As seen above, three different interfaces of the RACS make use of the Diameter protocol. It is worth noting that, even though Diameter is used for requesting resources, it is not the protocol used to provision resources into the transfer plan. However, from a service layer point of view, Diameter is the protocol to be used to interface with the transport layer, through the Gq′ interface in ETSI/TISPAN architecture (defined in ETSI TS 183017 [17]) or through the equivalent Gq interface in 3GPP architecture (defined in 3GPP TS 29209 [18], from which Gq′ has been defined).

The open model adopted by Diameter made it easy for 3GPP and ETSI to use Diameter their own way, without needing to ask IETF to work on that, by defining a specific application or adapted AVPs. Instead, 3GPP defined a new *Vendor-Specific* application for the Gq interface, which has been inherited by ETSI/TISPAN to build Gq′. ETSI TS 183017 [17] defined some particularities on how Diameter should be used, without violating RFC 3588 [4], but by using the offered possibilities for usage customization. Going through all these attributes, specificities and their usage would be fastidious and is not relevant without having the complete picture of interactions: the information given here is intended to give an example of customization and appropriation work that is going around Diameter. Here is a list of some NGN specificities when Diameter is used for the Gq′ interface in NGN architectures:

- Whereas Diameter can be used over TCP or SCTP, Gq' mandates to use SCTP only.
- The SPDF is the Diameter server, and the P-CSCF (AF) is the client.
- AF and SPDF will advertize their support of Gq in capability exchange messages (CER/CEA) by using the application ID 16777222 (*Auth-Application-Id* AVP). However, this AVP must be placed in the grouped AVP *Vendor-Specific-Application-Id*, along the *Vendor-Id* set to 10415 (corresponding to 3GPP). Two *Supported-Vendor-Id* AVPs will be included to mention the support of ETSI (vendor identifier 13019) and 3GPP (vendor identifier 10415);
- The accounting messages offered in RFC 3588 [4] are not used.

- For the case of Gq' (ETSI TS 183017 [17]), nine new AVPs are defined, and other AVPs are imported from various specifications:
  - 19 AVPs from the Gq interface (3GPP TS 29209 [18]);
  - two AVPs from the e4 interface (ETSI ES 283034 [19]);
  - 2 AVPs from NASREQ (RFC 4005 [7]).

There is now doubt at this time that Diameter promises to be widely used, as long as NGN architectures as defined by 3GPP and ETSI/TISPAN are adopted by network operators.

References

[1] IETF RFC 2989: 'Criteria for Evaluating AAA Protocols for Network Access'.
[2] IETF RFC 3411 (STD0062): 'An Architecture for Describing Simple Network Management Protocol (SNMP) Management Frameworks'.
[3] IETF RFC 2748: 'The COPS (Common Open Policy Service) Protocol'.
[4] IETF RFC 3588: 'Diameter Base Protocol'.
[5] IETF RFC 3127: 'Authentication, Authorization, and Accounting: Protocol Evaluation'.
[6] IETF RFC 4004: 'Diameter Mobile IPv4 Application'.
[7] IETF RFC 4005: 'Diameter Network Access Server Application'.
[8] IETF RFC 4006: 'Diameter Credit Control Application'.
[9] IETF RFC 4072: 'Diameter Extensible Authentication Protocol (EAP) Application'.
[10] IETF RFC 4740: 'Diameter Session Initiation Protocol (SIP) Application'.
[11] IETF RFC 4282: 'The Network Access Identifier'.
[12] IETF RFC 2475: 'An Architecture for Differentiated Services'.
[13] ETSI ES 282 001: 'Telecommunications and Internet Converged Services and Protocols for Advanced Networking (TISPAN); NGN Functional Architecture Release 1'.
[14] IETF RFC 3261: 'SIP: Session Initiation Protocol'.
[15] ITU-T H.248-1: 'Gateway Control Protocol'.
[16] IETF RFC 3525: 'Gateway Control Protocol'.
[17] ETSI TS 183 017: 'Resource and Admission Control: DIAMETER Protocol for Session Based Policy Setup Information Exchange Between the Application Function (AF) and the Service Policy Decision Function (SPDF); Protocol Specification'.
[18] 3GPP TS 29.209: 'Policy Control over Gq Interface'.
[19] ETSI ES 283 034: 'TISPAN; Network Attachment Sub-System (NASS); e4 Interface Based on the DIAMETER protocol'.

5

The Common Open Policy Service (COPS) Protocol

5.1 A New Scheme for Policy-based Admission Control

The Internet has become the privileged network infrastructure for the deployment and the operation of a wide range of IP service offerings, ranging from basic Internet access to advanced IPTV broadcast services. The dramatic evolution of such services through the development of value-added capabilities such as quality of service features has given rise to the need to accommodate the delivery of such services with the relevant levels of reliability, availability, quality and security.

From this perspective, service providers should now be able to monitor and control the use of the network and service resources they operate, according to a set of policies that would be derived from criteria such as traffic/bandwidth requirements that would encourage the dynamic enforcement of admission control policies adapted to the needs of customers and/or services.

Such policies would therefore consist in applying a set of rules to determine whether or not access to a specific resource (whatever such a resource may be – a trunk, a router, an access link, an optical wavelength, etc.) should be granted.

Within this context, the members of what used to be the resource allocation protocol (RAP [1]) working group of the IETF have specified the common open policy service (COPS) protocol (RFC 2748 [2], RFC 3084 [3]), which is described in detail in this chapter.

The initial motivation that yielded the chartering of the *rap* working group was related to the use of the resource reservation protocol (RSVP) (RFC 2750 [4]) as a means to dynamically reserve network resources along the path towards a given destination, by soliciting the participating routers that would be in charge of proceeding with the reservation requests and maintaining the corresponding states during the lifetime of the reserved resources.

In a typical RSVP-based resource admission control scheme, RSVP routers are supposed to make their own decisions according to their local vision of resource availability. Within

Service Automation and Dynamic Provisioning Techniques in IP/MPLS Environments C. Jacquenet, G. Bourdon and M. Boucadair

the context of a policy-based admission control as promoted by the *rap* WG, RSVP routers would delegate the decision-making process to a central PDP. Delegating decisions to a centralized PDP is commonly referred to as the 'outsourcing' mode.

This outsourcing approach, which yielded the specification of the COPS protocol, has several advantages:

- The centralized PDP maintains a network-wide, global and systemic view of available and unavailable resources, unlike RSVP routers.
- Policy-based admission control schemes facilitate the support for pre-emption, i.e. the ability to dynamically remove a previously installed state so that a new admission control request might be accepted.
- Policy-based admission control schemes also ease the monitoring of policy states and resource usage because of the aforementioned global, network-wide view of in-use and unused resources.

The COPS protocol has been designed to reliably convey the information that is exchanged between the PEP capability embedded in the network devices and the PDP capability entitled to make the policy decisions that will be applied by the corresponding PEP-embedding devices. Details on the COPS protocol and its machinery come next.

5.2 A Client–Server Architecture

The COPS protocol relies upon a client–server model where the PEP sends requests to the PDP and the PDP returns decisions back to the PEP. The PDP also has the ability to send unsolicited messages to the PEP. The connection between the PEP and the PDP is established over the transmission control protocol (TCP) (RFC 793 [5]), hence using a reliable transport mode. Figure 5.1 depicts the relationship between the PEP and PDP.

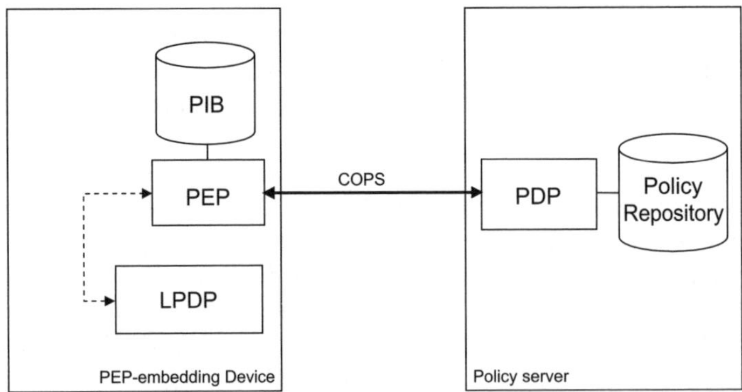

Figure 5.1 COPS client–server architecture

The PEP initiates the TCP connection with the PDP, and, once established, communication between the PEP and the PDP mostly consists of stateful requests and decision forwarding. The COPS protocol is stateful mostly because:

- Request and decision states are shared between the PEP and the PDP. This means that requests from the (client) PEP are installed and maintained by the PDP until they are explicitly deleted by the PEP. Likewise, decisions made by the (policy server) PDP can be generated asynchronously (by means of unsolicited messages) for any given installed request state.
- States that reflect various events (as request/decision pairs) may be interleaved. This means that the PDP may respond to new queries differently because of previously installed request/decision states that are related to the processing of the aforementioned queries and that may affect the decision-making process accordingly.

In a typical outsourcing fashion, the PEP outsources the decision-making process to the PDP. Although the PEP-embedding device can make local decisions with the LPDP, the final decision is made by the PDP (unless the latter is not accessible). There are three different types of outsourcing event that require a decision to be made by the PDP. Within the context of an RSVP-based policy admission control scheme, Figure 5.2 depicts the basic COPS chronology.

According to Figure 5.2:

- The RSVP incoming request is received by the PEP-enabled device ①, which causes the PEP to ask the PDP what the device should do with this incoming request (i.e. accept it or reject it).
- Once processed by the PDP, the decision made by the PDP is sent back to the PEP ②, and the PEP will then apply the corresponding decision by allocating the appropriate

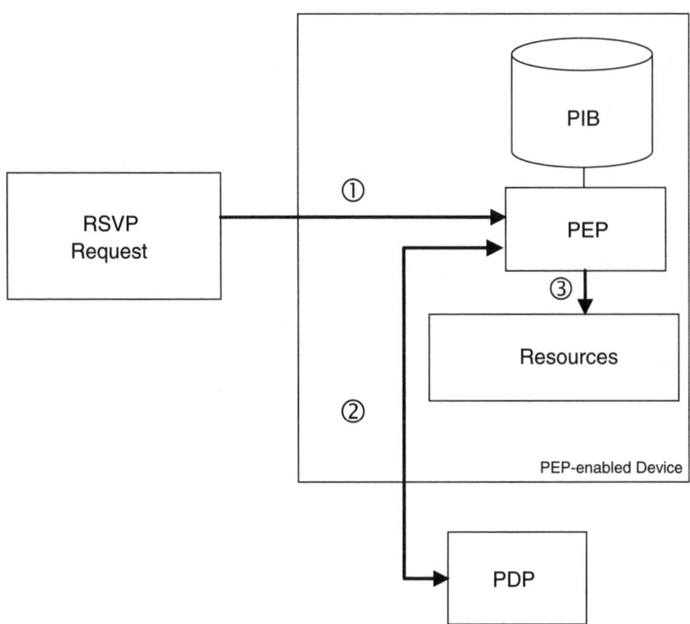

Figure 5.2 RSVP-based policy admission control scheme with COPS

resources ③, which will in turn allow the forwarding of an outgoing message as per the RSVP machinery.

5.3 The COPS Protocol

The COPS protocol has been designed so that COPS messages are self-identifying policy objects that contain policy elements. These policy elements can be defined as elementary pieces of information that depict the way policy rules need to be enforced. A single policy element may carry a user or an application identifier, while another policy element may carry user credentials.

By definition, policy elements are 'protocol agnostic', in the sense that they carry any kind of policy-formatted information that may be derived from the activation of QoS signaling protocols (such as RSVP), traffic engineering protocols (such as the path computation element communication protocol (PCEP) [6]) or routing protocols, such as the open shortest path first (OSPF) protocol (RFC 2328 [7]).

5.3.1 The COPS Header

COPS messages always begin with a common header, as depicted in Figure 5.3.

The corresponding fields of the COPS header message are defined as follows:

- The 4-bit encoded Version field provides information about the version of the COPS protocol, which is version 1 for the time being.
- The 4-bit encoded Flags field is meant to indicate the nature of the COPS message. Value 0x1 of the Flags field has been defined so far, to indicate that the message is solicited by another COPS message. This is the solicited message flag bit.
- The 8-bit encoded Op Code field denotes the nature of the COPS operation. Table 5.1 indicates the associated values.
- The 16-bit encoded Client Type field uniquely identifies the policy client, that is, the COPS client that relates to the enforcement of a specific policy. As a consequence, there will be as many client types as there are policies to enforce – quality of service, traffic engineering, security, addressing, routing, etc. Thus, the interpretation of all encapsulated objects is client type specific. Client types that set the most significant bit in the corresponding field are said to be 'enterprise specific' (these client types are identified within the 0x8000–0xFFFF range). For KA messages, the client type in the header is always set to '0', since the KA message is only used to check the COPS connection: this is not a per-client type session assessment operation.

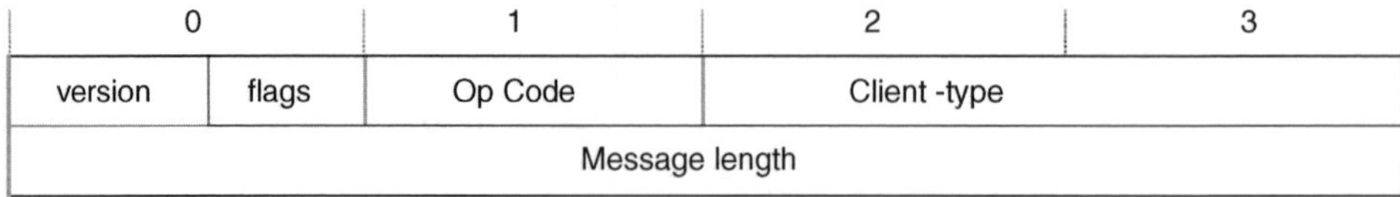

Figure 5.3 The COPS common header

Table 5.1 Op code values of COPS operations

Value	COPS operation	Abbreviated message
1	*Request*	REQ
2	*Decision*	DEC
3	*Report State*	RPT
4	*Delete Request State*	DRQ
5	*Synchronize State Request*	SSQ
6	*Client-Open*	OPN
7	*Client-Accept*	CAT
8	*Client-Close*	CC
9	*Keep-Alive*	KA
10	*Synchronization Complete*	SSC

- The last field of the header is the Message Length 32-bit encoded field, which indicates the size of the COPS message, expressed in bytes, taking into account the common header as well as the following policy elements. COPS messages need to be aligned on 4-byte intervals.

5.3.2 The COPS Message Objects

While the COPS header defines the nature of the COPS message [decision (DEC), request (REQ), etc.], the message itself is composed of several policy elements that provide information about the decisions that have been made by the PDP, among other information. The presence of a COPS object depends on the nature of the COPS message. The header for all COPS objects is depicted in Figure 5.4.

The corresponding fields of the COPS objects are defined as follows:

- The 16-bit encoded Length field indicates the number of bytes (header included) that compose the object. If the length does not fall into a 32-bit word boundary, padding will have to be added to the end of the object so that it is aligned with the next 32-bit word boundary before the message can be sent. On the receiving side, a subsequent object boundary can be located by rounding up the previous stated object length to the next 32-bit word boundary.
- The 8-bit encoded C-Num field designates the class of information contained in the object, while the 8-bit encoded C-Type field designates the subtype or version of the information contained in the object. As an example, the identification of IPv4 and IPv6 interfaces will

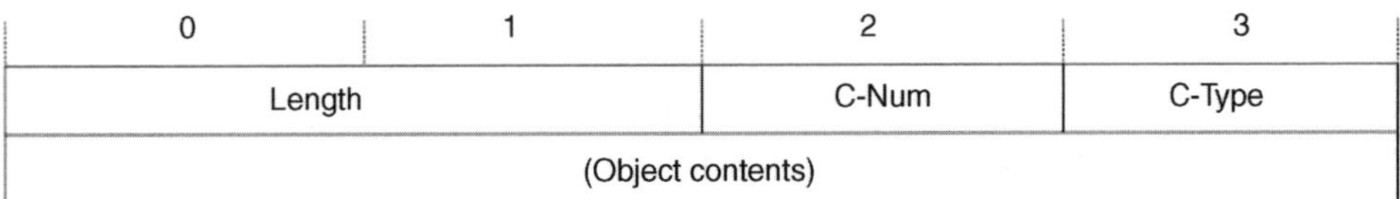

Figure 5.4 Format of the COPS object

rely upon (1) C-Num fields valued at '3' [or '4', depending on the nature of the interface – incoming (3) or outgoing (4)], and (2) a C-Type field valued at '1' (for IPv4) and '2' (for IPv6).

Table 5.2 summarizes the possible values of C-Num that have been specified so far.

The content of a COPS-specific object is of variable length depending on both the C-Num and C-Type fields. Messages exchanged between the PEP and the PDP are composed of such objects, and the following section provides details about these messages.

Table 5.2 C-Num values of COPS-specific objects

C-Num	Name	Explanation
1	Handle	The handle object encapsulates a unique value that identifies an installed state. This identification is used by most COPS operations.
2	Context	Specifies the type of event(s) that triggered the query. Required for request messages.
3	In Interface	This object is used to identify the incoming interface on which a particular request applies and the address where the received message originated.
4	Out Interface	This object is used to identify the outgoing interface to which a specific request applies and the address for where the forwarded message is to be sent.
5	Reason Code	This object specifies the reason why the request state was deleted. It appears in the delete request (DRQ) message.
6	Decision	Decision made by the PDP. Appears in replies.
7	LPDP Decision	Decision made by the PEP's local policy decision point (LPDP). May appear in requests.
8	Error	This is used to identify a particular COPS protocol error.
9	Client Specific Info	This contains client-type specific information, e.g. the contents of RSVP pathmessage, if the client is RSVP.
10	Keep-Alive Timer	Time given in seconds.
11	PEP Identification	The PEP identification object is used to identify the PEP client to the remote PDP.
12	Report Type	The type of report on the request state associated with a handle.
13	PDP Redirect Address	A PDP when closing PEP session for a particular client-type may optionally use this object to redirect the PEP to the specified PDP server address and TCP port number.
14	Last PDP Address	When a PEP sends a Client-Open message for a particular client-type the PEP SHOULD specify the last PDP it has successfully opened (meaning it received a Client-Accept) since the PEP last rebooted.
15	Accounting Timer	Optional timer value used to determine the minimum interval between periodic accounting-type reports.
16	Message Integrity	The integrity object includes a sequence number and a message digest useful for authenticating and validating the integrity of a COPS message.

5.4 COPS Messages

5.4.1 Client-Open (OPN)

The OPN message is sent by the PEP to the PDP. This message is used by the PEP to notify the PDP about the client types the PEP supports, and may also provide information about the PDP to which the PEP was last connected. The structure of the OPN message is depicted in Figure 5.5.

The PEPID is the identifier of the PEP, which is unique within an administrative domain. This identifier is encoded as an ASCII string and can be an IP address or a domain name system (DNS) name. The last PDP address denotes the last PDP for which the PEP is still caching decisions. The integrity object is used if security needs to be enforced, to make sure the PEP is entitled to establish the connection with the PDP. Note that the PEP can also send additional client-specific information by means of the client SI field.

```
<Client-Open>       ::= <Common Header>
                        <PEPID>
                        [<ClientSI>]
                        [<LastPDPAddr>]
                        [<Integrity>]
```

Figure 5.5 The *Client-Open* message

5.4.2 Client-Accept (CAT)

The CAT message is sent by the PDP to the PEP as a response to an OPN message if the PDP accepts the *Client-Open* request. The PDP will return a *Keep-Alive* (KA) timer value, which indicates the maximum time interval between KA messages. Optionally, the PDP can send information about the minimum allowed time interval between accounting report messages sent by the PEP (ACCT timer). Figure 5.6 depicts the format of the CAT message.

```
<Client-Accept>  ::= <Common Header>
                     <KA Timer>
                     [<ACCT Timer>]
                     [<Integrity>]
```

Figure 5.6 The *Client-Accept* message

5.4.3 Request (REQ)

The PEP establishes a *Request State Client Handle* for which the PDP may maintain state. The handle corresponds to the identification means that is used by the PDP to communicate with the PEP. Any change that is local to the PEP will yield a notification sent by the PEP to the PDP. Figure 5.7 depicts the format of the REQ message.

```
<Request Message> ::=  <Common Header>
                        <Client Handle>
                        <Context>
                        [<IN-Int>]
                        [<OUT-Int>]
                        [<ClientSI(s)>]
                        [<LPDPDecision(s)>]
                        [<Integrity>]

<ClientSI(s)> ::= <ClientSI> | <ClientSI(s)> <ClientSI>

<LPDPDecision(s)> ::= <LPDPDecision> |
                      <LPDPDecision(s)> <LPDPDecision>

<LPDPDecision> ::= [<Context>]
                    <LPDPDecision: Flags>
                    [<LPDPDecision: Stateless Data>]
                    [<LPDPDecision: Replacement Data>]
                    [<LPDPDecision: ClientSI Data>]
                    [<LPDPDecision: Named Data>]
```

Figure 5.7 The *Request* message

The context object describes the context where all other objects must be interpreted. The ClientSI information depicts the client-type-specific information, e.g. the contents of a RSVP *Request* message. LPDPDecision refers to the decisions made by the embedded LPDP, and which must be verified, completed or possibly overwritten by the PDP. Five different kinds of LPDP decision have been identified:

- The *Flags* decision denotes the normal request/decision process.
- The *Stateless Data* decision refers to a local decision which does not affect the state of the pending request.
- The *Replacement Data* decision replaces the existing data in a signaled message.
- The *ClientSI Data* decision is used to introduce additional decision types.
- The *Named Data* decision contains configuration information.

5.4.4 Decision (DEC)

Upon receipt of an REQ message, the PDP sends a DEC message back to the PEP. If the PDP has not responded to the REQ message within a given time limit, the PEP will remove the corresponding handle (see previous section) and create a new one – then try to get a response from the PDP again. Figure 5.8 depicts the format of the DEC message.

The PDP can send several decisions back to the PEP, as well as error messages if anything has gone wrong.

```
<Decision Message> ::= <Common Header>
                        <Client Handle>
                        <Decision(s)> | <Error>
                        [<Integrity>]

      <Decision(s)> ::= <Decision> | <Decision(s)> <Decision>

      <Decision> ::= <Context>
                     <Decision: Flags>
                     [<Decision: Stateless Data>]
                     [<Decision: Replacement Data>]
                     [<Decision: ClientSI Data>]
                     [<Decision: Named Data>]
```

Figure 5.8 The *Decision* message

5.4.5 Other COPS Messages

The previous sections elaborated on the most significant messages used by the COPS machinery. This section provides a brief description of the other messages that can be used.

5.4.5.1 *Report State* (RPT)

RPT messages are sent by the PEP to the PDP, and are used under the following conditions:

- To report to the PDP about the result of an action performed by the PEP (success, failure), as per a decision made by the PDP.
- To send unsolicited information to the PDP about accounting or state monitoring. Such RPT messages may contain client-specific information.

5.4.5.2 *Delete State Request* (DRQ)

DRQ messages are sent by the PEP to the PDP to indicate that the handle specified in the message is no longer applicable and must therefore be deleted. If the action is not performed by the PEP, the handle will be maintained by the PDP until the COPS connection is closed or the TCP connection is terminated.

DRQ messages contain a report object providing an indication of the reason for deletion, which is client-specific by nature.

5.4.5.3 *Synchronize State Request* (SSQ)

SSQ messages are sent by the PDP to the PEP, so as to make the PEP send the states related to the client type specified in the COPS header message to the PDP. This is a COPS synchronization mechanism that relies upon the resending of the requests that relate to the corresponding handle.

Whenever an unrecognized handle is specified in the SSQ message, the PEP must indicate to the PDP that this handle should be deleted. Once synchronization is completed, a *Synchronization State Complete* (SSC) message is sent by the PEP to the PDP.

5.4.5.4 *Synchronization State Complete* (SSC)

As per the previous section, SSC messages are sent by the PEP to the PDP to indicate that synchronization related to a specific handle (or all handles) is completed for the specified client type depicted in the COPS message header.

5.4.5.5 *Client-Close* (CC)

CC messages are sent either by the PEP or by the PDP to indicate to each other that a client type is no longer supported, the error object contained in the CC message providing an explanation for the closure. If the CC message is sent by the PDP, it may contain the identifier of another PDP that may support the client type specified in the message.

5.4.5.6 *Keep-Alive* (KA)

KA messages are sent either by the PEP or the PDP on a random basis (fractions of the minimum KA timer as described in CAT messages) to check the state of the connection. Upon receipt of a KA message by a PDP, the PDP must send back another KA message.

The Client Type field of the COPS common header associated with KA messages is set to ‘0’, since KA messages cover all the sessions that have been opened between the PEP and the PDP. Note that, if connection with the PDP is lost, the PEP is supposed to reach another PDP.

5.5 Summary of COPS Operations

Table 5.3 reflects the possible combinations of COPS messages with COPS operations, where ‘M’ denotes that the information must be provided in the message, while ‘O’ denotes that the information may be provided in the message.

Table 5.3 Matrix of COPS messages and operations

	REQ	DEC	RPT	DRQ	SSQ	OPN	CAT	CC	KA	SSC
Handle	M	M	M	M	O					O
Context	M									
In-interface	O									
Out-interface	O									
Reason				M						
Decision		M								
LPDP decision	O									
Error		M						M		
ClientSI	O		O			O				
KA timer							M			
PEP ID						M				
Report type			M							
PDP redirect address								O		
Last PDP address						O				
ACCT timer							O			
Integrity	O	O	O	O	O	O	O	O	O	O

5.6 Use of COPS in Outsourcing Mode

RFC 2748 [2] introduces two models for COPS usage: the outsourcing model and the configuration model. The latter has been further described in RFC 3084 [3] and will be the subject of Section 5.7. In the outsourcing model, the PEP delegates the decision-making process to the PDP, and thus most of the COPS operations are initiated by the PEP.

Within the context of the outsourcing mode, the following exchanges occur:

- The PEP sends an REQ message to the PDP with a specific handle, and the PDP reports back its decision for this handle by means of the relevant DEC message. The PEP then sends an RPT message back to the PDP that contains the result (success/failure) of the application of the decision.
- The PEP may request the PDP to delete a specific handle.
- The PDP sends an SSQ message to the PEP for synchronization purposes.

5.7 Use of COPS in Provisioning Mode

RFC-3084 [3] details the provisioning mode of COPS, which is referred to as 'COPS-PR'. The basic motivation for this mode is to encourage the use of unsolicited messages sent by the PDP to the PEP, to reflect the ability to initiate specific actions. As an example, the ability to proactively intervene on the network has become of utmost importance for service providers, as the deployment and the operation of QoS-demanding value-added IP service offerings need to accommodate: (1) incoming customers' QoS requirements, which may be dynamically negotiated with the service provider by means of service level specifications (SLS) [8]; (2) network planning policies, which aim to reflect the evolution of the overall usage of network resources.

Thus, any event (such as the number of SLS instances to be processed over a short period of time) may affect the way policies are defined by the PDP and then enforced by the PEP embedded in the network devices. Within this context, COPS REQ messages sent by the PEP refer rather to the capabilities that can be configured on the device that embeds the PEP than to incoming requests that may yield the delegation of the decision-making process to the PDP.

Figure 5.9 summarizes the COPS provisioning model.

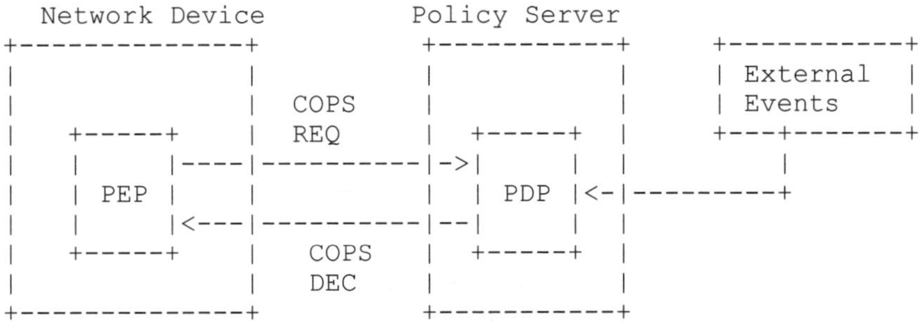

Figure 5.9 The COPS provisioning model

5.7.1 On the Impact of Provisioning Mode on COPS Operations

COPS-PR introduces additional operations. These new objects are conveyed within the COPS named client-specific information object and named decision data object. The format of these new objects is defined in Figure 5.10.

Length	S-Num	S-Type
Content of the Object (32-bit unsigned integer)		

Figure 5.10 Format of new COPS-PR objects

According to the Figure 5.10:

- The S-Num and S-Type fields are similar to the C-Num and C-Type fields that are used in the base COPS objects. The 8-bit encoded S-Num field identifies the purpose of the object, while the 8-bit encoded S-Type field provides information about the encoding (namely basic encoding rules (BERs), which is the encoding technique used for all of the COPS objects so far).
- The 16-bit encoded Length field denotes the size of the object (header included), expressed in bytes. Note that current technologies encourage the use of XML string-based encodings [9].
- The content of the object is indexed by the S-Num value, as per Table 5.4.

Table 5.4 S-Num values and new COPS-PR objects

S-Num	Object
1	Provisioning instance identifier (PRID). Uniquely identifies a particular PIB instance.
2	Provisioning instance identifier prefix. Provides a prefix that can be used when all PIB instances whose PRID begins with this prefix are associated with the same operation (namely withdrawal of data).
3	Encoded provisioning instance data (EPD). Denotes the encoded value of the PIB instance to which the operation applies.
4	Global provisioning error. This depicts general errors, and the format is similar to the COPS error object: a 16-bit encoded error code provides indication of the nature of the error, while a subsequent 16-bit encoded error subcode details the nature of the error.
5	Provisioning class provisioning error. This denotes errors related to a specific provisioning class (PRC) of the PIB. This assumes the presence of an error PRID OBJECT.
6	Error PRID. Uniquely identifies the PIB instance that is affected by a particular error.

The objects listed in Table 5.4 can be encapsulated in (1) the named ClientSI field of REQ messages, (2) the named ClientSI field of RPT messages or (3) the named Decision Data field of DEC messages.

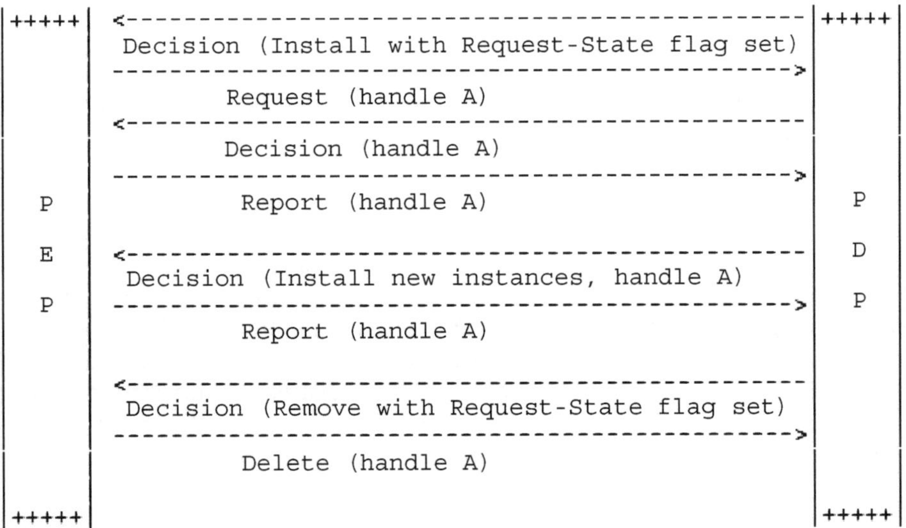

Figure 5.11 COPS-PR exchanges between a PEP and a PDP

5.7.2 On the Impact of Provisioning Mode on PEP–PDP Exchanges

COPS-PR messages are motivated by the (configuration) data stored and maintained in the PIB. Within the context of the provisioning model:

- The PDP can force the PEP to issue a new REQ message.
- The PDP can force the PEP to issue a new DRQ message.
- The PDP can send unsolicited DEC messages to the PEP that aim at installing additional PIB instances related to an existing request state.

Such new exchanges are summarized in Figure 5.11.

Upon receipt of the first REQ message, the PDP will provision the PEP back with the corresponding handle-specific policies, as stored in the PIB. The PDP can later update or remove some PIB instances that have been installed by the PEP as per the corresponding DEC message, or send another DEC message so that the PEP can install additional PIB instances. Such operations can be triggered by external events, as already discussed in Section 5.7.

COPS-PR makes the PIB a key component of the dynamic policy enforcement scheme: the PRC classes and PRI instances that are organized according to a tree structure (as introduced in Chapter 2) within the PIB therefore represent the core of the information that accurately depicts any given policy.

This implies that multiple yet independent PIB instances might be installed by the PEP, hence representing the various policy-specific configuration data that will be derived into (technology-specific) configuration information used by the network device to dynamically enforce any given set of policy rules.

5.8 Security of COPS Messages

OPN and CAT messages are used for security negotiation purposes if configured so. If the PEP is configured to use the security capabilities of COPS, the very first OPN message will include an integrity object that contains a 32-bit encoded Sequence Number field. This PEP-valued field will contain the initial sequence number, which the PEP expects the PDP to increment when the communication continues after the initial OPN/CAT message exchange, hence facilitating the prevention of replay attacks.

The integrity object contains two other fields – the 32-bit encoded KeyID identifier and the 96-bit encoded Keyed Message Digest. The KeyID field is used to identify the shared key that will be used by both the PEP and the PDP, as well as the cryptographic algorithm to be used by both entities, such as HMAC (Keyed-Hashing for Message Authentication) (RFC 2104 [10]). The digest is computed over every object and field of the message, with the notable exception of the aforementioned Keyed Message Digest field. This is the reason why the integrity object must always be the very last object contained in the COPS message.

References

[1] http://www.ieft.org/html.charters/OLD/rap-charter.html

[2] Boyle, J., Cohen, R., Durham, D., Herzog, S., Raja R. and Sastry A., 'The COPS (Common Open Policy Service) Protocol', RFC 2748, Proposed Standard, January 2000.

[3] Ho Chan, K., Durham, D., Gai, S., Herzog, S., McLoghrie, K., Reichmeyer, F., Seligson, J., Smith, A. and Yavatkar, R., 'COPS Usage for Policy Provisioning (COPS-PR)', RFC 3084, March 2001.

[4] Herzog, S., 'RSVP Extensions for Policy Control', RFC 2750, January 2000.

[5] Postel, J., "Transmission Control Protocol", RFC 793, September 1981.

[6] Vasseur, J.-P., Le Roux, J.-L. *et al.*, 'Path Computation Element (PCE) Communication Protocol (PCEP)', draft-ietf-pce-pcep-08.txt, Work in Progress, July 2007.

[7] Moy, J., 'OSPF Version 2', RFC 2328, April 1998.

[8] Goderis, D. *et al.*, 'Attributes of a Service Level Specification (SLS) Template', draft-tequila-sls-03.txt, Work in Progress, October 2003.

[9] World Wide Web Consortium (W3C), 'Extensible Markup Language (XML)',W3C Recommendation, February 1998, http://www.w3.org/TR/1998/REC-xml-19980210.

[10] Krawczyk, H., Bellare, M. and Canetti, R., 'HMAC: Keyed Hashing for Message Authentication', RFC 2104, February 1997.

6

The NETCONF Protocol

6.1 NETCONF at a Glance

6.1.1 Introduction

NETCONF (NETwork CONFiguration) protocol is a network management protocol through which not only individual network devices but also networks can be managed. By networks we mean a set of devices that are involved in the delivery of a given IP service offering. This service offering can be an IP connectivity service (aka layer-3 services including configuration of IP routing protocols or IP/MPLS VPNs) or high-level service [such as voice-over IP (VoIP)]. NETCONF is a building block that can be implemented within the context of an automated configuration system. NETCONF can be used jointly with other techniques, protocols and systems in order to offer a fully automated configuration solution. The NETCONF protocol specifications do not explicitly document how this service automation should be implemented, since it is up to protocol developers to design an overall automated system. In this overall system, NETCONF can be used to implement the configuration enforcement and reporting interface.

NETCONF protocol is structured functionally into several layers. Each layer is responsible for a well-defined functional part of the protocol operations. Moreover, NETCONF uses a remote procedure call (RPC) paradigm to exchange protocol messages between the *management application* (also called the NETCONF client) and the *managed device* (also called the NETCONF server). To avoid misusage of its operations, NETCONF protocol distinguishes *configuration* data from *state* data and implements methods dedicated to treat and manipulate each of these data types.

Communication between NETCONF clients and servers is performed through a request/response scheme, where all exchanged messages are encoded in XML (eXtended Markup Language). When enabling NETCONF between two peers, they can discover their capabilities and thereby adapt their respective behavior in order to take advantage of supported features.

Service Automation and Dynamic Provisioning Techniques in IP/MPLS Environments C. Jacquenet, G. Bourdon and M. Boucadair

Furthermore, NETCONF protocol is built above an underlying transport protocol which is responsible for establishment of secure and connection-oriented sessions. Security and reliability are managed by the underlying transport protocol and are not part of the NETCONF protocol itself. Even if NETCONF is transport independent, all NETCONF implementations must support at least secure shell (SSH) as a mandatory transport protocol. NETCONF can also be (optionally) activated over blocks extensible exchange protocol (BEEP) or over simple object access protocol (SOAP) since the IETF NETCONF working group has specified mapping with the aforementioned protocols. Other transport protocols can also be envisaged to convey NETCONF messages if and only if these protocols meet a set of transport requirements, mainly security and reliability.

One of the important features of the NETCONF protocol is its openness. Indeed, the base NETCONF protocol supports only a limited set of operations and additional operations and functionalities can be defined, specified and implemented to complement the core NETCONF functionalities. Associated NETCONF capabilities should be advertized to remote NETCONF peers so as to be used during a given NETCONF session. A template to define these new capabilities is provided in NETCONF [1] where the reader may find more information about this template.

This section aims to provide the reader with some preliminary information regarding the essence of NETCONF initiative, motivation and context, before getting into NETCONF core specification (i.e. supported functions, capabilities, etc.).

6.1.2 Motivations for Introducing NETCONF

Service providers, network providers, protocol designers and developers have gained experience in implementing, deploying and manipulating a large set of protocols and associated information required to manage networking infrastructure and services. Numerous data models have also been defined for network management purposes. Thus, several protocols have been standardized, such as the simple network management protocol (SNMP) (RFC 3410 [2]), the common open policy service (COPS) (RFC 2748 [3]) and COPS-PR (RFC 3084 [4]). Multiple data models have been defined and used by operators such as the core information model (CIM) [5], the directory-enabled network (DEN), SMI structure of management information (SMI) (RFC 2578 [6]), structure of policy provisioning information (SPPI) (RFC 3159 [7]), the management information base (MIB), and policy information management (PIB).

In spite of this relatively important amount of invested standardization effort within distinct standardization fora, some operators [both service providers (SPs) and network providers (NPs)] and standardization bodies address a negative report about the capacity of existing tools to deal with the operator's requirements related to network management and configuration operations. In order to understand these requirements, the Internet Architecture Board (IAB) held a dedicated workshop about network management in June 2002. This workshop was a continuation of the dialog initiated between some service providers, network providers and protocol designers. One of the major purposes of this workshop was to get feedback from the providers' community, and to drive a set of pertinent guidelines for future work in the network management field that would hopefully aim at addressing these providers' concerns. RFC 3535 [8] reflects the major discussions and conclusions of that workshop. This RFC encloses a set of requirements that should be satisfied by management protocols, such as:

- easiness of the management protocol (deployability, usability, etc.);
- implementation of a clear distinction between configuration data and state data;
- enabling operators to concentrate on the configuration of the network as a whole rather than individual devices;
- minimization of the impact caused by configuration changes on the network.

This RFC also lists a set of recommendations agreed during that workshop. Thus: 'The workshop recommends, with strong consensus from both protocol developers and operators, that the IETF focus resources on the standardization of configuration management mechanisms' (RFC 3535 [8]). Just after this workshop, The NETCONF initiative was launched within IETF.

6.1.3 NETCONF, an IETF Initiative

NETCONF protocol has been specified, designed and promoted within the NETCONF working group, a member of the IETF Operations and Management Area. This work was launched mid-2003 (the first NETONCF WG meeting was held at the 57th IETF meeting in Vienna). The working group has adopted as a starting document a proposal, edited by R. Enns from Juniper, suggesting the usage of XML as a means to encode configuration-related operations and associated data models. That proposal was called XMLCONF (XML CONFiguration). This document has been updated during the lifetime of the NETCONF WG and has now become the base NETCONF specification. Note that, at the time of launching this initiative, the RAP (resource allocation protocol) working group was active and the link between the two initiatives was ambiguous.

6.1.4 Missions of the IETF NETCONF Working Group

The NETCONF working group has been chartered by the IETF, more precisely inside the IETF Operations and Management Area, in order "to produce a protocol suitable for network configuration, with the following characteristics:

- Provides retrieval mechanisms which can differentiate between configuration data and non-configuration data.
- Is extensible enough that vendors will provide access to all configuration data on the device using a single protocol.
- Has a programmatic interface (avoids screen scraping and formatting-related changes between releases).
- Uses a textual data representation that can be easily manipulated using non-specialized text manipulation tools.

- Supports integration with existing user authentication methods.
- Supports integration with existing configuration database systems.
- Supports network wide configuration transactions (with features such as locking and rollback capability).
- Is as transport-independent as possible.
- Specify the protocol syntax and semantics of a notification message.
- Specify or select a notification content information model.
- Specify a mechanism for controlling the delivery of notifications during a session.
- Specify a mechanism for selectively receiving a configurable subset of all possible notification types."[1]

In order to meet its objective, the NETCONF working group has adopted as a starting document the XMLCONF proposal. This document has been enhanced during the lifetime of the working group. In the meantime, the working group has organized several interim meetings to speed up the progress on other critical issues such as the selection of transport protocol or the decision as to whether or not notification mechanisms should be part of the base specification document. These discussions have led to the delivery of an open protocol specification with a limited set of functionalities and methods. The protocol is simple and relies on a basic exchange scheme between involved entities.

6.1.5 NETCONF-related Literature[2]

This section provides a non-exhaustive list of recommended documents for readers who want to deeply understand NETCONF protocol and associated protocols, mainly SSH, SOAP and BEEP.

- NETCONF base documents:
  - NETCONF configuration protocol (RFC 4741 [1]);
  - Using the NETCONF configuration protocol over secure shell (RFC 4742 [9]);
  - Using the NETCONF protocol over blocks extensible exchange protocol (RFC 4744 [10]);
  - Using the network configuration protocol over the simple object access protocol (RFC 4743 [11]);
  - NETCONF event notifications (*draft-ietf-netconf-notification*);
  - Reporting schema for NETCONF protocol (*draft-adwankar-netconf-reporting*);
  - Data types for NETCONF data models (*draft-romascanu-netconf-datatypes*);
  - NETCONF architecture model (*draft-atarashi-netconfmodel-architecture*);
  - Requirements for efficient and automated configuration management (*draft-boucadair-netconf-req*).

[1]Quoted from the NETCONF charter, available at URL: http://www.ietf.org/html.charters/netconf-charter.html

[2]Several NETCONF implementations have been released, such as *YencaP* which is a NETCONF agent for Linux implemented in Python, distributed under GPL as part of the EnSuite (Extended NETCONF Suite) collection. Available at URL: http://madynes.loria.fr/ensuite

- SSH-related RFCs:
  - The secure shell protocol assigned mumbers (RFC 4250 [12]);
  - The secure shell protocol architecture (RFC 4251 [13]);
  - The secure shell authentication protocol (RFC 4252 [14]);
  - The secure shell transport layer protocol (RFC 4253 [15]);
  - The secure shell connection protocol (RFC 4254 [16]);
  - Using DNS to securely publish secure shell key fingerprints (RFC 4255 [17]);
  - Generic message exchange authentication for the secure shell protocol (RFC 4256 [18]);
  - The secure shell transport layer encryption modes (RFC 4344 [19]);
  - Secure shell session channel break extension (RFC 4335 [20]).
- BEEP-related RFCs:
  - BEEP core protocol definition (RFC 3080 [21]);
  - BEEP core protocol mapping onto TCP (RFC 3081 [22]);
  - On the design of application protocols (RFC 3117 [23]);
  - Reliable delivery for syslog (RFC 3195 [24]);
  - A definition to perform XML-RPC (RFC 3349 [25]);
  - An application layer proxy for BEEP peers (RFC 3620 [26]);
  - A protocol description to provide access to user security credentials (RFC 3767 [27]);
  - Using the Internet registry information service over the blocks extensible exchange protocol (RFC 3983 [28]);
  - Using SOAP on top of BEEP (RFC 4227 [29]).
- List of SOAP-related recommendations:
  - SOAP Version 1.2 Part 0: Primer;
  - SOAP Version 1.2 Part 1: Messaging framework;
  - SOAP Version 1.2 Part 2: Adjuncts;
  - SOAP Version 1.2 Specification assertions and test collection;
  - XML-binary optimized packaging;
  - SOAP message transmission optimization mechanism;
  - Resource representation SOAP header block.

6.1.6 What is In? What is Out?

The goal of the NETCONF section is not to describe XML, nor SSH, nor BEEP, nor SOAP, but only to describe the NETCONF protocol. Readers who want more detailed information about the aforementioned technologies are advised to refer to RFC 4254 [16], RFC 4253 [15], RFC 4252 [14], Gudgin *et al.* [30], RFC 3080 [21], RFC 3081 [22] and the references provided in Section 6.1.5.

This chapter is not a specification document; protocol implementers are advised to refer to base specification documents.

6.2 NETCONF Protocol Overview

This section describes an overview of NETCONF operations. Therefore, details about NETCONF messages, transport protocol and communication channels are elaborated in this section.

6.2.1 Some Words about XML

This book assumes that the reader is familiar with XML language and terminology. Nevertheless, we provide a brief description of some useful notions in order to ease the understanding of NETCONF protocol. For more information, it is recommended that the reader refer to the XML literature [31].

XML Document

XML (Extensible Markup Language) describes a class of data objects called XML documents and describes the behavior of applications that process them. An XML document contains several XML elements, which define the structure of the XML document. To illustrate this basic notion, Table 6.1 provides an example of an XML document that encloses **<RIB>**, **<Routes>** and **<Route>** elements.

XML Tag

An XML tag denotes what is enclosed between '<' and '>' in an XML document. An opening tag looks like **<tag>**, while a closing tag has a slash that is placed before the name of an XML element and looks like **</tag>**. Information belonging to an element is contained between the opening and closing tags of an element. For instance, in the above example, **<RIB>**, **<Routes>** and **<Route>** are opening tags and **</RIB>**,**</Routes>** and **</Route>** are closing ones.

XML Attribute

XML attributes are used to specify additional information related to a given XML element. In the example provided above, **IPADDR**, **NETMASK** and **NEXTHOP** are attributes of the **<Route>** element. These attributes provide information about the **<Route>** element (respectively the destination IP address, the network mask and the next hop to which packets destined for the IP address specified in the **IPADDR** tag are to be sent).

XML Comments

XML comments are included in an XML document to add a note or to provide explanation information outside the XML code. Comments are inserted in XML code to guide XML document readers (or viewers) and ease the comprehension of the XML documents. An XML comment is identified by a '!' character (this syntax is similar to HTML comment syntax). Two examples of comments embedded in an XML document are given in Tables 6.2 and 6.3.

XML Namespace

An XML namespace is a set of names that are used in XML documents as element types and attribute names. An XML namespace is identified by a uniform resource identifier (URI) as defined in RFC 2396 [32]. XML namespaces are used to avoid confusion and to make sure that the same data are unambiguously interpreted by all XML-enabled applications.

For a better understanding of the issue that motivated the introduction of namespaces, let us consider an XML document that contains the following elements.

The first element, called **<RIB>**, encloses information about the open shortest path first (OSPF) route information base entries as provided in Table 6.4.

The second element, also called **<RIB>**, encloses information about the border gateway protocol (BGP) routes as illustrated in Table 6.5.

Table 6.1 Example of XML elements

```
<RIB>
  <Routes>
    <Route>
      <IPADDR>12.12.3.44</IPADDR>
      <NETMASK>24</NETMASK>
      <NEXTHOP>31.54.67.8</NEXTHOP>
    <Route>
    <Route>
      <IPADDR>1.2.3.4</IPADDR>
      <NETMASK>24</NETMASK>
      <NEXTHOP>123.6.6.7</NEXTHOP>
    </Route>
    <Route>
      <IPADDR>51.15.26.96</IPADDR>
      <NETMASK>24</NETMASK>
      <NEXTHOP>123.6.6.7</NEXTHOP>
    </Route>
  </Routes>
</RIB>
```

Table 6.2 Example of XML comments outside XML tags

```
<!-
Structure of the RIB
->
<RIB>
  <Routes>
    <Route>
      <IPADDR>12.12.3.44</IPADDR>
      <NETMASK>24</NETMASK>
      <NEXTHOP>31.54.67.8</NEXTHOP>
    <Route>
    <Route>
      <IPADDR>1.2.3.4</IPADDR>
      <NETMASK>24</NETMASK>
      <NEXTHOP>123.6.6.7</NEXTHOP>
    </Route>
  <Route>
      <IPADDR>51.15.26.96</IPADDR>
      <NETMASK>24</NETMASK>
      <NEXTHOP>123.6.6.7</NEXTHOP>
    </Route>
</Routes>
</RIB>
```

Table 6.3 Example of XML comments inside XML tags

```
<RIB>
  <!- List of routes installed in the RIB ->
  <Routes>
    <!-
      First route entry in the RIB
      ->
    <Route>
      <IPADDR>12.12.3.44</IPADDR>
      <NETMASK>24</NETMASK>
      <NEXTHOP>31.54.67.8</NEXTHOP>
    <Route>
    <!-
      Second route entry in the RIB
      ->
    <Route>
      <IPADDR>1.2.3.4</IPADDR>
      <NETMASK>24</NETMASK>
      <NEXTHOP>123.6.6.7</NEXTHOP>
    </Route>
    <!-
      Third route entry in the RIB
      ->
    <Route>
      <IPADDR>51.15.26.96</IPADDR>
      <NETMASK>24</NETMASK>
      <NEXTHOP>123.6.6.7</NEXTHOP>
    </Route>
  </Routes>
</RIB>
```

The above XML codes enclose an element called **<RIB>**. Nevertheless, the definition of this element is not the same for the two pieces of XML code. The first XML code refers to an OSPF RIB and the second one refers to a BGP RIB (for readers who are familiar with OSPF and BGP, these RIBs are not formal descriptions of OSPF and BGP tables but only a lightweight version of these tables). If these two codes are inserted together in the same document, it would induce a conflict because both documents contain the **<RIB>** element and XML parsers would not know which **<RIB>** schema to use: OSPF or BGP.

Table 6.4 OSPF **<RIB>** element

```
<RIB>
  <OSPF>
    <Area_ID>5</Area_ID>
    <Route_ID>2</Route_ID>
  </OSPF>
</RIB>
```

Table 6.5 BGP **<RIB>** element

```
<RIB>
  <BGP>
    <Local_AS>5511</Local_AS>
    <Prefix>132.12.34.5</Prefix>
    <Next_Hop>21.12.34.5</Next_Hop>
  </BGP>
</RIB>
```

In order to resolve this conflict, it is recommended that a reference be added to the namespace where these tables are defined.

If OSPF elements are differentiated from BGP elements by defining two distinct namespaces, **http://serviceauto.org/routing/ospf** and **http://serviceauto.org/routing/bgp**, then the previous two XML codes become as shown in Tables 6.6 and 6.7.

These two **<RIB>** elements are distinct (even if they have the same name) since they are not declared in the same namespace. As a consequence, there is no conflict anymore between these elements when quoted in the same XML document. Therefore, the XML document shown in Table 6.8 will be parsed unambiguously.

Table 6.6 Updated OSPF RIB element

```
<!-
    the XML namespace is identified by the URI:
    http://serviceauto.org/routing/ospf
->
<RIB xmlns="http://serviceauto.org/routing/ospf">
  <OSPF>
  <Area_ID>5</Area_ID>
    <Route_ID>2</Route_ID>
  </OSPF>
</RIB>
```

Table 6.7 Updated BGP RIB element

```
<!-
    the XML namespace is identified by the URI:
    http://serviceauto.org/routing/bgp
->
<RIB xmlns="http://serviceauto.org/routing/bgp">
  <BGP>
    <Local_AS>5511</Local_AS>
    <Prefix>132.12.34.5</Prefix>
    <Next_Hop>21.12.34.5</Next_Hop>
  </BGP>
</RIB>
```

Table 6.8 Merged **<RIB>**

```
<RIB xmlns="http://serviceauto.org/routing/ospf">
  <OSPF>
    <Area_ID>5</Area_ID>
    <Route_ID>2</Route_ID>
  </OSPF>
</RIB>
<RIB xmlns="http://serviceauto.org/routing/bgp">
  <BGP>
    <Local_AS>5511</Local_AS>
    <Prefix>132.12.34.5</Prefix>
    <Next_Hop>21.12.34.5</Next_Hop>
  </BGP>
</RIB>
```

6.2.2 NETCONF Terminology

The following terms are used within this chapter:

- *Management application/application management*: denotes an application that is NETCONF enabled and that performs/enforces NETCONF operations on a managed device. It is also denoted as *NETCONF client* or *client*.
- *Managed device* denotes a node that implements NETCONF protocol and has access to management instrumentation [33]. This is also known as *NETCONF server* (denoted only as *server*) or *managed entity*.

6.2.3 NETCONF Layer Model

As specified by the NETCONF working group, the NETCONF protocol can be partitioned functionally into several layers. At least four layers can be conceptually distinguished in the NETCONF model. Note that the usage of 'layer' terminology within NETCONF is misleading since it should be read as 'sublayer of OSI application layer'. This term should not be confused with OSI layers. For instance, the 'transport layer' should not be understood to mean the same as that of the OSI model. In the rest of this section, the terms 'layer' and 'sublayer' are used interchangeably.

NETCONF layers are defined below and illustrated in Figure 6.1.

- *Transport layer.* Within the context of NETCONF, this layer provides the required functions for enabling communication between a client (aka configuration client) and a server (aka configuration server). NETCONF can use any transport protocol satisfying the set of requirements listed in Section 6.4.1. At the current stage of NETCONF specifications, only secure shell (SSH) (RFC 4254 [16], RFC 4253 [15], RFC 4252 [14]) transport protocol is mandatory to be supported by any NETCONF implementation. In addition, two other alternatives, the blocks extensible exchange protocol (BEEP) (RFC 3080 [21], RFC 3081 [22]) and the simple object access protocol (SOAP) [30] have been identified,

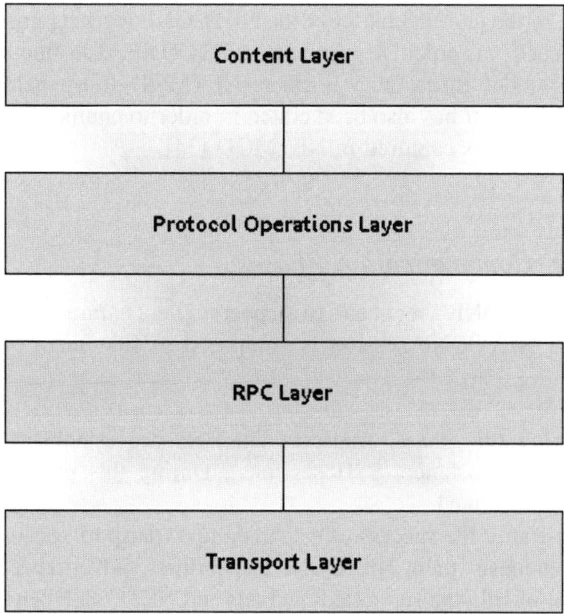

Figure 6.1 NETCONF layered model

and the mappings have been adopted as NETCONF working group documents (these documents have also been adopted by the IETF in the standard track and not in the informational track). Section 6.4.2 provides more information regarding how to activate NETCONF over BEEP [10], SSH [9] or SOAP [11].

- *RPC layer.* This layer provides a transport-independent mechanism (note that 'transport' is to be understood as the 'NETCONF transport layer') for encoding and decoding RPC messages such as **<rpc>** and **<rpc-reply>** elements. These elements are used to convey requests to the managed device, or to acknowledge the correct processing of a NETCONF request, by issuing an **<rpc-reply>** element. The RPC layer elements are used as framing means for NETCONF requests such as **<get>** and **<copy-config>**, since NETCONF operations are enclosed in RPC elements. Because of this close relationship, the RPC layer can be merged with the 'operations layer' and forms only one functional layer responsible for the framing of NETCONF methods.
- *Operations layer.* This layer groups a limited set of basic NETCONF operations. These operations are also called RPC methods. The encoding of these operations follows strict XML schemas as defined in NETCONF specifications. As an example of these operations, the **<get>** method can be invoked to retrieve the whole or only a portion of the configuration data, the **<copy-config>** method can be invoked to copy a part or the whole configuration data and **<delete-config>** can be invoked to delete configuration data. These methods and additional ones are described in Section 6.2.8. NETCONF protocol allows new methods and their associated (new) capabilities to be defined.
- *Content layer.* This layer is used to denote the configuration data manipulated using RPC methods, members of the NETCONF operations layer. This layer is clearly identified

outside the scope of the current charter of the NETCONF working group. Some initiatives have been launched in order to promote NETCONF data models, especially the NETCONF data model Birds Of a Feather (NETMOD BOF) held at the 60th IETF meeting. An Internet draft has also been edited in order to enumerate issues related to the NETCONF data model as captured in NETMOD [34].

6.2.4 NETCONF Communication Phases

When enabling NETCONF between two peers (i.e. *management application* and *managed device*), a NETCONF session is composed of four main phases as illustrated in Figure 6.2:

- *First phase*. During this phase, transport channels are established between the two NETCONF peers (for instance, SSH sessions). During this phase, authentication and key exchange are performed.
- *Second phase*. Just after the success of establishing a transport session, NETCONF peers exchange simultaneously their NETCONF capabilities, advertizing to their peer their capabilities and implicitly the supported methods and NETCONF behavior with regards to basic NETCONF operations. Section 6.2.5 provides more information about capability exchange.
- *Third phase*. When capabilities are exchanged, NETCONF peers can send their RPC methods and issue NETCONF operations such as `<get>` or `<copy-config>` methods.
- *Fourth phase*. The NETCONF session can be terminated by sending appropriate NETCONF methods such as `<kill-session>` or `<close-session>`.

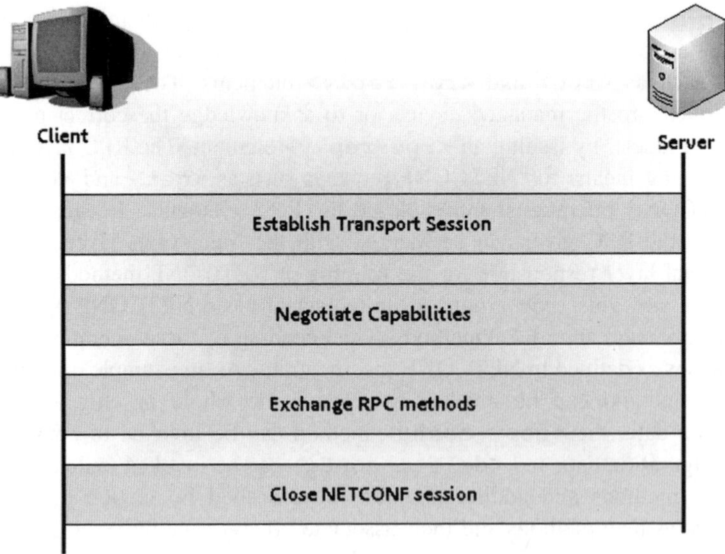

Figure 6.2 Main NETCONF steps

6.2.5 NETCONF Data

6.2.5.1 Configuration Data and State Data

The NETCONF protocol manipulates two types of data that can be enforced or retrieved from NETCONF-enabled devices such as routers, switches or any other device embedding a NETCONF stack:

- *Configuration data.* This category of data groups all writable data that contribute to transforming a given system from one state into another state (or its current state). These data can be conveyed to a NETCONF-enabled device either to be enforced on or be retrieved from this device in order to have more information about its status. NETCONF protocol introduces dedicated methods for manipulating data of this type (such as **<get-config>**, refer to Section 6.3 for more information about NETCONF operations).
- *State data.* The set of non-configuration data. These data can be, for instance, read-only, like status information such as accounting or reporting information. Data of this type can be retrieved by invoking dedicated NETCONF methods like **<get>**, to be notified by the managed entity like notification events introduced in Chisholm *et al.* [33]. As defined by Chisholm *et al.* [33], an event is 'something' that happens and that interests the management application (e.g. a configuration change, a fault, a status change, crossing a threshold, etc.). Several classes of events can be distinguished as listed below:
  1. *Fault events.* These events are generated when an error or a warning occurs. This can lead the managed device to generate alarms that are of several types, such as communications alarms, a processing error alarm, a quality of service (QoS) alarm or a threshold crossing event.
  2. *Configuration events.* These events are used to notify the management application that a configuration change has occurred (e.g. add/delete hardware, create/modify/delete a service, etc.). Events of this type can be notified, for instance, when invoking copy configuration, delete configuration, or the edit configuration operations (refer to Section 6.3 for more information about these NETCONF operations).
  3. *Audit events.* These events indicate that some specific actions have been performed by a given managed device.
  4. *Data dump events.* This class of events covers asynchronous events containing information about a given managed device in terms of its configuration, its state and other appropriate information.
  5. *Maintenance events.* These events inform the management application about the processing or the termination of an action executed on a given managed device.
  6. *Metrics events.* These events contain the performance metrics related to a given managed device.
  7. *Heartbeat events.* These events are sent periodically to evaluate/check/assess the status of the communications channel between the application manager and the managed device.

Because of their read-only characteristic, manipulating *state data* can induce some problems. An example of these problems is trying to write 'read-only' data. In order to avoid problems of this kind, NETCONF has defined two methods: the **<get-config>** operation responsible for retrieving exclusively *configuration data*, and the **<get>** method responsible for retrieving both *configuration and state data.*

6.2.5.2 NETCONF Configuration Datastores

Within the context of NETCONF, a configuration datastore is a set of *configuration data* that is enforced to get a device from a given state to a different one. Configuration datastores should not contain any *state data*. Several configuration datastores can be distinguished:

- *Running datastore*. This datastore is the complete configuration currently active on a given managed device. This type of datastore is always present, and only one configuration running datastore exists on a given NETCONF managed device. The running configuration datastore is the unique datastore present in the base NETCONF model. Additional configuration datastores may be defined. In the remaining part of this chapter, we refer to this datastore as the **<running>** element.
- *Candidate datastore*. This datastore is used as the workplace for creating and manipulating, i.e. modifying, deleting and adding, configuration data without impacting upon the current configuration of the managed device. This latter, if it supports this type of datastore, should advertize the **:candidate** capability (refer to Section 6.5 for more information about NETCONF capabilities) to its NETCONF peers. In the remaining part of this chapter, we refer to this datastore as **<candidate>**.
- *Startup datastore*: The startup configuration datastore contains the enforced configuration during the initialisation of a given Managed Device. This configuration datastore can be distinct from running configuration datastores. A Managed Device supporting this type of datastores should advertize the **:startup** capability (refer to Section 6.5 for more information about NETCONF capabilities) to its NETCONF peers. Within the NETCONF context, operations executed on the running configuration will not affect the startup configuration unless an explicit NETCONF operation, e.g. **<copy-config>** (see Section 6.3 for more information about this method), is invoked. In the remaining part of this chapter, we refer to this datastore as **<startup>**.

6.2.6 NETCONF Capability Exchange

After establishing a transport session, NETCONF peers proceed to exchange their respective capabilities. This section details this procedure.

6.2.6.1 NETCONF XML Namespace

For the correct parsing and interpretation of XML data, NETCONF protocol uses a dedicated namespace that must be declared in all NETCONF XML documents. The base NETCONF namespace is defined by the following URN: **urn:ietf:params:xml:ns:netconf:base:1.0**. Additional namespaces can be declared in a NETCONF document if new capabilities and methods are introduced (refer to Section 6.2.6.2 for more namespace URNs)

6.2.6.2 Capability Exchange Framework

In order to discover the capabilities of a NETCONF node, either a client or a server, NETCONF protocol introduces a capability exchange procedure that aims to advertize to a remote NETCONF peer the capabilities supported by the local node. Owing to this

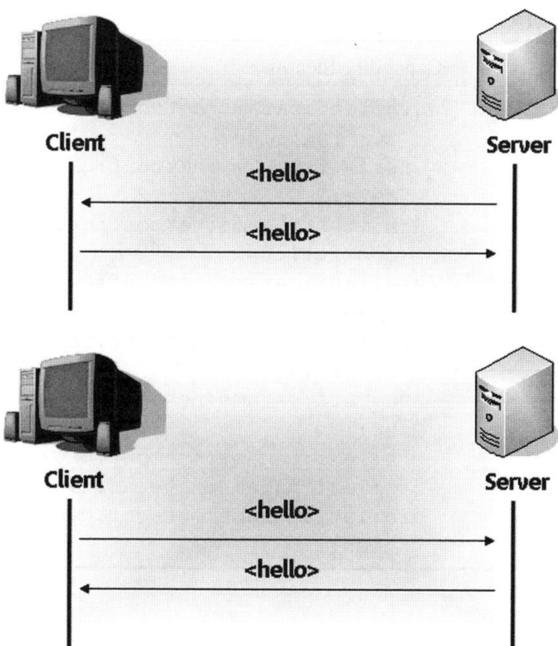

Figure 6.3 NETCONF capability exchange

procedure, a NETCONF node is aware of the methods supported by its remote peers. When receiving a set of capabilities from a remote peer, only the set of capabilities supported by the local peer are taken into account and may be used in the future for NETCONF purposes. However, the capabilities that are not understood by the local peer are ignored.

NETCONF capabilities are enclosed in a dedicated element called **<hello>**. This message is exchanged just after the NETCONF session is opened and must be sent by both communication parties as illustrated in Figure 6.3. No recommendation regarding the order of issuing this message is specified by NETCONF.

Within the context of NETCONF, capabilities are defined as URN references and should follow the format **urn:ietf:params:netconf:capability:{name}:1.0**, where **{name}** is the name of the corresponding capability. In the remaining part of this chapter, we use interchangeably **:{name}** or **<name>** to refer to a given NETCONF capability.

Table 6.9 lists the capabilities introduced by NETCONF protocol (these capabilities are described in Section 6.5). Additional capabilities can be defined in the future by other extension documents.

NETCONF protocol demands that each peer send the base NETCONF capability identified by the URN (**urn:ietf:params:netconf:base:1.0**) to its NETCONF peers.

The example in Figure 6.4 shows a client advertizing the base NETCONF capability, the **:startup** capability, which is defined in the base NETCONF specification, and a server advertizing to the client the base NETCONF capability, the **:startup** and the **:candidate** capabilities which are defined in the base NETCONF protocol. This

Table 6.9 List of NETCONF capability namespaces

Index	Capability Identifier
:writable-running	urn:ietf:params:netconf:capability:writable-running:1.0
:candidate	urn:ietf:params:netconf:capability:candidate:1.0
:confirmed-commit	urn:ietf:params:netconf:capability:confirmed-commit:1.0
:rollback-on-error	urn:ietf:params:netconf:capability:rollback-on-error:1.0
:validate	urn:ietf:params:netconf:capability:validate:1.0
:startup	urn:ietf:params:netconf:capability:startup:1.0
:url	urn:ietf:params:netconf:capability:url:1.0
:xpath	urn:ietf:params:netconf:capability:xpath:1.0
:notification	urn:ietf:params:netconf:capability:notification:1.0

process will lead both peers to use **:startup** capability only. **:candidate** capability will be ignored by the client.

The format of the **<hello>** message sent by the NETCONF server is slightly different from the one issued by the client since it encloses an additional attribute called the **session-id** attribute. This attribute identifies the transport session established between the client and the server. Such an attribute is not present in the **<hello>** message issued by a NETCONF client.

6.2.7 RPC Layer

Once the capability exchange process has been achieved, NETCONF peers may issue RPC methods. These methods carry NETCONF operations and associated answers. At least two types of RPC can be defined: *one-way RPCs* and *two-way RPCs.*

6.2.7.1 One-way RPC

NETCONF messages are generally two-way. However, in the context of some applications, only messages to be generated from the server side or only from the client side are required. In order to support this model, NETCONF WG introduced the concept of the one-way RPC message represented as an **<rpc-one-way>** element. The one-way RPC message is similar to the two-way RPC message, except that no response is expected to the enclosed method, as illustrated in Figure 6.5.

For instance, in the context of event notifications, RPCs will be originated from the NETCONF server and not from the NETCONF client. Figure 6.6 shows an example of usage of this RPC.

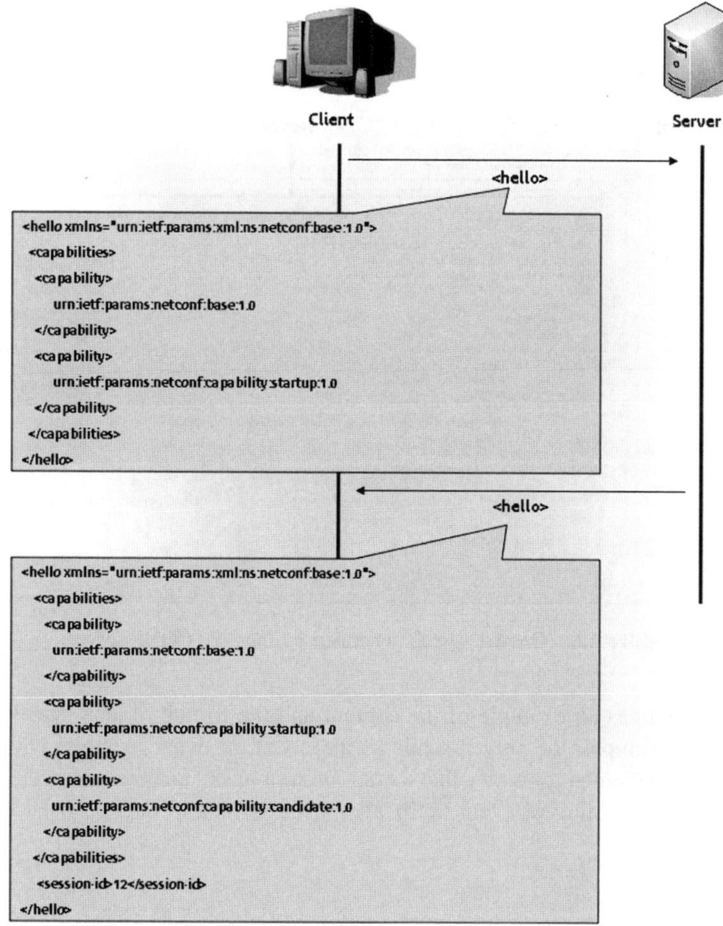

Figure 6.4 NETCONF capability exchange – hello XML format

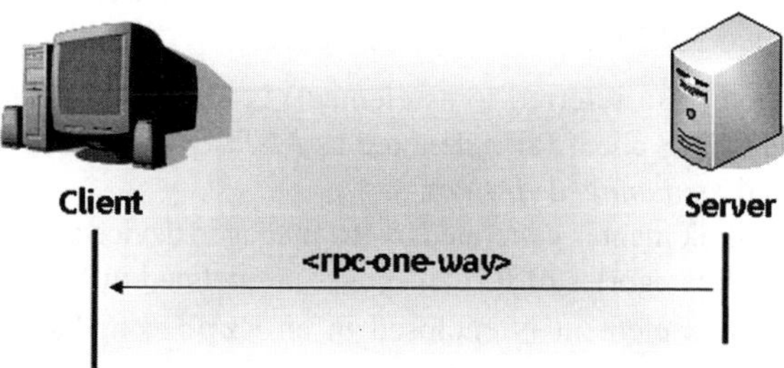

Figure 6.5 One-way RPC

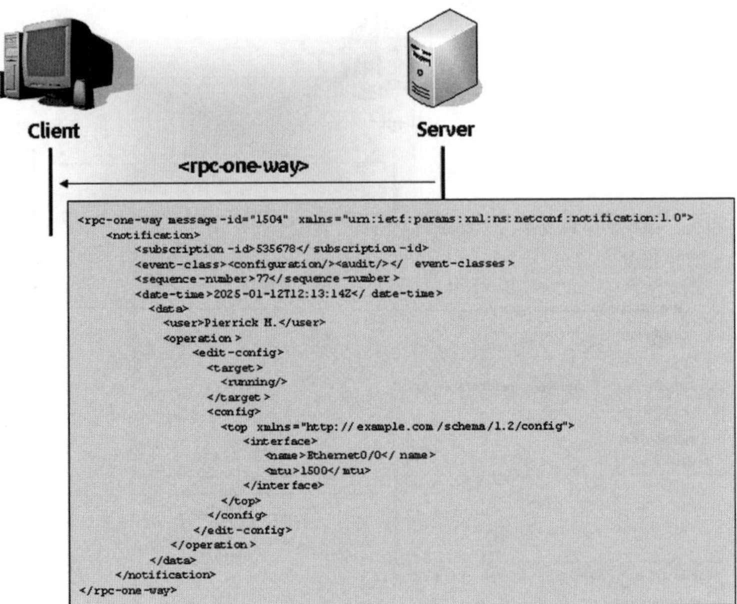

Figure 6.6 One-way RPC generated by the NETCONF server

Figure 6.6 illustrates an example of the content and the format of an **<rpc-one-way>** element. In this example, the server sends to the client an event message enclosed in an **<rpc-one-way>** element to notify that a configuration task has been enforced. Additional information on the structure of event notifications is provided in Section 6.3.13.

6.2.7.2 Two-way RPC

NETCONF-enabled nodes use dedicated methods to provide transport-protocol-independent framing for NETCONF requests and responses between clients and servers. This procedure is commonly denoted within NETCONF terminology as the RPC communication model. RPC methods allow NETCONF protocol to be independent of the underlying transport protocol used to convey communication messages.

The base NETCONF specification documents make use of an RPC scheme that is two-way, i.e. an RPC request must be answered by an RPC response, as shown in Figure 6.7.

In such a model, two RPC methods can be invoked:

- **<rpc>**. This method (also referred to as element) is used to carry a NETCONF request issued from a NETCONF client (also denoted as *NETCONF manager*) and destined for a server (also denoted as *managed device*).
- **<rpc-reply>**. This element is generated by the managed device as an answer to a recei ved **<rpc>** element. Two categories of RPC response are defined in NETCONF protocol:
  - **<rpc-error>**: this element is enclosed in an **<rpc-reply>** element and is sent back to the NETCONF manager if an error or a warning has occurred when processing an **<rpc>** element or when enforcing required actions by the managed device;

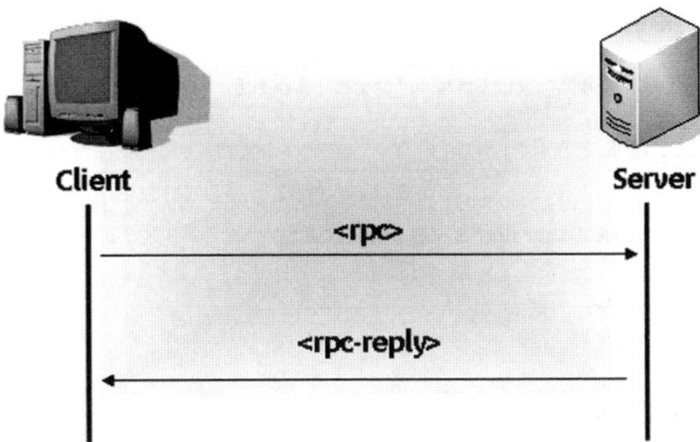

Figure 6.7 Two-way RPC

– **<ok>**: this element is enclosed in an **<rpc-reply>** element and is sent back to the client if no errors and no warnings have occurred during the processing of a received **<rpc>** element and during the enforcement of required actions by the managed device.

More information about these elements and their usage provided below through several examples.

6.2.7.3 <rpc> element

An **<rpc>** element is used to carry a NETCONF request generated by a client and destined for its attached server. This element, i.e. **<rpc>**, encloses at least the **message-id** attribute. The value of this attribute is an arbitrary string assigned by the client. Usually, this value is a monotonically increasing integer. The XML schema of the **message-id** attribute is provided in Table 6.10.

Figure 6.7 provides an example of the content of a basic **<rpc>** element and its XML format. The RPC request indicates the URN of the supported NETCONF namespace; in this example this is indicated by the xmlns line: **xmlns="urn:ietf:params:xml:ns:netconf:base:1.0"**.

In this example (Table 6.11), the **<rpc>** element has a **message-id** equal to '8'. Only one method is contained in this element. The core of this method is empty. For readers unfamiliar with XML notations, the '**!**' symbol is used at the beginning of an XML line to identify comments. These comments are not part of the definition of the elements, nor are they considered as attributes, but they provide clarifications and additional useful information for the reader of XML documents.

Table 6.10 XML schema of the **message-id** attribute

```
<xs:simpleType name="messageIdType">
  <xs:restriction base="xs:string">
    <xs:maxLength value="4095"/>
  </xs:restriction>
</xs:simpleType>
```

Table 6.11 Example of an **<rpc>** element

```
<rpc message-id="8" xmlns="urn:ietf:params:xml:ns:netconf:
base:1.0">
  <method-example>
    <!-
      List of method-example's parameters
    ->
    <!- Parameter 1 ->
    <!- Parameter 2 ->
    <!- Parameter 3 ->
    <!- Parameter 4 ->
        . . .
        . . .
    <!- Parameter i ->
    <!- Parameter i+1->
  </ method-example >
</rpc>
```

6.2.7.4 **<rpc-reply>** element

Upon receipt of an **<rpc>** element, the server saves the value of the **message-id** attribute and uses the corresponding value in any **<rpc-reply>** message generated as a response to the received **<rpc>** element. In addition, if other attributes are present in an **<rpc>** element, a NETCONF server returns them in the resulting **<rpc-reply>** elements without any modifications, i.e. the values of these attributes are copied by the server and returned back to the requesting client. If several **<rpc>** elements have been received by a NETCONF server, the server processes them serially. Then, the server sends **<rpc-reply>** elements in the order the requests were received.

An example of an **<rpc-reply>** element is shown in Table 6.12.

The **<data>** tag identifies the part of the **<rpc-reply>** that contains the data requested by the method enclosed in a request received by the server. An **<rpc-reply>** element can also contain other elements, mainly the **<ok>** or **<rpc-error>** elements. These elements are described in the subsections below.

Table 6.12 Example of an **<rpc-reply>** element

```
<rpc-reply message-id="8" xmlns="urn:ietf:params:xml:ns:netconf:
base:1.0">
  <data>
    <!- Body of data enclosed in the rpc-reply ->
    <tag1>value1</tag1>
    <tag2>value2</tag2>
    <tag3>value3</tag3>
  </data>
</rpc-reply>
```

Table 6.13 Example of <ok> element

```
<rpc-reply message-id="8" xmlns="urn:ietf:params:xml:ns:netconf:
base:1.0">
    <ok/>
</rpc-reply>
```

<ok> Element

In order to inform a remote peer (i.e. one with which a NETCONF session has been established) that no errors or no warnings have occurred during the processing of a received **<rpc>** request, a NETCONF server issues an **<ok>** element. Table 6.13 provides an example of the content of an **<ok>** element.

<rpc-error> Element

When an error (or a warning in some NETCONF implementations) has occurred during the processing of a received **<rpc>** request, a NETCONF server issues an **<rpc-reply>** element enclosing an **<rpc-error>** element. If multiple errors have been encountered by the server during the processing of the **<rpc>** request, the server can enclose one or several **<rpc-error>** elements in the same **<rpc-reply>** element. NETCONF implementations could opt not to send any **<rpc-error>** element if one or many warnings have occurred during the processing of the request.

An **<rpc-error>** element may include the attributes listed in Table 6.14. Optional attributes can be omitted.

Table 6.15 lists NETCONF errors as defined in the base NETCONF document. For all these errors, the error severity attribute value is equal to 'error' (the corresponding error column is not represented in the table). Only mandatory attributes are described in Table 6.15.

Tables 6.16 and 6.17 provide two examples illustrating the format of issued **<rpc-error>** elements. The first example (Table 6.16) illustrates the response sent by the server in the case where the **<rpc>** element does not enclose a **message-id** attribute.

The second example (Table 6.17) shows an **<rpc-error>** element indicating that the IP address assigned by the client to one of the server's interfaces is not valid.

Table 6.14 <rpc-error> attributes

error-type	Indicates the type of error that has occurred. The following values can be set: TRANSPORT, RPC, PROTOCOL or APPLICATION.
error-tag	A string identifying the error condition.
error-severity	Contains a string identifying the error severity, as determined by the device. This can take the values: error or warning.
error-app-tag	Contains a string identifying the data-model-specific or implementation-specific error condition. This attribute is optional.
error-path	Contains an identification of the element path to the node that is associated with the error being reported in a particular **<rpc-error>**. This attribute is optional.
error-message	This attribute describes the error condition.
error-info	Contains protocol-specific or data-model-specific error content.

Table 6.15 List of NETCONF errors

Error-tag	Error-type	Error-info	Description
IN_USE	• PROTOCOL • APPLICATION	None. None.	The request requires a resource that is already in use.
INVALID_VALUE	• PROTOCOL • APPLICATION	None.	The request specifies an unacceptable value for at least one parameter.
TOO_BIG	• TRANSPORT • RPC • PROTOCOL • APPLICATION	None.	The request or response is too large.
MISSING_ATTRIBUTE	• RPC • PROTOCOL • APPLICATION	• **<bad-attribute>**: name of the missing attribute; • **<bad-element>**: name of the element that should contain the missing attribute.	An expected attribute is missing.
BAD_ATTRIBUTE	• RPC • PROTOCOL • APPLICATION	• **<bad-attribute>**: name of the attribute with the bad value; • **<bad-element>**: name of the element that contains the attribute with the bad value.	An attribute value is not correct.
UNKNOWN_ATTRIBUTE	• RPC • PROTOCOL • APPLICATION	• **<bad-attribute>**: name of the unexpected attribute; • **<bad-element>**: name of the element that contains the unexpected attribute.	An unexpected attribute is present.
MISSING_ELEMENT	• RPC • PROTOCOL • APPLICATION	**<bad-element>**: name of the. missing element	An expected element is missing.
BAD_ELEMENT	• RPC • PROTOCOL • APPLICATION	• **<bad-element>**: name of . the element	An element value is not correct.

UNKNOWN_ELEMENT	• RPC • PROTOCOL • APPLICATION	• **<bad-element>**: name of the unexpected element.	An unexpected element is present.
UNKNOWN_NAMESPACE	• RPC • PROTOCOL • APPLICATION	Name of the unexpected namespace.	An unexpected namespace is present.
ACCESS_DENIED	• RPC • PROTOCOL • APPLICATION	None.	Access to the requested RPC, protocol operation or data model is denied because of authorization failure.
LOCK_DENIED	PROTOCOL	• **<session-id>**: session ID of the session holding the requested lock, or zero to indicate a non-NETCONF entity holding the lock.	Access to the requested lock is denied because it is currently held by another NETCONF entity.
RESOURCE_DENIED	• TRANSPORT • RPC • PROTOCOL • APPLICATION	None.	Request could not be completed because of insufficient resources.
ROLLBACK_FAILED	• PROTOCOL • APPLICATION	None.	Request to roll back some configuration change was not completed for some reason.
DATA_EXISTS	• APPLICATION	None.	Request could not be completed because the relevant data model content already exists.
DATA_MISSING	• APPLICATION	None.	Request could not be completed because the relevant data model content does not exist.

(Continued)

Table 6.15 (*Continued*)

Error-tag	Error-type	Error-info	Description
OPERATION_NOT_SUPPORTED	• RPC • PROTOCOL • APPLICATION	None.	Request could not be completed because the requested operation is not supported by this implementation.
OPERATION_FAILED	• RPC • PROTOCOL • APPLICATION	None.	Request could not be completed because the requested operation failed for some reason not covered by any other error condition.
PARTIAL_OPERATION	APPLICATION	• **<ok-element>**: identifies an element in the data model for which the requested operation has been completed for that node and all its child nodes; • **<err-element>**: identifies an element in the data model for which the requested operation has failed for that node and all its child nodes; • **<noop-element>**: identifies an element in the data model for which the requested operation was not attempted for that node and all its child nodes.	Some part of the requested operation failed.

Table 6.16 First example of an **<rpc-error>**

```
<rpc-reply message-id =``8'' xmlns=``urn:ietf:params:xml:ns:netconf:
base:1.0''>
  <error-type>rpc</error-type>
    <error-tag>MISSING_ATTRIBUTE</error-tag>
    <error-severity>error</error-severity>
    <error-info>
      <bad-attribute>message-id</bad-attribute>
      <bad-element>rpc</bad-element>
    </error-info>
</rpc-error>
</rpc-reply>
```

Table 6.17 Second example of an **<rpc-error>**

```
<rpc-reply message-id =``8'' xmlns=``urn:ietf:params:xml:ns:netconf:
base:1.0''>
<rpc-error>
  <error-type>rpc</error-type>
    <error-tag>INVALID_VALUE</error-tag>
    <error-severity>error</error-severity>
    <error-message>Invalid IP Address</error-message>
  </rpc-error>
</rpc-reply>
```

6.2.8 *NETCONF Filtering*

6.2.8.1 Terminology

NETCONF protocol allows an application manager to specify which portion of the *configuration* and/or *state data*, as maintained by the managed evice, to be included in a given **<rpc-reply>** element. This feature is termed *XML subtree filtering* and allows more flexibility to manipulate configuration data (Table 6.18).

NETCONF introduced the following terminology to denote the components that may be present in a given subtree filter:

Table 6.18 Example of an XML subtree filter

```
<filter type=``subtree''>
    <network xmlns=``http://serviceauto.org/filters/''>
      <routers/>
    </network>
</filter>
```

Table 6.19 Content match node

```
<filter type=``subtree''>
  <network xmlns=``http://serviceauto.org/filters/''>
    <routers>
      <router>
        <routerID>ASBR_157485<routerID>
      </router>
    </routers>
  </network>
</filter>
```

- *Namespace selection.* The output when applying this filter is the list of all nodes defined within the specified namespace. Note that at least one node must be specified when enclosing this filter in a given NETCONF request. All child nodes will be listed in the response to be sent back to the application manager. In the example in Table 6.18, all child elements of **<network>** defined in the namespace **http://serviceauto.org/filters/** must be listed in a response to a request enclosing this filter.
- *Containment node.* A containment node is a node that contains child elements within a filter. A containment node can also contain other containment nodes, etc. When only one containment node is specified in a given filter included in a NETCONF request, all child nodes must be listed in the response. In the example in Table 6.18, **<network>** is a containment node. A request including this filter should be answered by a response containing all **<routers>** data.
- *Selection node.* A leaf element of a subtree filter is called a selection node. In the example in Table 6.18, the **<routers>** element is a selection node. When specifying this filter in a given NETCONF request, the corresponding reply should include all **<routers>** data.
- *Content match node.* Any leaf element with simple type content is called a content match node. More concretely, in the following example, **<routerID>** is a content match node. The response to this filter must include only data related to a router with an identifier equal to **ASBR_157485**.

6.2.8.2 Examples

In order to illustrate the usage of filters within NETCONF, this section lists numerous examples (Tables 6.20 and 6.21) and related output when applying the corresponding filter to a given configuration. The XML document in Table 6.20 is used for all listed examples. Note that the data model used for these examples is not part of the NETCONF specification itself.

Table 6.20 Example of an XML configuration


```
<data>
  <network xmlns=``http://serviceauto.org/filters/''>
    <routers>
      <router>
        <routerID<ASBR_157485
        <interfaces>
          <interface>
            <interfaceName<Eth0>/interfaceName>
            <interfaceBW<100M>/interfaceBW>
          </interface>
          <interface>
            <interfaceName<Eth1>/interfaceName>
            <interfaceBW<10M>/interfaceBW>
          </interface>
          <interface>
            <interfaceName<Eth3>/interfaceName>
            <interfaceBW<100M>/interfaceBW>
          </interface>
        </interfaces>
      </router>
      <router>
        <routerID<ASBR_787878
        <interfaces>
          <interface>
            <interfaceName<Eth0>/interfaceName>
            <interfaceBW<10M>/interfaceBW>
          </interface>
          <interface>
            <interfaceName<Eth1>/interfaceName>
            <interfaceBW<10M>/interfaceBW>
          </interface>
        </interfaces>
      </router>
    </routers>
```


6.3 NETCONF Protocol Operations

The NETCONF base protocol specifies nine operations. These operations are invoked to retrieve, install or modify configuration data of a given managed device. These basic operations are as follows:

- **<get>**. This method is invoked in order to retrieve both *configuration* and *state data* from a managed device.
- **<get-config>**. This method is similar to the **<get>** operation. However, it is used to retrieve *configuration data* only.

Table 6.21 Filter examples

Filter

```
<filter type="subtree">
</filter>
```

This filter is empty. No selection criterion is specified.

Filter output

```
<data>
</data>
```

Returned data is empty since no criterion has been specified in the filter.

Filter

```
 <filter type="subtree">
      <network xmlns="http://serviceauto.org/filters/">
            <routers/>
      </network>
 </filter>
```

This filter selects all <routers> elements.

Filter output

```
<data>
 <network xmlns="http://serviceauto.org/filters/">
  <routers>
    <router>
      <routerID>ASBR_157485<routerID>
      <interfaces>
        <interface>
            <interfaceName>Eth0</interfaceName>
            <interfaceBW>100M</interfaceBW>
        </interface>
        <interface>
            <interfaceName>Eth1</interfaceName>
            <interfaceBW>10M</interfaceBW>
        </interface>
        <interface>
            <interfaceName>Eth3</interfaceName>
            <interfaceBW>100M</interfaceBW>
        </interface>
      </interfaces>
     </router>
     <router>
       <routerID>ASBR_787878<routerID>
       <interfaces>
         <interface>
            <interfaceName>Eth0</interfaceName>
            <interfaceBW>10M</interfaceBW>
          </interface>
          <interface>
            <interfaceName>Eth1</interfaceName>
            <interfaceBW>10M</interfaceBW>
          </interface>
```

```
   </interfaces>
  </router>
 </routers>
 </network>
</data>
```

All configured routers are listed in the reply to a request containing the filter.

```
<filter type="subtree">
     <network xmlns="http://serviceauto.org/filters/">
          <routers>
               <router>
                    <routerID>ASBR_157485<routerID>
               </router>
          </routers>
     </network>
</filter>
```

This filter selects only the router identified by a router ID equal to **ASBR_157485**

```
<data>
 <network xmlns="http://serviceauto.org/filters/">
  <routers>
   <router>
    <routerID>ASBR_157485<routerID>
    <interfaces>
     <interface>
       <interfaceName>Eth0</interfaceName>
       <interfaceBW>100M</interfaceBW>
     </interface>
     <interface>
       <interfaceName>Eth1</interfaceName>
       <interfaceBW>10M</interfaceBW>
     </interface>
     <interface>
       <interfaceName>Eth3</interfaceName>
       <interfaceBW>100M</interfaceBW>
     </interface>
    </interfaces>
   </router>
  </routers>
 </network>
</data>
```

Only **ASBR_157485** related elements are enclosed in the response.

```
<filter type="subtree">
     <network xmlns="http://serviceauto.org/filters/">
          <routers>
               <router>
                    <routerID/>
               </router>
```

```
<data>
 <network xmlns="http://serviceauto.org/filters/">
  <routers>
   <router>
    <routerID>ASBR_157485<routerID>
   </router>
```

(*Continued*)

Table 6.21 (*Continued*)

```
            </routers>
      </network>
</filter>
```

This filter requests retrieving all routers' IDs of all configured routers.

```
      <router>
        <routerID>ASBR_787878<routerID>
      </router>
    </routers>
  </network>
</data>
```

All routers'IDs are listed in the response.

```
<filter type="subtree">
      <network xmlns="http://serviceauto.org/filters/">
            <routers>
                  <router>
                        <routerID>ASBR_157485<routerID>
                           <interfaces>
                               <interface>
                                  <interfaceName/>
                               <interface>
                           </interfaces>
                  </router>
            </routers>
      </network>
</filter>
```

This filter selects showing only all interfaces names of **ASBR_157485.**

```
<data>
 <network xmlns="http://serviceauto.org/filters/">
  <routers>
    <router>
      <routerID>ASBR_157485<routerID>
      <interfaces>
        <interface>
            <interfaceName>Eth0</interfaceName>
        </interface>
        <interface>
            <interfaceName>Eth1</interfaceName>
        </interface>
        <interface>
            <interfaceName>Eth3</interfaceName>
        </interface>
      </interfaces>
     </router>
    </routers>
 </network>
</data>
```

All configured interfaces names of **ASBR_157485** are shown.

- **<copy-config>**. This operation is invoked in order to create or to replace all or portion of the *configuration data*.
- **<edit-config>.** This operation is invoked to load all or portions of the *configuration data*.
- **<delete-config>**. This method is invoked to delete a configuration datastore.
- **<lock>** This method is invoked in order to lock a configuration datastore so as to avoid misconfiguration and inconsistent operations.
- **<unlock>**. This method is invoked to release locks hold by a NETCONF client on a configuration datastore.
- **<close-session>**. This method is used to close gracefully a NETCONF session.
- **<kill-session>**. This method forces the termination of a NETCONF session.

In addition to the base NETCONF protocol operations, NETCONF peers can invoke additional methods if appropriate capabilities are supported:

- **<validate>**. This method is invoked to validate the contents of the specified configuration.
- **<commit>**. This method is invoked by a client to instruct a server to implement the candidate configuration data.
- **<discard-changes>**. This operation is invoked by a NETCONF client if the candidate configuration should not be committed.

And finally, NETCONF methods introduced in Chisholm *et al.* [33] in order to support notification events can also be invoked by NETCONF speakers:

- **<subscribe-notification>**. This method is issued when a management application wants to subscribe to a NETCONF notification service.
- **<notification>**. This method is sent to a management application by a server to enclose events of interest to the management application.

For each operation, a definition table is provided with the following information:

- *Parameters*: describes input of the NETCONF method, such as source, target or filter;
- *Positive response*: describes the output of the NETCONF method to acknowledge the correct processing of the request;
- *Negative response*: describes the output of the NETCONF method to indicate that an error has occurred during the processing of the request.

6.3.1 Retrieve Configuration Data

To retrieve the whole or only part of the configuration datastore of a given managed device, NETCONF protocol defines a dedicated method called **<get-config>**. This method is issued by a management application to the managed device. More information about this method is captured in Table 6.22.

To illustrate the usage of this method, the example in Figure 6.8 shows a NETCONF exchange that occurs between a client and a server when retrieving configuration information.

In this example, the client issues a **<get-config>** operation embedded in an **<rpc>** element and destined for the managed device. The core body of this operation encloses a filter specifiying which portion of the configuration data is to be retrieved from the managed

Table 6.22 **<get-config>** method

Parameter: source	The source element specifies the name of the configuration datastore being queried.
Filter	The filter element identifies the portions of the managed device configuration to retrieve. If no filter is specified, the entire configuration is returned.
Positive response	If the device can satisfy the request, the server sends an **<rpc-reply>** element containing a **<data>** element with the results of the query.
Negative response	An **<rpc-error>** element is included in the **<rpc-reply>** errors, or problems have been encountered when processing the request by the managed device.

device. Indeed, the embedded filter, shown in this example, selects retrieving all configured interfaces in the router (i.e. server).

As a response to this RPC request, the managed device generates an **<rpc-reply>** response carrying the whole list of configured interfaces. For all listed interfaces, all leaf

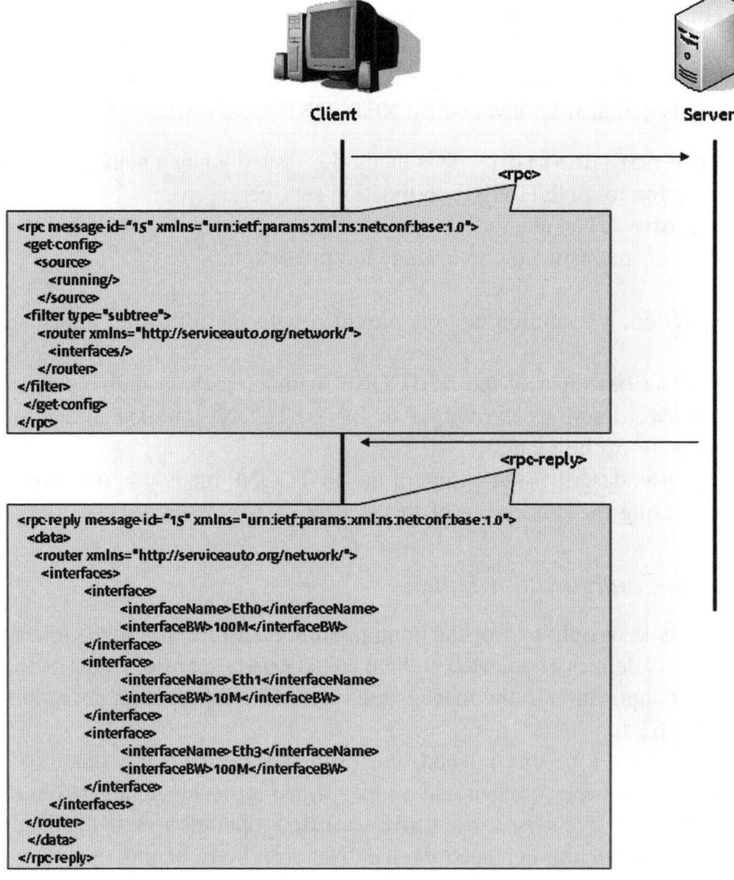

Figure 6.8 Example of the **<get-config>** operation

Table 6.23 Definition of the **<get>** operation

Filter	This parameter specifies the portion of the managed device configuration and state data to retrieve. If this parameter is empty, all configuration and state data are returned.
Positive response	If the device can satisfy the request, the server sends an **<rpc-reply>** element containing a **<data>** element with the results of the query.
Negative response	An **<rpc-error>** element is included in the **<rpc-reply>** errors, or problems have been encountered when processing the request by the managed device.

configuration elements, such as interface name (**interfaceName**) and interface bandwidth (**interfaceBW**), are also retrieved from the managed device. Three interfaces (**eth0, eth1** and **eth3**) are configured in the managed device. Interfaces **eth0** and **eth3** are 100 MB/s Ethernet cards, and **eth1** is a 10 MB/s Ethernet card.

6.3.2 Get

Unlike the previous method, the **<get>** operation is used to retrieve both running *configuration* and device *state* data. **<get>** and **<get-config>** have been introduced by NETCONF to avoid potential problems that can arise when trying to enforce some unhallowed operation such as writing 'read-only' data. Table 6.23 gives more information about the **<get>** method.

Figure 6.9 provides an example illustrating the invocation of the **<get>** method. Therefore, the client issues an **<rpc>** element including the **<get>** method. A filter is also specified in this request to select the portion of the configuration and state data to be retrieved from the managed device (server). Within this example, the enclosed filter requests retrieving information related to the interface whose **interfaceName** is equal to **Eth0**. As a response to his request, the server issues an **<rpc-reply>** element embedding data related to the interface **Eth0**. Therefore, the client is notified that **Eth0** has a capacity of 100 MB/s. Moreover, this interface has received 155 456 bytes of traffic and sent 871 423 bytes of traffic, and 45 bytes have been dropped.

6.3.3 Delete Configuration Data

In order to delete a configuration datastore, except the **<running>** configuration one, a NETCONF client can issue the **<delete-config>** method. This method is characterized as in Table 6.24.

To illustrate the usage of the **<delete-config>** operation, the example in Figure 6.10 shows the invocation of this method by the client to delete the **<startup/>** datastore.

As shown in Figure 6.10, the client sends an **<rpc>** element specifying the datastore to delete as indicated inside the **<target>** tag. Upon receipt of this request, the server enforces appropriate actions so as to delete the **<startup/>** datastore. A positive answer is then sent to the client to notify the successful enforcement of the requested action, using the **<ok>** method.

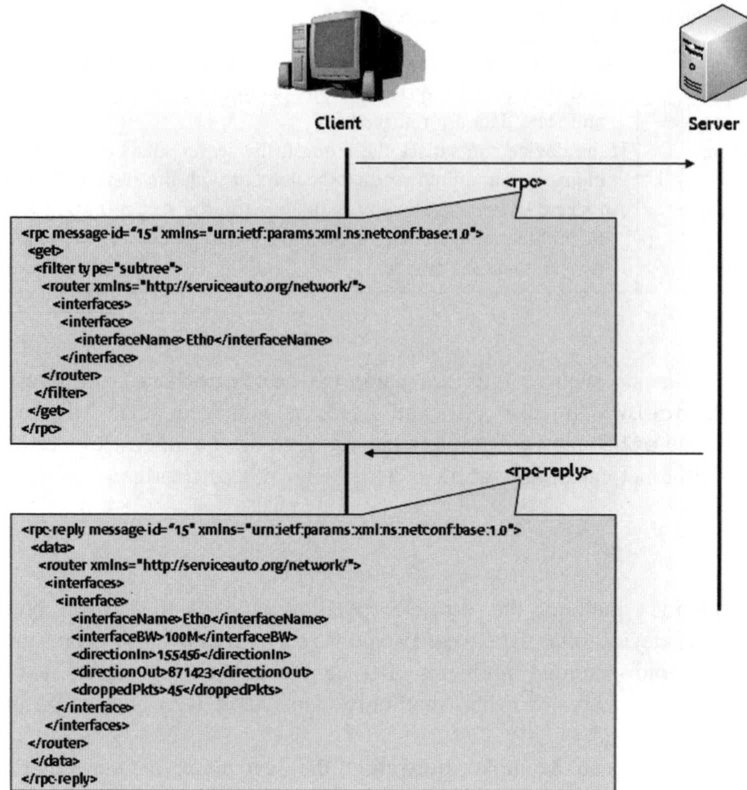

Figure 6.9 Example of the `<get>` operation

6.3.4 Copy Configuration

The `<copy-config>` operation is invoked to create or to replace an entire configuration datastore, specified as a target, with another one, specified as a source. If no configuration datastore exists, the device creates a new one when invoking the `<copy-config>` method. Table 6.25 provides more information about the `<copy-config>` operation.

Figure 6.11 shows a second example of the `<copy-config>` operation where the `<running>` datastore is replaced by a file located in a given URL. The source file is

Table 6.24 `<delete-config` definition

Target	Target element specifies the name of the configuration datastore to delete.
Positive response	If the device was able to satisfy the request, an `<rpc-reply>` including an `<ok>` element is sent to the application manager.
Negative response	An `<rpc-error>` element is included in the `<rpc-reply>` errors, or problems have been encountered when processing the request by the managed device.

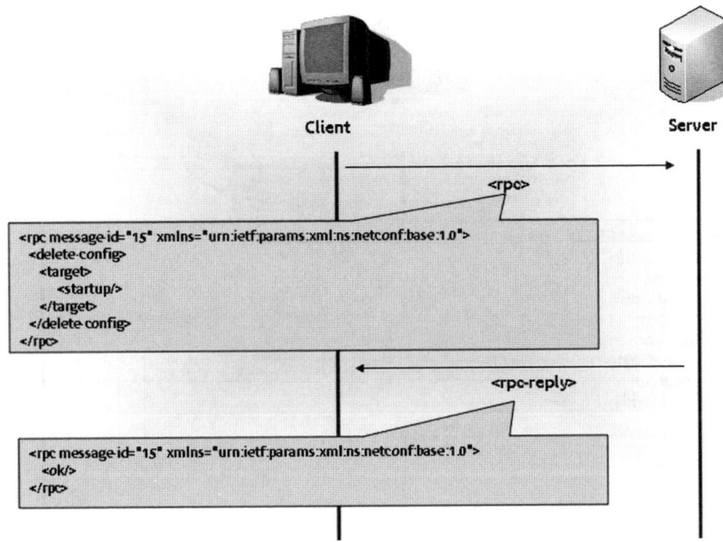

Figure 6.10 Example of the **<delete-config>** operation

specified in the **<source/>** element and the target datastore is enclosed in the **<target/>** element, as shown in Figure 6.11. This operation is embedded in an **<rpc>** element issued by the client to the managed device. Upon receipt of this request, and once the operation is successfully enforced by the server, the latter sends back a positive answer to the client by invoking the **<ok>** method which is embedded in an **<rpc-reply>** element.

6.3.5 Edit Configuration Data

The **<edit-config>** method is responsible for loading all or portions of the configuration data. The retrieved configuration data are identified by the **<config>** attribute. Several operations can be performed on the specified configuration data as listed below:

Table 6.25 Definition of the **<copy-config>** operation

Target	This element specifies the name of the configuration datastore to use as the destination of the copy operation.
Source	This element specifies the name of the configuration datastore to use as thesource of the copy operation or the **<config>** element containing theconfiguration subtree to copy.
Positive response	If the device was able to satisfy the request, an **<rpc-reply>** including an **<ok>** element is sent to the application manager.
Negative response	An **<rpc-error>** element is included in the **<rpc-reply>** errors, or problems have been encountered when processing the request by the managed device.

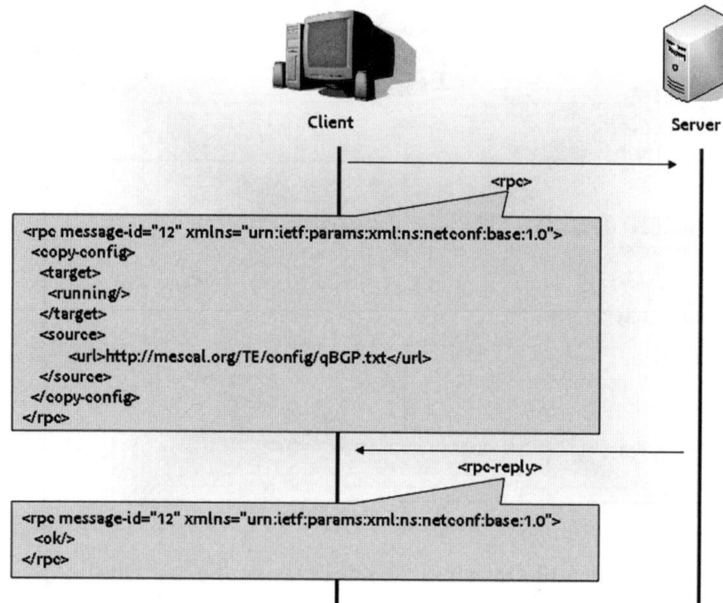

Figure 6.11 Example of the **<copy-config>** operation

- *Merge*. The configuration data are merged with the configuration at the corresponding level in the configuration datastore identified by the target parameter (the target parameter is defined in Table 6.26).
- *Replace*. The configuration data replace any related configuration in the configuration datastore identified by the target parameter. Unlike a **<copy-config>** operation, which replaces the entire target configuration, only the configuration actually present in the **<config>** parameter is affected.
- *Create*. The configuration data are added to the configuration if and only if the configuration data do not already exist on the managed device.
- *Delete*. The configuration data are deleted in the configuration datastore identified by the target parameter.

Note that the default behavior is the 'merge' operation.

Table 6.26 provides additional information about the **<edit-config>** method.

Figure 6.12 shows an example of the **<edit-config>** method with the default operation **merge**. The management application sends an **<edit-config>** message so as to set the IP address of the **Eth0** interface to **1.2.3.4**. Upon receipt of this message, the managed device enforces the new configuration and proceeds to setting the IP address of the **Eth0** interface to the new assigned IP address by the management application. Once all related actions are enforced, the managed device issues a response to the management application, acknowledging that the new configuration changes have been taken into account and are successfully enforced.

Table 6.26 Definition of the **<edit-config** operation

Parameters	Target	Name of the configuration datastore to use as the destination of the copy operation.
	Default operation	Selects the default operation for the **<edit-config>** request. The default value for the default operation parameter is 'merge'.
	Test option	The test option element may be specified only if the device advertizes the **:validate** capability. The test option element has one of the following values: • ***test-then-set***: perform a validation test before attempting to set. If validation errors occur, do not perform the **<edit-config>** operation. This is the default test option; • ***set***: perform a set without a validation test.
	Error option	The error option element can take one of the following values: • ***stop-on-error***: abort the **<edit-config>** operation on first error. This is the default error option. • ***continue-on-error***: continue to process configuration data on error. The error is recorded and a negative response is generated if any errors occur. • ***rollback-on-error***: if an error occurs, the server will stop processing the **<edit-config>** operation and restore the specified configuration to its complete stateat the start of this **<edit-config>** operation. This option requires the managed device to support the **:rollback-on-error** capability.
	Config	A hierarchy of configuration data as defined by one of the managed device's data models.
Positive response		If the device was able to satisfy the request, an **<rpc-reply>** including an **<ok>** element is sent to the application manager.
Negative response		An **<rpc-error>** element is included in the **<rpc-reply>** errors, or problems have been encountered when processing the request by the managed device.

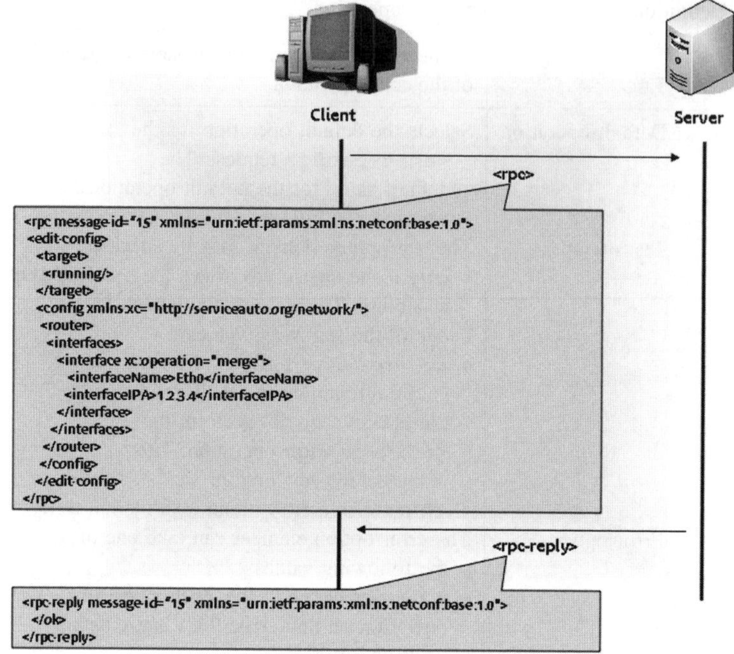

Figure 6.12 Example of the **`<edit-config>`** operation

6.3.6 Close a NETCONF Session

The method **`<close-session>`** is invoked in order to terminate gracefully a NETCONF session. Upon receipt of a **`<close-session>`** request by a NETCONF server, the latter releases all existing locks and resources associated with the corresponding session. This method is defined in Table 6.27.

Figure 6.13 provides an example of the usage of the **`<close-session>`** method.

Within this example, the client sends an **`<rpc>`** element requesting the closure of NETCONF communication between the managed device and the server. Upon receipt of this request, the server gracefully closes all alive sessions between the server and client and releases associated resources. All received NETCONF messages, after treating the **`<close-session>`** operation, are ignored by the server.

Table 6.27 Definition of the **`<close-session>`** operation

Parameters	None.
Positive response	If the device was able to satisfy the request, an **`<rpc-reply>`** including an **`<ok>`** element is sent to the application manager.
Negative response	An **`<rpc-error>`** element is included in the **`<rpc-reply>`** errors, or problems have been encountered when processing the request by the managed device.

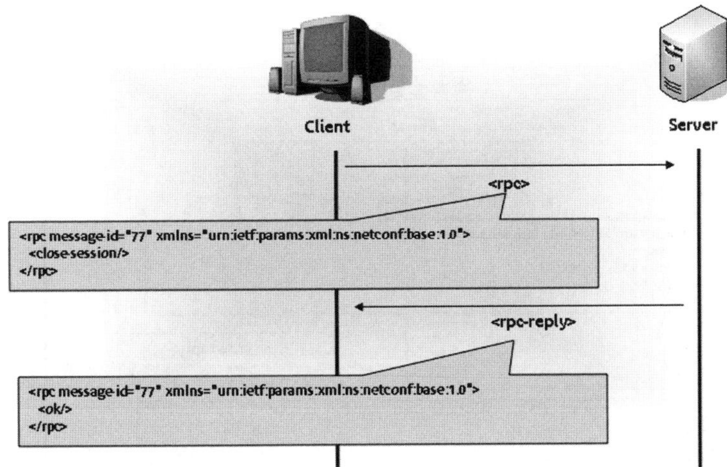

Figure 6.13 Example of the **<close-session>** operation

6.3.7 Kill a Session

The method **<kill-session>** is used to force terminate a NETCONF session. Upon receipt of a **<kill-session>** request by a NETCONF peer, the later will abort any running operation and will release all existing locks and resources associated with this session. This method is defined in Table 6.28.

If a NETCONF server receives a **<kill-session>** request while processing a confirmed commit, it restores the configuration to its state before the confirmed commit was issued. Figure 6.14 illustrates how a session can be closed by using the **<kill-session>** method.

The example in Figure 6.14 shows a client issuing an **<rpc>** element embedding a **<kill-session>** operation. The purpose of this operation is to kill the session identified by a **session-id** equal to 15. Upon receipt of this request, the server proceeds to appropriate actions to handle this request. An **<rpc-reply>** element is sent back to the client to notify that no problems have been encountered when processing this request.

Table 6.28 Definition of the **<kill-session>** method

Parameters	Session identifier of the NETCONF session to be terminated. If this value is equal to the current session ID, an 'invalid value' error is returned.
Positive response	If the device was able to satisfy the request, an **<rpc-reply>** including an **<ok>** element is sent to the application manager.
Negative response	An **<rpc-error>** element is included in the **<rpc-reply>** errors, or problems have been encountered when processing the request by the managed device.

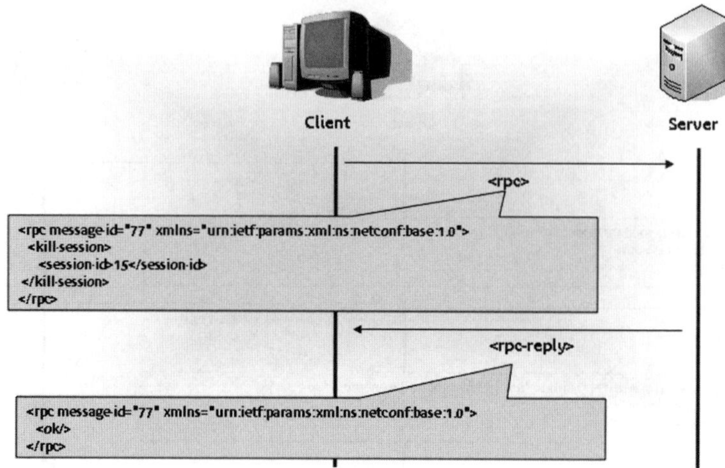

Figure 6.14 Example of the **<kill-session>** operation

6.3.8 Lock NETCONF Sessions

NETCONF protocol uses the **<lock>** operation in order to lock a configuration datastore and deny the modification of the configuration to other NETCONF clients. A server accepting a lock operation should allow only the client holding the lock to modify the configuration data and should send back appropriate error messages to clients trying to modify the locked configuration datastore.

Table 6.29 provides more information about the definition of this method.

A first example illustrating the usage of the **<lock>** operation is provided in Figure 6.15. This method is carried inside an **<rpc>** element issued by the client and destined for the managed device. This request specifies the datastore to be locked. Upon receipt of this request, the server checks if the **<candidate>** datastore is not locked by another client. In this example, the **<running>** datastore is not locked; therefore, the server accepts the request, proceeds to enforcing appropriate actions to lock the **<running>** datastore and

Table 6.29 Definition of the **<lock>** operation

Target	This element specifies the name of the configuration datastore to lock.
Positive response	If the device was able to satisfy the request, an **<rpc-reply>** including an **<ok>** element is sent to the application manager.
Negative response	An **<rpc-error>** element is included in the **<rpc-reply>** errors, or problems have been encountered when processing the request by the managed device. If the lock is already held by another client, the **<error-tag>** element will be set to 'lock denied', and the **<error-info>** element will include the **<session-id>** of the lock owner. If the lock is held by a non-NETCONF entity, a **<session-id>** of 0 is included.

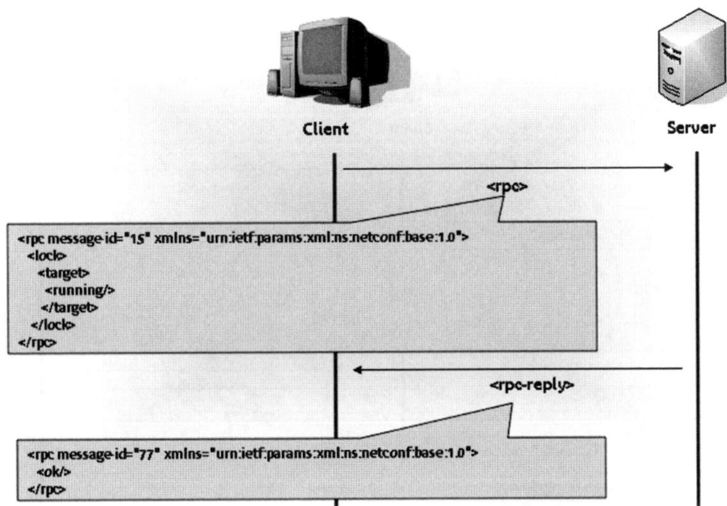

Figure 6.15 Example of the **<lock>** operation

assigns the lock to the requesting client. An **<rpc-reply>** including an **<ok>** element is then sent to the client.

A second example (Figure 6.16) provides an overview of the NETCONF exchange that occurs when errors are encountered when processing a lock request. Upon receipt of the request by the server, the latter proceeds to appropriate checking regarding the locks associated with the **<candidate>** datastore. In this example, the **<running>** datastore is already locked by another client. Therefore, the server issues an rpc-reply> with an appropriate error message to notify the client that its request is denied.

6.3.9 Unlock NETCONF Sessions

In order to release a configuration lock, the **<unlock>** operation can be invoked by a given management application. A client is only allowed to unlock a configuration datastore it has previously locked.

Table 6.30 provides additional information about the **<unlock>** operation.

A first example illustrating a successful **<unlock>** operation is shown in Figure 6.17. In this example, the client issues a request including an **<unlock>** operation together with the targeted datastore. This request is received by the server. The latter handles the received request by enforcing appropriate actions and checking operation. After verifying that this client holds a lock on the **<running>** datastore, the server releases this lock and sends back an **<rpc-reply>** embedding an **<ok>** method to notify the client about the successful handling of its request. Consequently, the **<running>** datastore is no longer locked and is a candidate to be locked by any other management application.

The second example (Figure 6.18) illustrates the exchange of NETCONF messages that occurs when a client tries to unlock a configuration datastore that has been locked by another NETCONF client. The server notifies the client about the failure of its request and lists the reasons for its failure in the response embedded in a **<rpc-reply>** element.

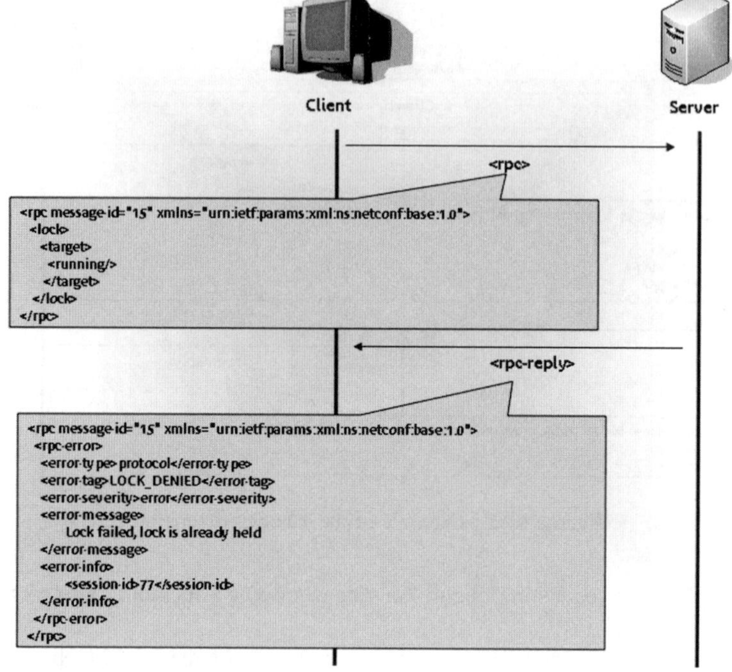

Figure 6.16 Example of the **<lock>** operation with an error message

6.3.10 Validate Configuration Data

To validate the content of a given configuration, a managed application can issue the **<validate>** method specifying the targeted configuration datastore to be validate. This method is defined in Table 6.31.

Figure 6.19 provides an example illustrating the usage of the **<validate>** operation. In this example, the client sends an **<rpc>** request containing a **<validate>** operation. The target datastore is the candidate configuration one. Upon receipt of this request, the server proceeds to appropriate actions and enforces them so as to validate the candidate configuration. Once these actions are successfully enforced, the server issues an **<rpc-reply>** element destined for the client, including an **<ok>** method. This response acknowledges to the client that validation-related actions have been executed on the candidate configuration datastore.

Table 6.30 Definition of the **<unlock>** operation

Target	This element specifies the name of the configuration datastore to unlock.
Positive response	If the device was able to satisfy the request, an **<rpc-reply>** including an **<ok>** element is sent to the application manager.
Negative response	An **<rpc-error>** element is included in the **<rpc-reply>** errors, or problems have been encountered when processing the request by the managed device.

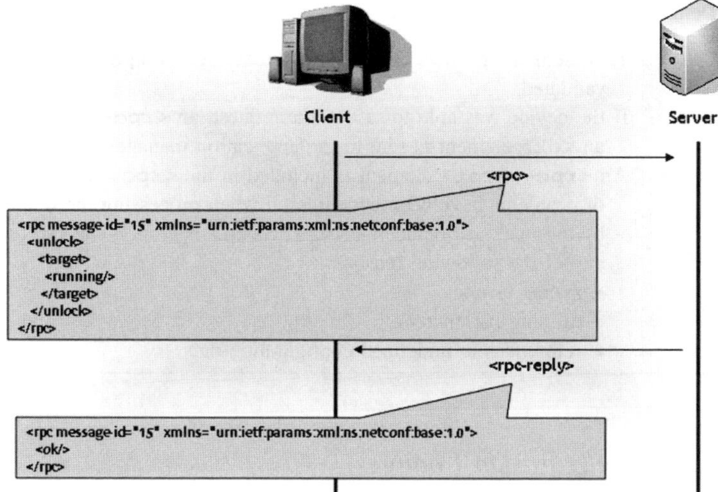

Figure 6.17 Example of the **<unlock>** operation

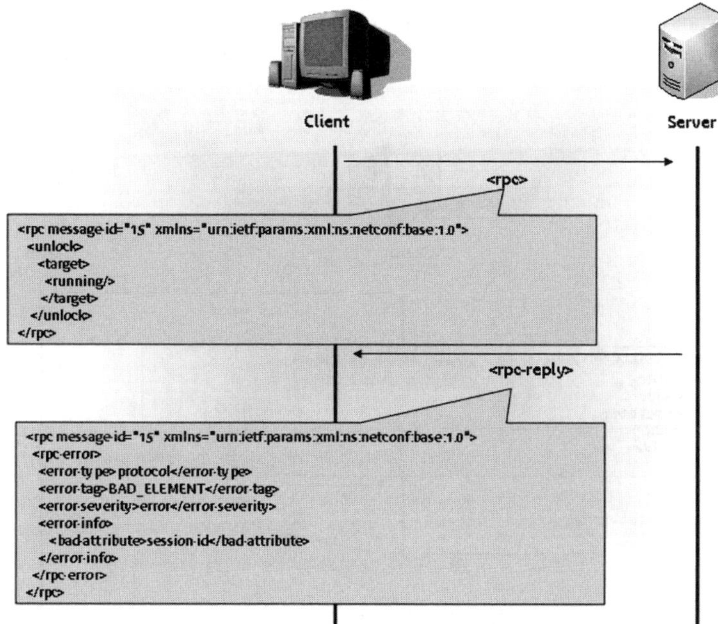

Figure 6.18 Example of the **<unlock>** operation with errors

Table 6.31 Definition of **<validate>** operation

Source	This element specifies the name of the configuration datastore being validated.
Positive response	If the device was able to satisfy the request, an **<rpc-reply>** including an **<ok>** element is sent to the application manager.
Negative response	An **<rpc-error>** element is included in the **<rpc-reply>** errors, or problems have been encountered when processing the request by the managed device. A **<validate>** operation can fail for any of the following reasons: • syntax errors; • missing parameters; • references to undefined configuration data.

6.3.11 Commit Configuration Changes

A NETCONF client instructs a server to implement the candidate configuration data by invoking the **<commit>** method. This method is allowed only if the **:candidate** capability is supported by both NETCONF peers. The running configuration datastore must remain unchanged if the NETCONF managed device was unable to commit all the changes in the candidate configuration datastore.

Table 6.32 defines the **<commit>** operation.

Figure 6.20 shows an example of the usage of the **<commit>** method. This method is enclosed in an **<rpc>** element sent by the client to the server. Once this request is received by the server, the latter implements the changes that have been added to the candidate configuration and issues a response to the client acknowledging that all commit-related operations have been achieved and changes implemented with the managed device.

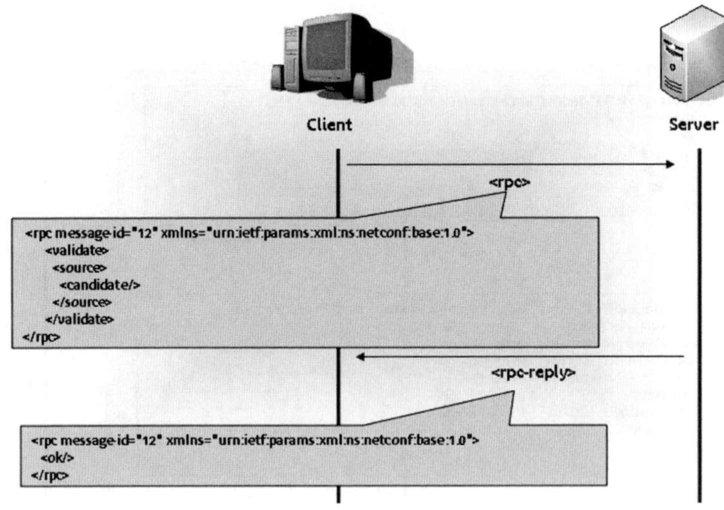

Figure 6.19 Example of the **<validate>** operation

Table 6.32 Definition of the **<commit>** operation

Parameters	None.
Positive response	If the device was able to satisfy the request, an **<rpc-reply>** including an **<ok>** element is sent to the application manager.
Negative response	An **<rpc-error>** element is included in the **<rpc-reply>** errors, or problems have been encountered when processing the request by the managed device.

6.3.12 Discard Changes of Configuration Data

A **<discard-changes>** method can be invoked by a given NETCONF client if the candidate configuration should not be committed. Therefore, the candidate configuration is reverted to the current running configuration. This operation discards any uncommitted changes by resetting the candidate configuration with the content of the running configuration. This method is only invoked when the **:candidate** capability is supported by both NETCONF peers (i.e. management application and managed device).

Figure 6.21 provides an example of the usage of the **<discard-changes>** method. Once the **<discard-changes>** method is received by the server, the latter will proceed to revert the content of the candidate configuration to the content of the running configuration datastore. An **<rpc-reply>** response including an **<ok>** method is then sent to the client as illustrated in Figure 6.21.

6.3.13 NETCONF Notification Procedure

Within the NETCONF architecture, managed devices can send notification messages when requested by NETCONF management applications. The NETCONF notification procedure is illustrated in Figure 6.22.

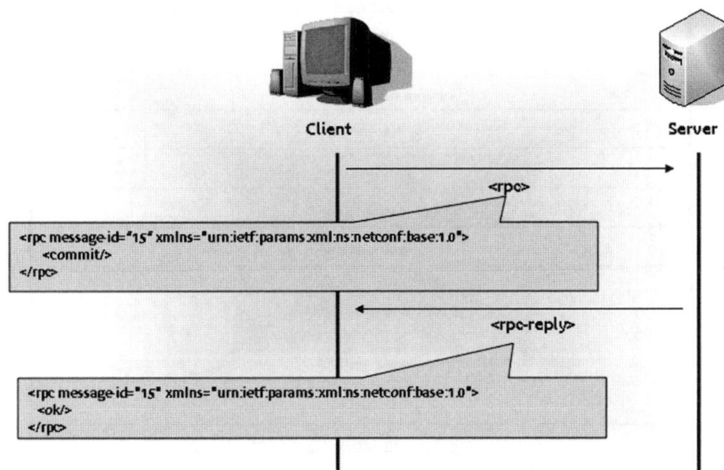

Figure 6.20 Example of the **<comit>** operation

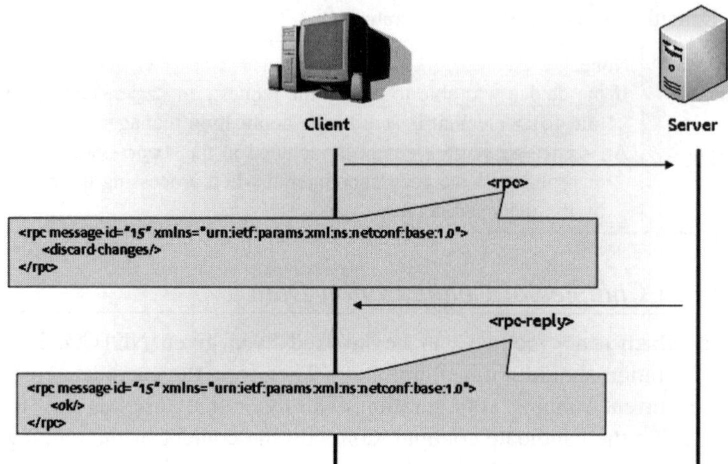

Figure 6.21 Example of the **<discard-changes>** operation

Firstly, the transport session is established and capabilities are exchanged between two NETCONF peers. Then, the client sends a **<create-subscription>** to the server to specify the event class of interest. When an event belonging to the class to which the client has subscribed occurs, the server sends **<notification>** messages to the client. These

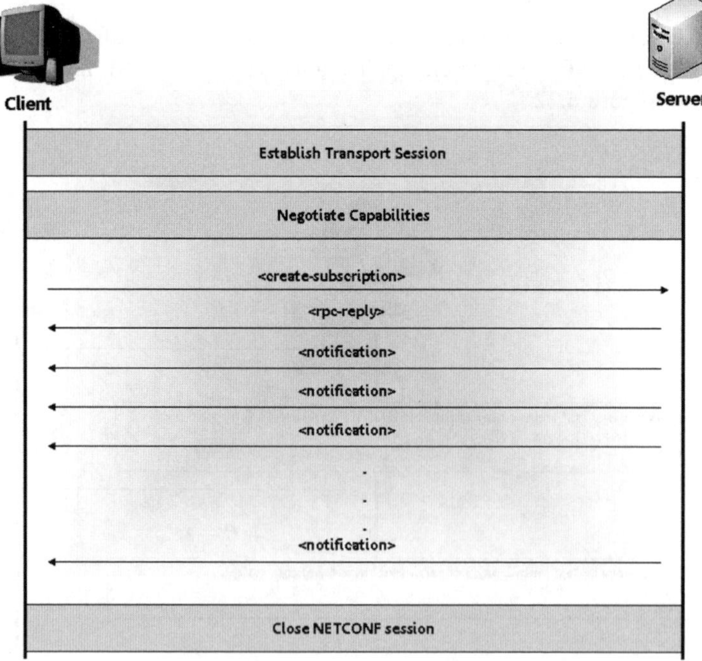

Figure 6.22 Notification procedure

Table 6.33 Definition of the **<create-subscription>** operation

Stream	Indicates which event classes are of interest to the client. If not present, events of all classes will be sent to the client.
Positive response	If the device was able to satisfy the request, an **<rpc-reply>** including an **<ok>** element is sent to the application manager.
Negative response	An **<rpc-error>** element is included in the **<rpc-reply>** errors, or problems have been encountered when processing the request by the managed device.

notification messages enclose information about the occurring event. In the first version of the document [33], notification subscriptions can be modified during the lifetime of a NETCONF session owing to the invocation of a **<modify-subscription>** method. This feature has been dropped in the latest version of that document.

6.3.13.1 Subscribing to Event Notifications

In order to subscribe to an event notification, which consists in receiving asynchronous event notifications from the server, a NETCONF client sends a **<create-subscription>** method to a managed device. The client indicates to the server the type of event notification it wishes to receive through the 'stream' attribute.

This method is defined in Table 6.33.

Figure 6.23 shows the format of a **<create-subscription>** method. In this example, the client sends a subscription request to be notified of events belonging to a given event class. The server accepts the request and sends an **<ok>** element to the client to indicate that the subscription is accepted.

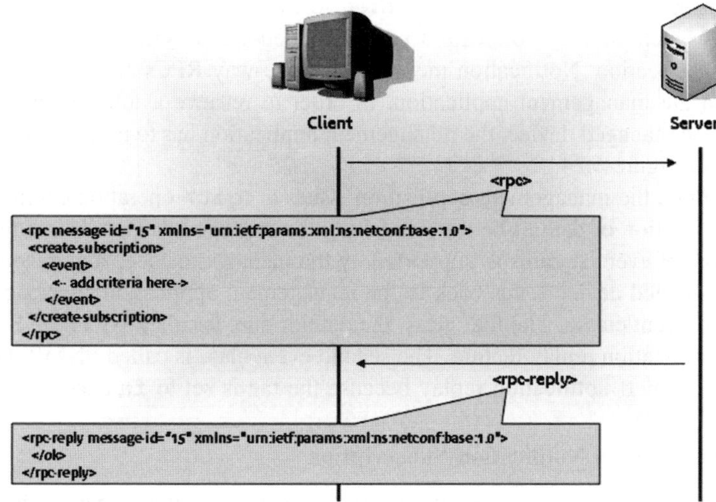

Figure 6.23 Example of **<create-subscription>**

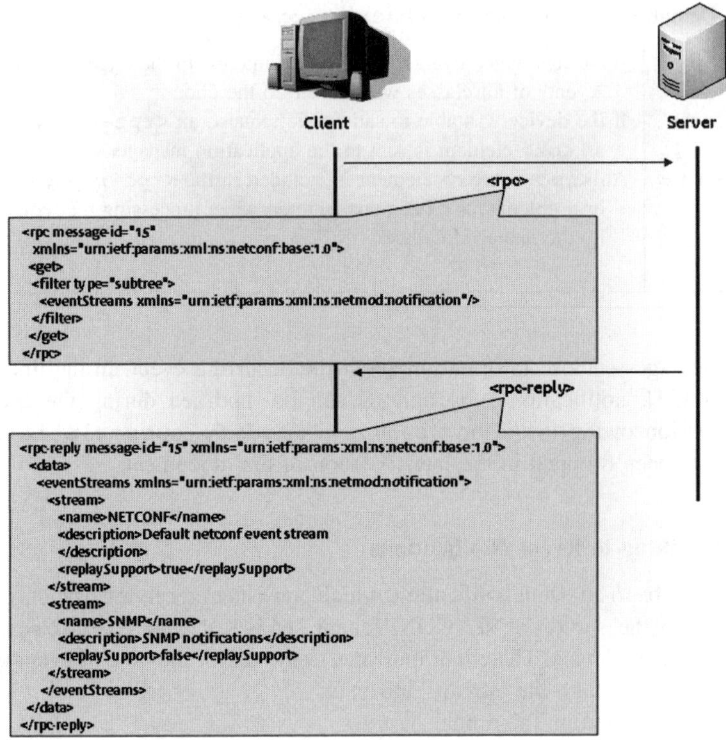

Figure 6.24 Retrival of event classes

6.3.13.2 Sending an Event Notification

Once a management application has subscribed to event classes of interest, the managed device sends asynchronously notification messages, denoted as **<notification>**, to that management application. Notification messages are one-way RPCs and do not require any response from the management application. In order to retrieve available event classes as supported by the managed device, the management application has to issue a **<get>** method as illustrated in Figure 6.24.

In this example, the management application issues a **<get>** operation enclosing a filter specifying the portion of data to be retrieved from the managed device. This filter specifies showing the list of event streams as supported by the managed device. As a response to this request, the managed device sends back to the management application an **<rpc-reply>** containing two event classe. The first class, the default one, is called NETCONF. This class supports the notification replay feature. The second event class is called SNMP. This second class does not support notification replay because the tag is set to **false**.

6.3.13.3 Terminating a Notification Subscription

To terminate a subscription to an event notification, a management application should issue a **<kill-session>** method or terminate the transport session. Under normal condition, i.e.

the NETCONF session is active, the subscription notification terminates when the 'stop time' as indicated in the `<create-subscription>` message is reached.

6.4 NETCONF Transport Protocol

6.4.1 NETCONF as Transport-independent Protocol

NETCONF protocol is a transport-independent protocol. Therefore it can be activated using any transport protocol that satisfies a set of requirements, which are listed below:

- *Connection oriented transport protocol.* NETCONF is connection oriented, requiring a persistent connection between NETCONF peers.
- *Authentication and integrity mechanisms should be supported by the transport protocol.* NETCONF protocol requires secure channels to convey sensitive data and means to validate the identity of the remote NETCONF peer. NETCONF protocol recommends the support of RADIUS the remote access dial-in user server (RADIUS) (RFC 2865 [35]), transport layer security (TLS) (RFC 2246 [36]) and SSH.

Nevertheless, NETCONF requires that any NETCONF implementation must support the SSH transport protocol mapping as a transport protocol. NETCONF implementations can support any additional transport protocol that meets the aforementioned requirements.

In Section 6.4.2, we provide more information about the mapping of using BEEP, SSH and SOAP as NETCONF transport protocols. Note that the purpose of this section is not to provide technical details about SSH, BEEP and SOAP but only to describe how NETCONF is invoked over these transport protocols.

6.4.2 Transport Protocol Alternatives

6.4.2.1 SSH[3]

This section describes how NETCONF is mapped to the SSH connection protocol (RFC 4254 [16]) over the SSH transport protocol (RFC 4253 [15]). We assume that the reader is familiar with the SSH system. For readers who want to have detailed knowledge about SSH, please refer to Refs [9], [12] and [14] to [16].

Where is Secure Shell Being Standardized?

The first version of SSH (aka SSH1) protocol was released in 1995. This first version provides secure means for interactive applications such as Telnet, RSH, REXEC or RLOGIN. This version has been updated within the IETF on account of some security failures. This work has been conducted by the SECSH working group (http://www.ietf.org/html.charters/secsh-charter.html). Several RFCs have been edited by this working group, especially the second version of SSH (aka SSH2).

[3]The reference SSH website is available at www.openssh.org

The RFCs published by the SECSH working group are listed below:

- The secure shell protocol assigned numbers (RFC 4250 [12]);
- The secure shell protocol architecture (RFC 4251 [13]);
- The secure shell authentication protocol (RFC 4252 [14]);
- The secure shell transport layer protocol (RFC 4253 [15]);
- The secure shell connection protocol (RFC 4254 [16]);
- Using DNS to securely publish secure shell key fingerprints (RFC 4255 [17]);
- Generic message exchange authentication for the secure shell protocol (RFC 4256 [18]);
- The secure shell transport layer encryption modes (RFC 4344 [19]);
- Secure shell session channel break extension (RFC 4335 [20]).

The following IETF Drafts has been adopted by the SECSH working group:

- SSH file transfer protocol (*draft-ietf-secsh-filexfer*);
- GSSAPI authentication and key exchange for the secure shell protocol (*draft-ietf-secsh-gsskeyex*);
- SSH public key file format (*draft-ietf-secsh-publickeyfile*);
- Diffie–Hellman group exchange for the SSH transport layer protocol (*draft-ietf-secsh-dh-group-exchange*);
- Uniform resource identifier scheme for secure file transfer protocol (SFTP) and secure shell (*draft-ietf-secsh-scp-sftp-ssh-uri*);
- Secure shell public-key subsystem (*draft-ietf-secsh-publickey-subsystem*);
- X.509 authentication in SSH (*draft-ietf-secsh-x509*).

What is SSH?
SSH is a security protocol used for secure remote login over a public/shared network. The SSH system is structured as follows:

- *The transport layer.* This layer provides server authentication, confidentiality and integrity. The transport layer is usually used over TCP. An example of SSH transport message exchange is provided in Figure 6.25. The main steps that occur during an SSH transport session are:
  - *TCP connection setup*. The client initiates the connection to the server which listens on port 22.
  - *SSH version string exchange*. Both sides send a version string. The current SSH protocol version is 2.0. This negotiation is used to indicate the capabilities of an implementation and to trigger compatibility extensions.
  - *SSH key exchange*. The encryption algorithm is negotiated. SSH2 supports several algorithms such as **`3des-cbc`**, **`blowfish-cbc`**, **`twofish256-cbc`**, **`twofish128-cbc`**, **`aes192-cbc`** and **`serpent192-cbc`**. Note that the encryption algorithm can be different for each direction.
  - After this stage, SSH data can be exchanged between two SSH peers.
  - SSH sessions can be closed by invoking appropriate SSH messages.
- *User authentication protocol*. This protocol is responsible for authenticating the client to the server. It runs over the transport layer protocol. An example of user authentication

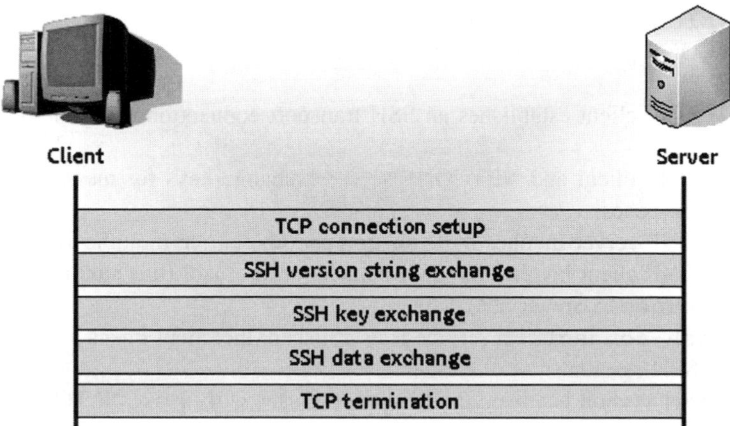

Figure 6.25 SSH Transport connection setup

message exchange is provided in Figure 6.26. Definitions of these functions are not provided in this book; for more information about the content of exchanged SSH messages, please refer to SSH base documents. This service is also denoted as **`ssh userauth`**.

- *The connection protocol.* This protocol multiplexes the encrypted tunnel into several logical channels. This layer runs over the user authentication protocol. Note that channels are identified by channel numbers at both ends of the connection and that channel numbers for the same channel at the client and server sides may differ.

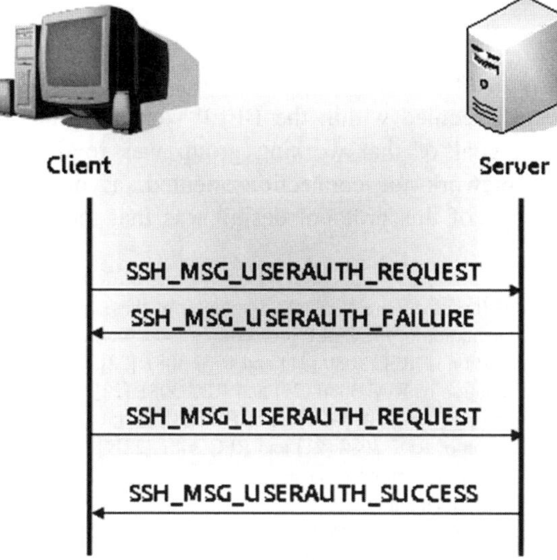

Figure 6.26 SSH user authentication protocol

Initiating NETCONF Sessions over SSH

To run NETCONF over SSH, the following steps should be followed:

1. The NETCONF client establishes an SSH transport connection using the SSH transport protocol.
2. The NETCONF client and NETCONF server exchange keys for message integrity and encryption purposes.
3. The NETCONF server invokes the **`ssh userauth`** service to authenticate the client.
4. The NETCONF client invokes the SSH connection protocol (this step is also denoted as the **`ssh-connection`** service).
5. After the **`ssh-connection`** service is established, the client opens a channel that will result in a SSH session.
6. Once the SSH session has been established, the user will invoke NETCONF as an SSH subsystem.

Exchanging NETCONF Methods over SSH

Once a NETCONF session over SSH has been established, NETCONF methods can be exchanged between NETCONF peers as described in previous sections. Nevertheless, the special character sequence '**`]]<]]>`**' is sent by the client and the server after each XML document. This character sequence is used to identify the end of an XML document. The motivation behind the use of this 'illegal' XML character is to ease synchronisation of the client and the server in the case of an error.

Figures 6.27 and 6.28 provide two examples of NETCONF exchange messages. In the first example, Figure 6.27 illustrates the capability exchange that occurs after NETCONF session establishment. The second example (Figure 6.28) provides an example of a NETCONF operation (in this example the **`<copy-config>`** method) sent over SSH. These NETCONF elements are ended by '**`]]<]]>`**' characters.

6.4.2.2 BEEP

Where was BEEP Designed?[4]

The BEEP protocol was specified within the BEEP working group, which disbanded in March 2002. The main goal of that working group was to develop a standards-track application protocol framework for connection-oriented, asynchronous request/response interactions. The challenge of this protocol design was that the designed framework had

[4]Several implementations of BEEP protocol are available. Examples of these implementations are as follows:

- *wodBeep:* an ActiveX implementation of the BEEP protocol;
- *beepcore-c:* C/C++ implementation of RFC 3080 [21] and RFC 3081 [22];
- *beepcore-java:* Java implementation of RFC 3080 [21] and RFC 3081 [22];
- *beepcore-tcl:* Tcl implementation of RFC 3080 [21] and RFC 3081 [22];
- *BEEPy:* a python implementation of RFC 3080 [21] and RFC 3081 [22];
- *IBM BeepLite:* BEEP implementation written in Java;
- *Net::BEEP::Lite:* a Perl BEEP implementation;
- *PermaBEEP-Java:* a toolkit for writing Java-based applications that use the BEEP framework;
- *RoadRunner BEEP Framework:* open source BEEP implementation written in C;
- *JsBEEP:* JavaScript implementation of the BEEP protocol.

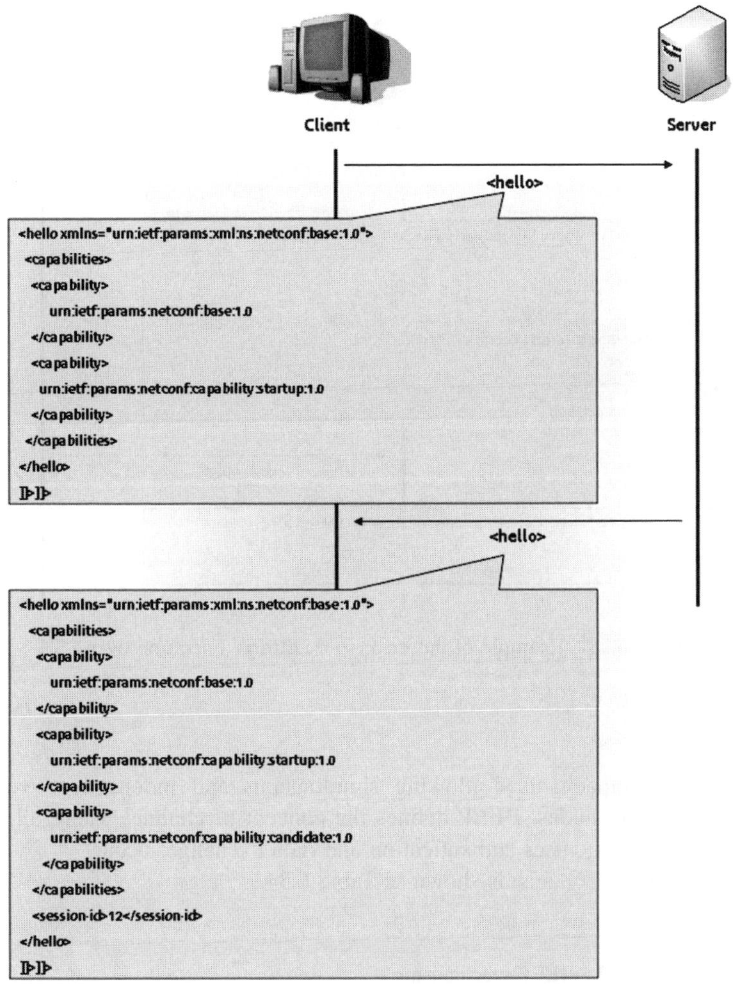

Figure 6.27 Example of capability exchange over SSH

to permit multiplexing of independent request/response streams over a single transport connection and had to support both textual and binary messages. The BEEP working group produced two RFCs:

- The blocks extensible exchange protocol core (RFC 3080 [21]);
- Mapping the BEEP core onto TCP (RFC 3081 [22]).

We also recommend reading the following RFC, which is not an output of the BEEP WG but provides much more information about application protocols:

- On the design of application protocols (RFC 3117 [23]).

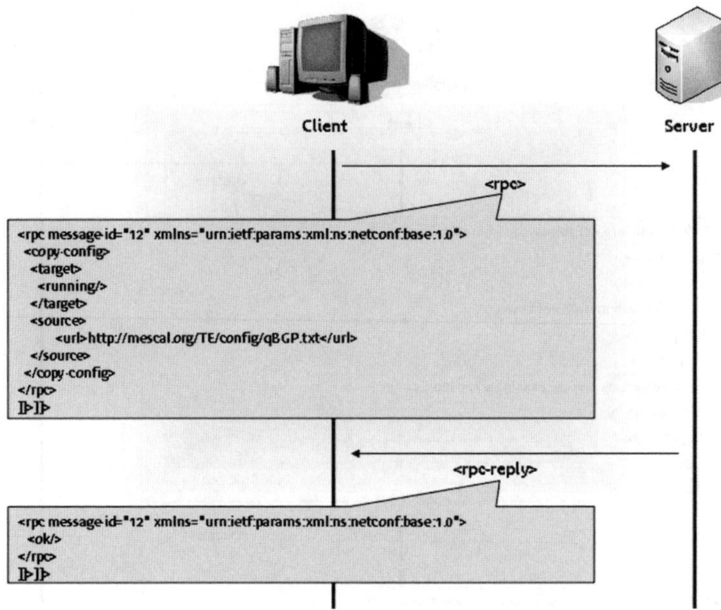

Figure 6.28 Example of the **`<copy-confing>`** operation over SSH

What is BEEP?[5]

BEEP is a framing mechanism allowing simultaneous and independent exchanges of messages between two nodes. BEEP defines the concept of channel where all exchanges, such as transport security, user authentication and data exchange, occur.

BEEP supports the mechanisms shown in Table 6.34.

Table 6.34 BEEP mechanisms

Mechanism	BEEP
Framing	Counting with a trailer
Encoding	MIME or text/xml
Reporting	Three-digit and localized textual diagnostic
Asynchrony	Channels
Authentication	SASL (RFC 2222 [37])
Privacy	SASL or TLS (RFC 2246 [36])

[5]The original name of the BEEP protocol was BXXP (the Blocks eXtensible eXchange Protocol).

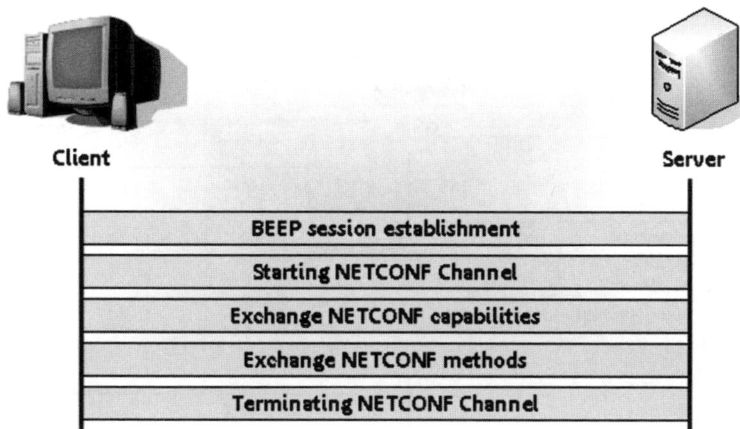

Figure 6.29 NETCONF over BEEP

BEEP defines the following messages:

- *MSG*. This message is used to send a request asking the server to perform a given task.
- *RPY*. This message is sent back by the server to inform the client that the requested task has been completed.
- *ERR*. This message is sent back by the server to inform the client that an error has occurred during the execution of the requested task.

NETCONF over BEEP

To enable NETCONF over BEEP, the following steps should be followed (see Figure 6.29):

1. A BEEP session is established between the NETCONF client and the server.
2. The NETCONF client advertizes the NETCONF profile to the server, as illustrated in Figure 6.30.
3. NETCONF capabilities are exchanged between the client and the server, as illustrated in Figure 6.31. Each NETCONF peer sends its capabilities, which should be positively acknowledged by the remote peer.
4. Once capabilities are exchanged, the NETCONF client can send NETCONF methods to the server. These methods are encoded in RPC elements.

For more information about BEEP-related messages, refer to the BEEP base specifications.

6.4.2.3 SOAP

Where is SOAP Standardized?

SOAP is a recommendation of W3C (World Wide Web Consortium, URL: *http://www.w3.org*). W3C initiated the SOAP initiative in 1999. When the SOAP 1.0 version was released, it

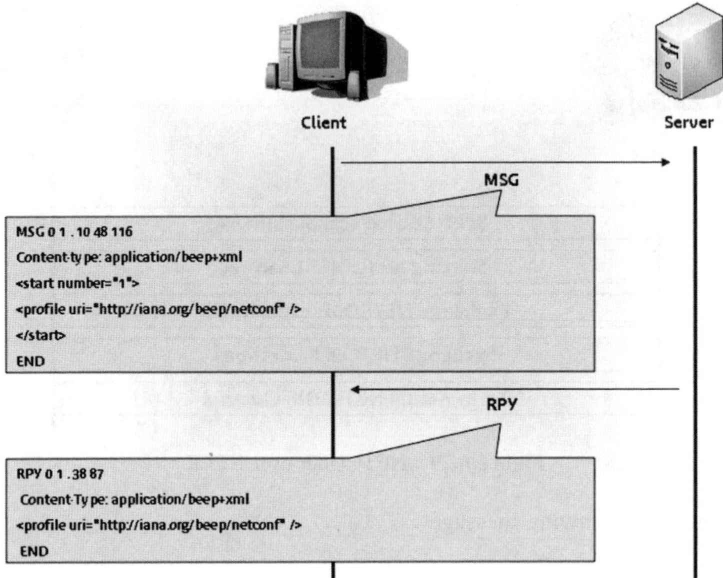

Figure 6.30 NETCONF over BEEP: starting a BEEP channel

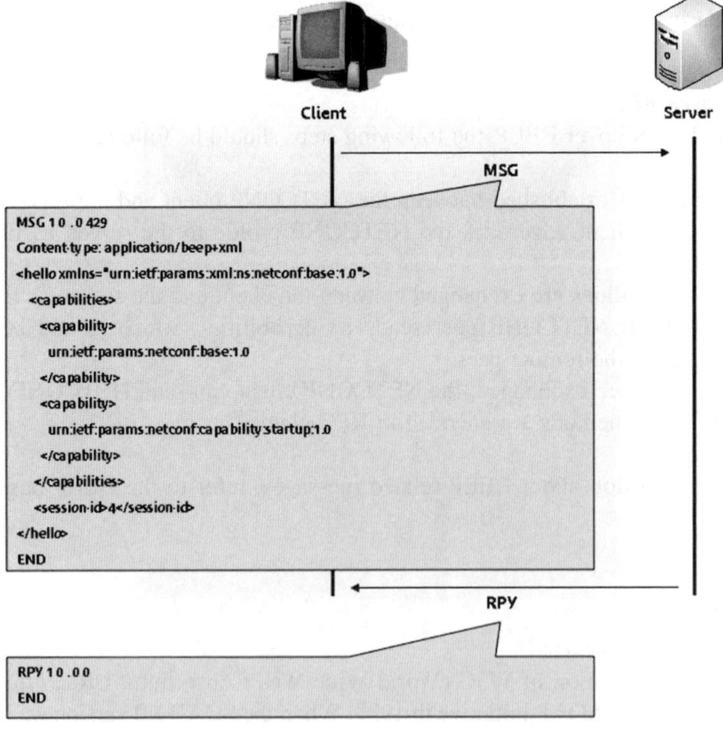

Figure 6.31 NETCONF over BEEP: exchange of capabilities

was based only on HTTP. However, in May 2000, SOAP 1.1 was released and included other transport protocols. Then, the first draft of SOAP 1.2 was presented in July 2001 and promoted to a 'recommendation'. The following URLs provide a list of literature about SOAP and XML-related effort:

- http://www.w3.org/TR/soap12-part0/;
- http://www.w3.org/TR/soap12-part1/;
- http://www.w3.org/TR/soap12-part2/;
- http://www.w3.org/2000/09/XML-Protocol-Charter;
- http://www.w3.org/2002/ws/Activity.html;
- http://www.w3.org/TR/xmlp-reqs/;
- http://www.w3.org/TR/xmlp-am/;
- http://www.w3.org/TR/xmlp-scenarios/.

What is SOAP?

SOAP provides means for exchanging XML-encoded information. It defines a framework for messages that can be exchanged over a variety of underlying protocols such as HTTP or BEEP. The format of SOAP messages is shown in Figure 6.32.

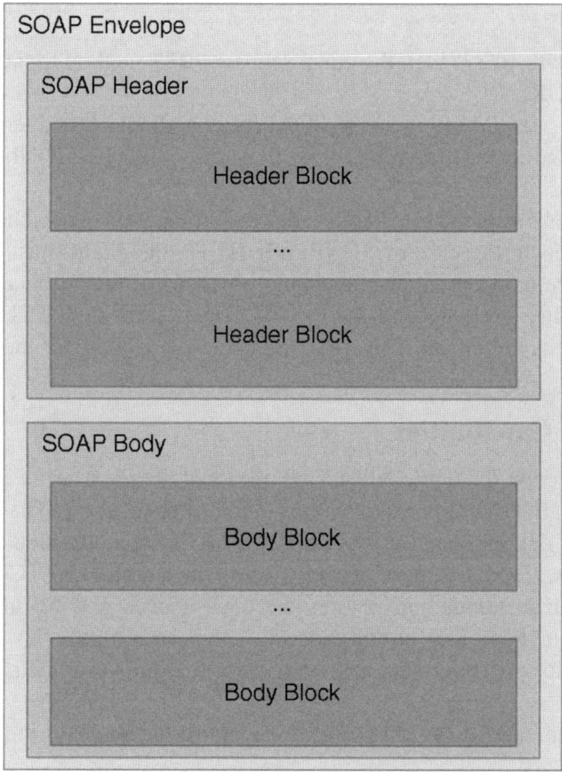

Figure 6.32 Format of SOAP message

Table 6.35 Example of a SOAP message

```
<env:Envelope xmlns:env="http://www.w3.org/2003/05/soap-envelope">
<env:Header>
<n:alertcontrol xmlns:n="http://example.org/alertcontrol">
<n:priority>1</n:priority>
<n:expires>2025-12-21T03:34:34-05:00</n:expires>
</n:alertcontrol>
</env:Header>
<env:Body>
<m:alert xmlns:m="http://example.org/alert">
<m:msg>EXAMPLE EXAMPLE EXAMPLE EXAMPLE
</m:alert>
</env:Body>
</env:Envelope>
```

Table 6.35 provides an example of the content of a SOAP message.

NETCONF over SOAP

Two alternatives can be taken into account in order to initiate a NETCONF session over SOAP:

- *If SOAP is used with HTTP*, then the client sends a **`POST`** method to the NETCONF server identified by an URI.
- *If SOAP is used over BEEP*, then the SOAP profile should be advertized over a BEEP session as specified in the above BEEP section. Then the NETCONF can be initiated.

Once a NETCONF session is established, capabilities can be exchanged between the client and the server. In the case of SOAP over HTTP, the client must send its **`<hello>`** message first. Figure 6.33 provides an example of the capability exchange that occurs.

After the capability exchange phase, the client can send a NETCONF request to the server. These requests are encoded in RPC and enclosed in a SOAP message.

6.5 NETCONF Capabilities

As mentioned in Section 6.2.6, the NETCONF protocol defines a set of capabilities that can be supported by a NETCONF speaker, either a client or a server. The implementation of these capabilities is optional, and, if supported, a NETCONF node should advertize them to its NETCONF peers. Upon receipt of this advertisement, a remote NETCONF peer uses only the capabilities it understands and ignores the ones it does not. In this section, a list of capabilities and their definitions are provided.

For each capability presented, the following information is provided:

- *Dependencies*: indicates if the support of a capability depends on the support of another capability;
- *Capability identifier*: provides the identifier of the capability;

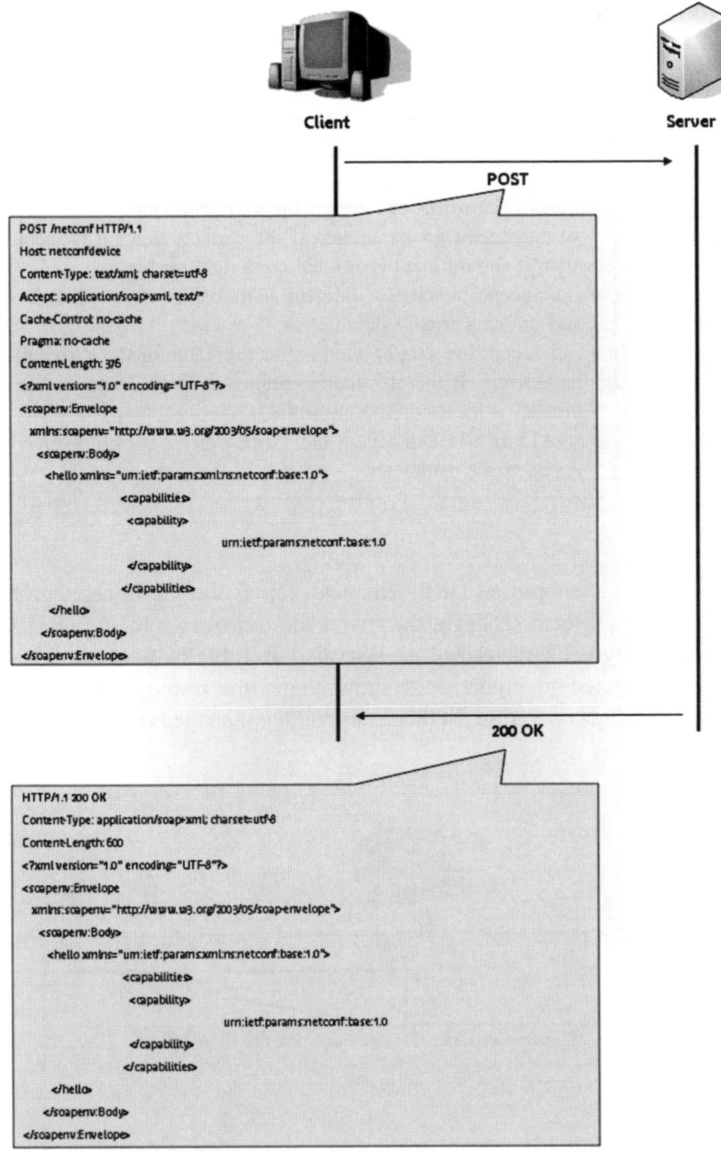

Figure 6.33 Example of capabilities exchange over SOAP

- *Modifications to existing operations*: lists the modifications to NETCONF operations when the capability is supported.

6.5.1 URL Capability

A NETCONF device advertizing the support of **`:url`** capability has the ability to accept the **`<url>`** element in **`<source>`** and target> parameters. The capability is identified by URL

Table 6.36 URL capability

Dependencies	None.
Capability identifier	**urn:ietf:params:netconf:capability:url:1.0?scheme={name,...}** where the 'scheme' argument is assigned to a comma-separated list of scheme names supported by the NETCONF peer such as ftp, http, etc.
Modifications to existing operations	• **<edit-config>** can accept the **<url>** element as an alternative to the **<config>** parameter. If the **<url>** element is specified, then it should identify a local configuration file; • can accept the **<url>** element as the value of the **<source>** and the **<target>** parameters; • can accept the **<url>** element as the value of the **<target>** parameters. If this parameter contains an URL, then it should identify a local configuration file; • **<validate>** can accept the **<url>** element as the value of the **<source>** parameter.

arguments indicating the supported URL schemes such as the file transfer protocol (FTP), the hypertext transfer protocol (HTTP), the trivial file transfer protocol (TFTP), etc.

The **:url** capability is characterized as described in Table 6.36.

The example illustrated in Figure 6.34 showsshows the usage of the **<url>** element inside a **<copy-config>** operation. In this example, the managed device is asked to replace

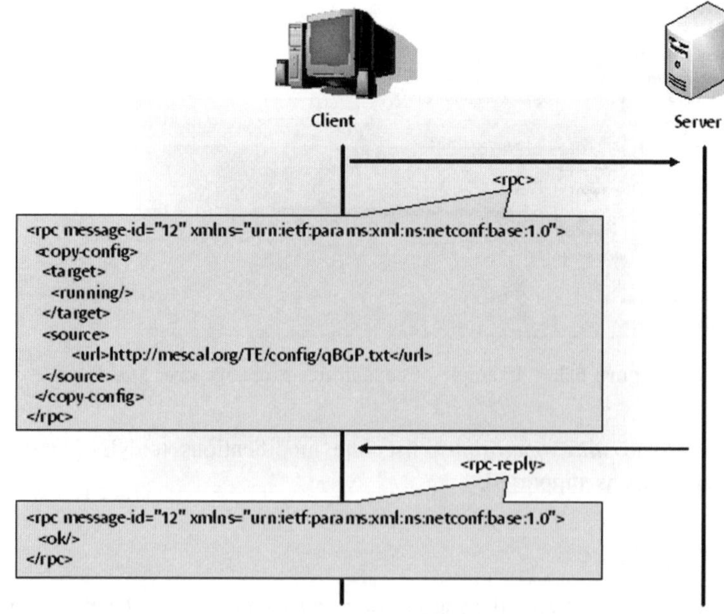

Figure 6.34 Example of **<copy-config>** with **<url>** element

Table 6.37 XPath capability

Dependencies	None.
Capability identifier	**urn:ietf:params:netconf:capability:xpath:1.0**
Modifications to existing operations	**<get-config>** and **<get>** can accept the value **xpath** in the type attribute of the filter element. When the type attribute is set to **xpath**, the contents of the filter element will be treated as an XPath expression and used to filter the returned data.

its running configuration datastore with the content of the file located in the indicated URL.

6.5.2 XPath Capability

A NETCONF device advertising the support of **:xpath** capability ensures its ability to accept the use of XPath expression [38] in the filter element. The XPath expression is evaluated in a context where the context node is the root node, and the set of namespace declarations are those in scope of the filter element, including the default namespace.

This capability is described in Table 6.37.

The following example shows an example of usage of the XPath expression in the filter element of the **<get>** operation. The example illustrated in Figure 6.35 captures a NETCONF call flow between a client and a server. The server is asked to retrieve both state and configuration data related to the session initiation protocol (SIP) channel as

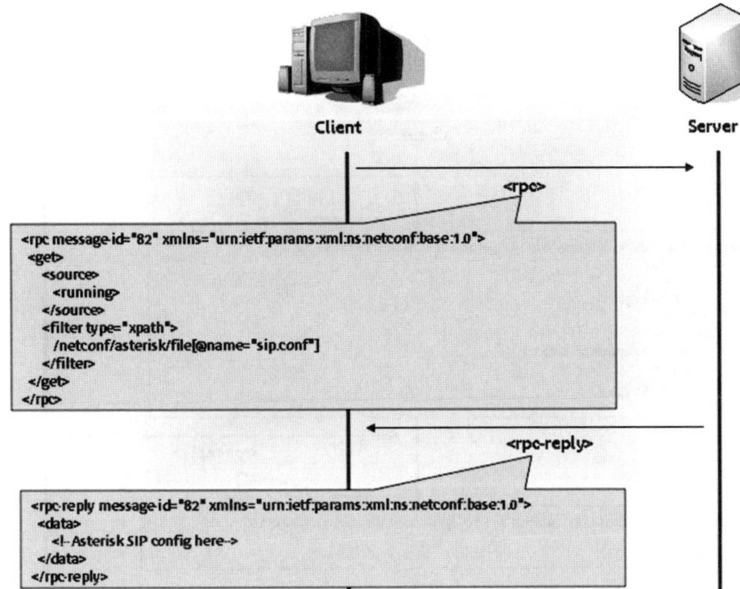

Figure 6.35 Example of **<get>** with Xpath in the filter element

Table 6.38 Writable-running capability

Dependencies	None.
Capability identifier	**urn:ietf:params:netconf:capability: writable-running:1.0**
Modifications to existing operations	**<edit-config>** and **<copy-config>** operations can accept the **<running>** element as a **<target>**.

configured in an asterisk VoIP server (www.asterisk.org). The managed device returns the content of the 'SIP.conf' file and associated state data related to the SIP channel.

6.5.3 Writable-Running Capability

A NETCONF device can advertize its support of **:writable-running** capability if this device supports 'write' operations on the **<running>** configuration datastore.

Table 6.38 provides more information about this capability.

As indicated in Table 6.38, a device that supports the **:writable-running** capability can invoke the **<edit-config>** and **<copy-config>** operations with **<running>** configuration as target (e.g. see Figure 6.36 see which illustrates an example of the **<edit-config>** operation invoked with **<running>** as a target configuration datastore).

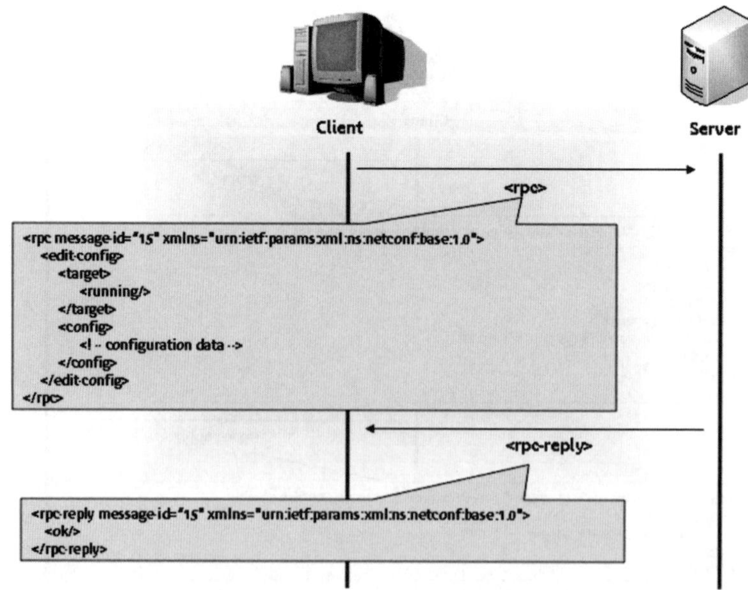

Figure 6.36 Example of **<copy-config>** with **<running>** as a target

Table 6.39 Candidate capability

Dependencies	The : **confirmed-commit** capability is only relevant if the :candidate capability is also supported.
Capability identifier	**urn:ietf:params:netconf:capability: candidate:1.0**
Modifications to existing operations	• The candidate configuration can be used as a source or target of any **<get-config>**, **<edit-config>**, **<copy-config>** or **<validate>** operations. • The candidate configuration can be locked or unlocked using the **<lock>**/**<unlock>** operation with the **<candidate>** element as the **<target>** parameter.

6.5.4 Candidate Configuration Capability

A NETCONF device can advertize that it supports the **:candidate** capability if a **<candidate>** configuration datastore is available on this device. NETCONF operations can be invoked with the **<candidate>** configuration datastore as **<target>** or **<source>** elements. A **<commit>** operation may be performed in order to set the device's running configuration to the value of the candidate configuration. The client can discard any uncommitted changes to the candidate configuration by executing the **<discard-changes>** operation. Note that a **<candidate>** configuration datastore may be shared among several NETCONF sessions. However, it is suitable for a client to lock the **<candidate>** configuration datastore before modifying it.

Table 6.39 provides more information about this capability.

Figure 6.37 provides an example of the usage of **<candidate>** as a target for **<lock>** and **<unlock>** operations.

6.5.5 Confirmed Commit Capability

The **:confirmed-commit** capability indicates that the server supports the **<confirmed>** and **<confirm-timeout>** parameters which can be enclosed in a **<commit>** operation. The timeout period can be adjusted with the **<confirm-timeout>** element. The default value of this timeout is 600 seconds.

The NETCONF server must restore the configuration to its state before the confirmed commit was issued in the following cases:

- The session issuing the confirmed commit is terminated before the confirm timeout expires.
- The managed device reboots for any reason before the confirm timeout expires.
- A confirming commit is not issued.

Table 6.40 provides more precise details about this capability.

Figure 6.38 shows an example of usage of the **<commit>** operation. In this example, the management application sends an **<rpc>** request in order to set the commit timeout to 300 seconds.

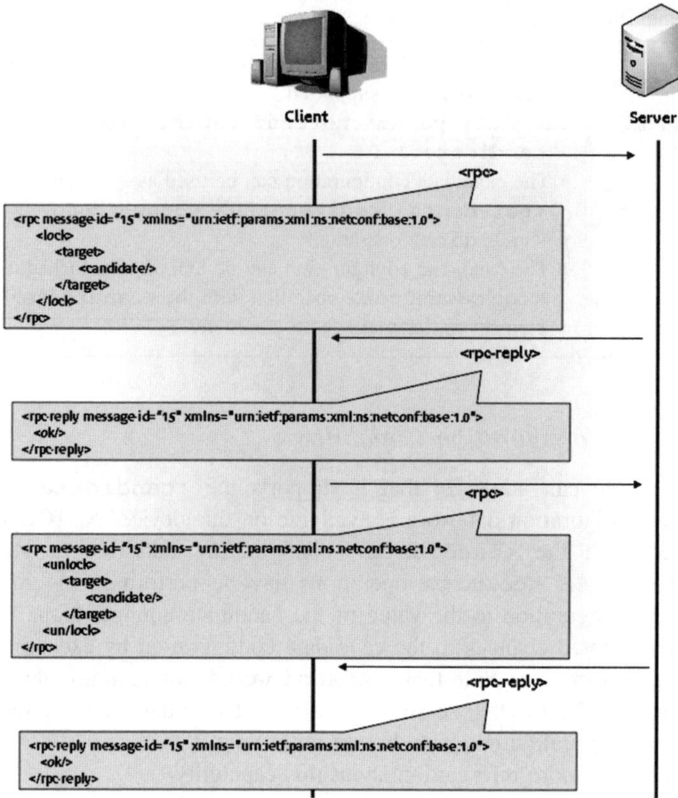

Figure 6.37 Example of **`<lock>`** and **`<unlock>`** operations with **`<candidate>`** as a target

6.5.6 Validate Capability

The support of the **`:validate`** capability means that a NETCONF node can perform checking operations on a candidate configuration for syntactical and semantic errors before the enforcement of the configuration. If this capability is supported, the device supports the **`<validate>`** operation and checks at least for syntax errors.

Table 6.41 provides more precise details about this capability.

Refer to Section 6.3 for an example of the usage of the **`<validate>`** operation.

Table 6.40 Confirmed commit capability

Dependencies	The **`:confirmed-commit`** capability is only relevant if the **`:candidate`** capability is also supported.
Capability identifier	**`urn:ietf:params:netconf:capability:confirmed-commit:1.0`**
Modifications to existing operations	The **`:confirmed-commit`** capability allows two additional parameters to the **`<commit>`** operation: • confirmed: perform a confirmed commit operation; • confirm timeout: timeout period for confirmed commit, in seconds. If unspecified, the confirm timeout defaults to 600 seconds.

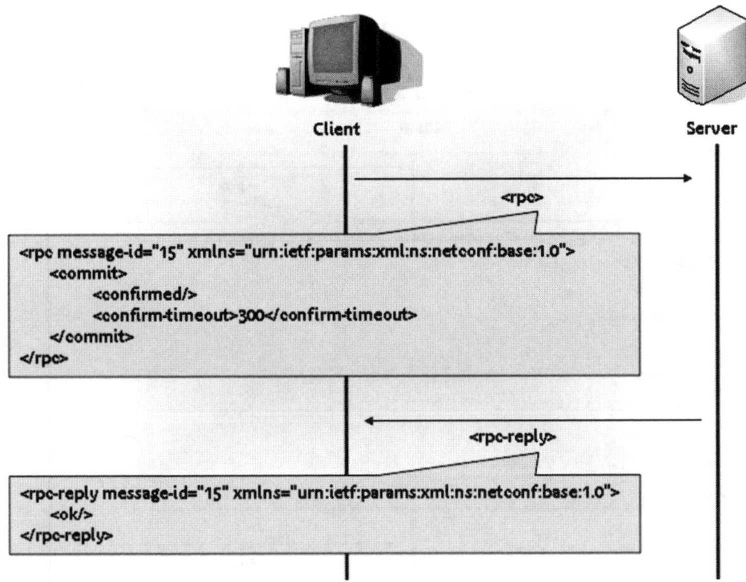

Figure 6.38 Example of **<commit>** with **<confirm-timeout>** as a target

6.5.7 Distinct Startup Capability

A NETCONF device can advertize that it supports the **:startup** capability if a **<startup>** configuration datastore is available on this device. In order to update the **<startup>** configuration datastore to the current **<running>** configuration, a **<copy-config>** operation must be issued. Table 6.42 provides more information about this capability.

Table 6.41 Validate capability

Dependencies	None.
Capability identifier	**urn:ietf:params:netconf:capability:validate:1.0**
Modifications to existing operations	None.

Table 6.42 Distinct startup capability

Dependencies	None.
Capability identifier	**urn:ietf:params:netconf:capability:startup:1.0**
Modifications to existing operations	The **:startup** capability adds the **<startup/>** configuration datastore to arguments of several NETCONF operations. The server must support **<startup>** being used as a **<source>** parameter for **<get-config>** and also for **<validate>** if **:validate** capability is supported, as **<target>** for **<lock>** and **<unlock>** and as both **<source>** and **<target>** for **<copy-config>**.

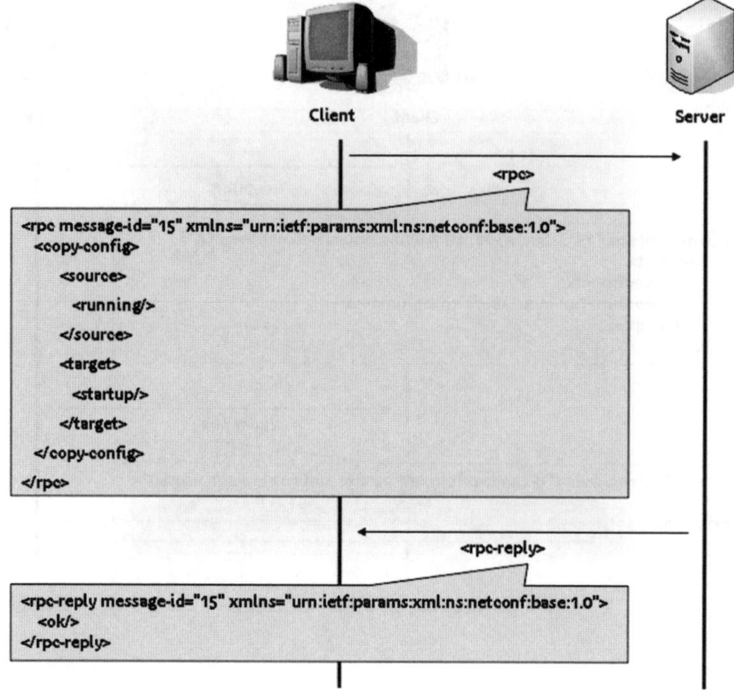

Figure 6.39 Example of **<copy-config>** with **<startup>** as a target

Figure 6.39 provides an example of invoking a **<copy-config>** method to update the content of the **<startup>** configuration with the content of the running configuration.

6.5.8 Rollback on Error Capability

A NETCONF device advertizing the support of **:rollback-on-error** capability can indicate a **rollback-on-error** value in the **<error-option>** parameter response to received **<edit-config>** operations.

Table 6.43 provides some information about this capability.

Table 6.43 Rollback on error capability

Dependencies	None.
Capability identifier	**urn:ietf:params:netconf:capability: rollback-on-error:1.0**
Modifications to existing	**<edit-config>** can have a **rollback-on-error** value to the **<error-option>** parameter

Table 6.44 Notificiation capability

Dependencies	None.
Capability identifier	**urn:ietf:params:netconf:capability: notification:1.0**
Modifications to existing operations	**<subscribe-notification>** and **<notification>** are supported by the managed device.

6.5.9 Notification Capability

The support of notification capability by a given managed device denotes its ability to send event information, statistical data and other changes to applications that subscribed to this service. Other criteria may be enclosed in the notification subscription in order to select event classes of interest to the management application.

Table 6.44 provides some information about this capability.

6.6 Configuring a Network Device

This section provides some guidelines when configuring a network device. It also describes NETCONF operations that should be invoked for correct configuration behaviors. Figure 6.40 presents a NETCONF client communicating with a network device via a NETCONF channel over one of the available transport protocols (SSH, BEEP or SOAP).

In order to configure this network device, the following tasks should be performed by the management application, in this order:

1. *Acquiring the configuration lock.* In order to avoid configuration inconsistency, it is recommended, for a NETCONF client desiring to configure a network device, to acquire the lock on the configuration datastore. This is achieved by invoking the **<lock>** method, as illustrated in example 1 presented in Figure 6.41.
2. *Loading the update.* Then the new configuration can be loaded onto the network device, as illustrated in example 2 of Figure 6.41. If the **:candidate** capability is supported, it is recommended that the candidate datastore be used as a target parameter of NETCONF operations, as illustrated in example 1 of Figure 6.42.
3. *Validating the incoming configuration.* When the configuration is loaded or the candidate datastore is modified, the client is invited to invoke the **<validate>** method, as illustrated in example 3 of Figure 6.41 or in example 2 of Figure 6.42.

Figure 6.40 Configuration use case

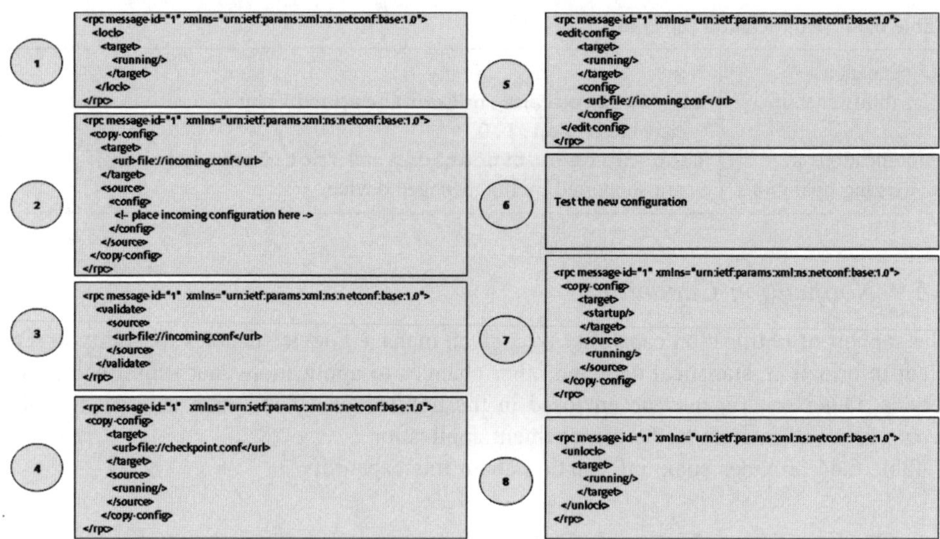

Figure 6.41 Set of NETCONF operations to configure a network device (**:candidate** and **:confirmed-commit** capabilities are not supported)

4. *Saving the running configuration.* In order to avoid misconfiguration, it is recommended that the running datastore be saved before enforcing the new configuration so as to be able to regress to the previous device operating state. This operation can be achieved by invoking the **<copy-config>** method, as illustrated in example 4 of Figure 6.41.

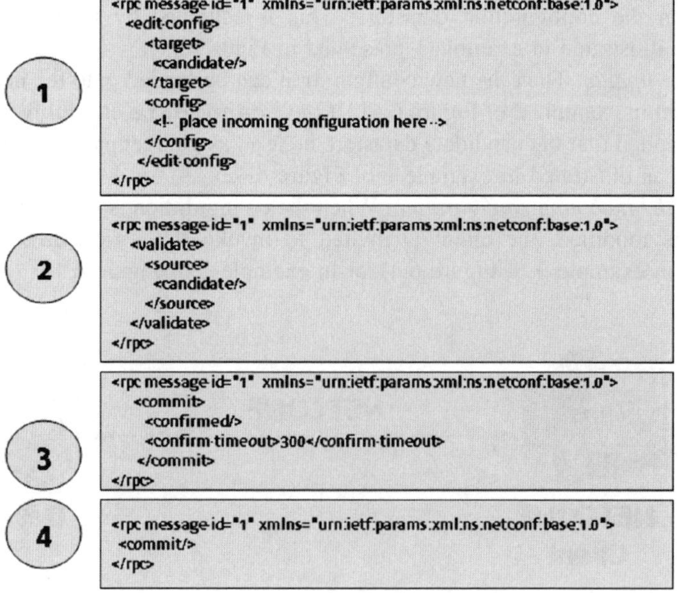

Figure 6.42 Set of NETCONF operations to configure a network device

5. *Changing the running configuration.* Once the running configuration datastore is saved, the client is invited to invoke the **<edit-config>** so as to apply the new configuration or to promote the candidate datastore to the running datastore via a **<commit>** method. See example 5 in Figure 6.41 or example 3 in Figure 6.42.
6. *Testing the new configuration.* When the new configuration is promoted to running state, the manager should test the behavior of the device and conclude if the new configuration is valid or not.
7. *Making the change permanent.* When performed tests have confirmed that the behavior of the device is as expected, the running datastore is copied to the startup datastore, as shown in example 7 in Figure 6.41 or in example 4 in Figure 6.42.
8. *Releasing the configuration lock.* Once all configuration tasks have been performed, the lock should be released. See example 8 in Figure 6.41.

6.7 NETCONF Content Layer

The NETCONF protocol separates protocol operations from data models. Data model design is clearly beyond the scope of the NETCONF working group (at least the current charter of the WG). The design of data models that will be conveyed in NETCONF messages should be developed by application developers themselves. Within the IETF, an initiative to promote NETCONF data models has been launched. At least two NETMOD BOF have been held during previous IETF meetings so as to gather the community interest and to argue the need for clear definitions of data models, but no working group has been created yet.

References

[1] Enns, R. *et al.*, 'NETCONF Configuration Protocol', RFC 4741, December 2006.
[2] Case, J. *et al.*, 'Introduction and Applicability Statements for Internet Standard Management Framework', RFC 3410, December 2002.
[3] Boyle, J., Cohen, R., Durham, D., Herzog, S., Raja, R. and Sastry, A., 'The COPS (Common Open Policy Service) Protocol', RFC 2748, January 2000.
[4] Chan, K. *et al.*, 'COPS Usage for Policy Provisioning (COPS-PR)', RFC 3084, March 2001.
[5] Distributed Management Task Force, 'Common Information Model (CIM) Specification Version 2.2', DSP 0004, June 1999.
[6] McCloghrie, K., Perkins, D. and Schoenwaelder J., 'Structure of Management Information Version 2 (SMIv2)', STD 58, RFC 2578, April 1999.
[7] McCloghrie, K., Fine, M., Seligson, J., Chan, K., Hahn, S., Sahita, R., Smith, A. and Reichmeyer, F., 'Structure of Policy Provisioning Information (SPPI)', RFC 3159, August 2001.
[8] Schoenwaelder, J., 'Overview of the 2002 IAB Network Management Workshop', May 2003.
[9] Wasserman, M. and Goddard, T., 'Using the NETCONF Configuration Protocol over Secure Shell (SSH)', RFC 4742, December 2006.
[10] Lear, E. and Crozier, K., 'Using the NETCONF Protocol over Blocks Extensible Exchange Protocol (BEEP)', RFC 4744, December 2006.
[11] Goddard, T., 'Using the Network Configuration Protocol (NETCONF) over the Simple Object Access Protocol (SOAP)', RFC 4743, December 2006.
[12] Lehtinen, S. and Lonvick, C., 'The Secure Shell (SSH) Protocol Assigned Numbers', RFC 4250, January 2006.

[13] Lonvick, C. *et al.*, 'The Secure Shell (SSH) Protocol Architecture', RFC 4251, January 2006.
[14] Ylonen, T. and Lonvick, C., 'The Secure Shell (SSH) Authentication Protocol', RFC 4252, January 2006.
[15] Ylonen, T. and Lonvick, C., 'The Secure Shell (SSH) Transport Layer Protocol', RFC 4253, January 2006.
[16] Ylonen, T. and Lonvick, C., 'The Secure Shell (SSH) Connection Protocol', RFC 4254, January 2006.
[17] Schlyter, J. and Griffin, W., 'Using DNS to Securely Publish Secure Shell (SSH) Key Fingerprints', RFC 4255, January 2006.
[18] Cusack, F. and Forssen, M., 'Generic Message Exchange Authentication for the Secure Shell Protocol (SSH)', RFC 4256, January 2006.
[19] Bellare, M., Kohno, T. and Namprempre, C., 'The Secure Shell (SSH) Transport Layer Encryption Modes', RFC 4344, January 2006.
[20] Galbraith, J. and Remaker, P., 'The Secure Shell (SSH) Session Channel Break Extension', RFC 4335, January 2006.
[21] Rose, M., 'The Blocks Extensible Exchange Protocol Core', RFC 3080, March 2001.
[22] Rose, M., 'Mapping the BEEP Core onto TCP', RFC 3081, March 2001.
[23] Rose, M., 'On the Design of Application Protocols', RFC 3117, November 2001.
[24] New, D. and Rose, M., 'Reliable Delivery for syslog', RFC 3195, November 2001.
[25] Rose, M., 'A Transient Prefix for Identifying Profiles under Development by the Working Groups of the Internet Engineering Task Force', RFC 3349, July 2002.
[26] New, D., 'The TUNNEL profile', RFC 3620, October 2003.
[27] Farrell, S. *et al.*, 'Securely Available Credentials Protocol', RFC 3767, June 2004.
[28] Newton, A. and Sanz, M., 'Using the Internet Registry Information Service (IRIS) over the Blocks Extensible Exchange Protocol (BEEP)', RFC 3983, January 2005.
[29] O'Tuathail and Rose, M., 'Using the Simple Object Access Protocol (SOAP) in Blocks Extensible Exchange Protocol (BEEP), RFC 4227, January 2006.
[30] Gudgin, M., Hadley, M., Moreau, J.J. and Nielsen, H., 'SOAP Version 1.2 Part 1: Messaging Framework', W3C Recommendation REC-soap12-part1-20030624, June 2002, http://www.w3.org/TR/soap12-part1/.
[31] Yergeau, F., Bray, T., Paoli, J., Sperberg-McQueen, C.M. and Maler, E., 'Extensible Markup Language (XML) 1.0 (Third Edition)', W3C Recommendation, 4 February 2004.
[32] Berners-Lee, T., Fielding, R. and Masinter, L., 'Uniform Resource Identifiers (URI): Generic Syntax', RFC 2396, August 1998.
[33] Chisholm, S. *et al.*, 'NETCONF Event Notifications', draft-ietf-netconf-notification-08.txt, July 2007 (work in progress).
[34] Chisholm, S. and Adwankar, S., 'Framework for NETCONF Data Models', ID draft-chisholm-netconf-model-06, January 2007.
[35] Rigney, C., Willens, S., Rubens, A. and Simpson, W., 'Remote Authentication Dial-In User Service (RADIUS)', RFC 2865, June 2000.
[36] Dierks, T. and Allen, C., 'The TLS Protocol Version 1.0', RFC 2246, January 1999.
[37] Myers, J., 'Simple Authentication and Security Layer (SASL)', RFC 2222, October 1997.
[38] Clark, J. and DeRose, S., 'XML Path Language (XPath) Version, 1.0', W3C REC REC-xpath-19991116, November 1999.

7

Control and Provisioning of Wireless Access Points (CAPWAP)

In recent years, radio access technologies have rapidly evolved, often deprecating the well-known RJ45-cable widely used to access companies' network first, or to reach the Internet gateway at home. Another usage of radio technology has also emerged: extending network access infrastructure for network operators, to provide Internet access to roaming and mobile users. Control and provisioning of wireless access points (CAPWAP) intends to address access point management challenges, especially when massive deployment of this equipment exists. So far, home access is not affected, but enterprises and network operators have to deal with the difficulties of a well-managed, secured, wireless access infrastructure composed of hundreds or thousands of access points. Whereas one might only think of the widespread 802.11 'WiFi' technology, CAPWAP is also open to the 802.16 'WiMax' standard, taking into consideration that 802.11 might not be suitable for network operators to provide an acceptable coverage in public areas without a considerable investment effort to deploy access points.

At the time of writing this book, CAPWAP is not a standardized protocol yet, and the ongoing work for standardizing this protocol reflects the efforts done on management, control and provisioning of access points (APs), aiming to have interoperating APs developed from various vendors. The main goal of CAPWAP is to provide an interoperable solution to centralize authentication and policy enforcement functions. This effort is part of the IETF, coordinating with IEEE 802 on this specific task.

This chapter will present wireless LAN (WLAN) operational challenges as identified by CAPWAP, as well as specific CAPWAP concepts and terminology. CAPWAP objectives will then be presented, followed by a brief presentation of candidate protocols that have been reviewed to address CAPWAP requirements, and a more detailed review of the lightweight

Service Automation and Dynamic Provisioning Techniques in IP/MPLS Environments C. Jacquenet, G. Bourdon and M. Boucadair

access point protocol (LWAPP), which has been chosen and modified to be the CAPWAP protocol, and renamed accordingly.

7.1 CAPWAP to Address Access Point Provisioning Challenges

As a matter of fact, WLAN access technologies bring along a set of new issues that have to be addressed specifically. These issues are presented in RFC 3990 [1] and are the foundation of CAPWAP developments which will be detailed later in this chapter. Basically, four major problems have been identified in managing wireless access devices:

- The limited coverage of radio technology used (especially with 802.11) requires a massive deployment of access points. All these devices are network elements that require monitoring, management and control, on a large scale. Often, device configurations are equivalent but not always identical: the problem lies in the need for a very rigorous management framework.
- The large number of APs deployed by a network administrator makes its job very difficult during the configuration phase: maintaining a consistent configuration among all access points constituting the access infrastructure is a challenge. Beyond the time required for configuration (often implying service disruption), the network remains in an inconsistent state for a certain period of time.
- The WLAN technology being used is very sensitive and often requires specific tweaks depending on the AP location: there may be radio interferences, or too many users sharing the same AP. All these difficulties require a specific consideration of each network access point. Periodic and rigorous monitoring is required, as well as time and competences to manage the access infrastructure properly.
- Securing the network is essential: access points are supposed to be deployed in every public place, and it is important for a network administrator to prevent malicious users from plugging in to install a rogue access point, or to steal security parameters that would have been configured in the access device.

These problems are well known for network operators and access point vendors. Of course, proprietary solutions and implementations already exist to address these issues. But here comes the fifth and transversal management challenge: interoperability. It is nowadays impossible for network administrators to have a plain and consistent view of their wireless access infrastructure, except if all devices are provided by the same vendor.

7.2 CAPWAP Concepts and Terminology

One of the first CAPWAP tasks was to provide an architecture taxonomy of various WLAN accesses, in order to give a structure and a common vocabulary to concepts that would be manipulated to build an interoperable management system for access points. This work is documented in RFC 4118 [2] and proposes the grouping of concepts adopted by AP vendors, extracted from a survey conducted among a dozen of them. Furthermore, this document provides a specific terminology related to the developed concepts.

WLAN architectures are composed of various elements (see Figure 7.1): stations (STA) and access points (APs) are elements that any network engineer can easily figure out. Other concepts have a specific designation, such as the Basic Service Set (BSS), usually identified in station configuration wizards as the SSID. The Extended Service Set (ESS) represents a common BSS usage made by a set of APs connected by the Distribution System (DS). The DS provides network resources for several APs to communicate between each other. Using these communication facilities, APs and the DS make a wide wireless network identified as a large BSS: the ESS.

The AP by itself is an important concept used in CAPWAP. It is not only a network element that enables a station to reach a network, it also breaks down into functional elements that may be spread at different places within the DS, depending on the architecture:

- The Wireless Termination Point (WTP) is the physical device in charge of transmitting data from and to the station.
- The Access Controller (AC) (also called the WLAN controller) implements WLAN logical functions. More generally, it is the equipment that centralizes authentication and policy enforcement functions for wireless networks.

CAPWAP introduces the notion of CAPWAP functions and 802.11 functions. 802.11 functions can be defined as mechanisms used to make possible 'over-the-air' communications between two entities. Among the numerous 802.11 functions, one can mention association and authentication phases, but also 802 frame transmission and retransmission

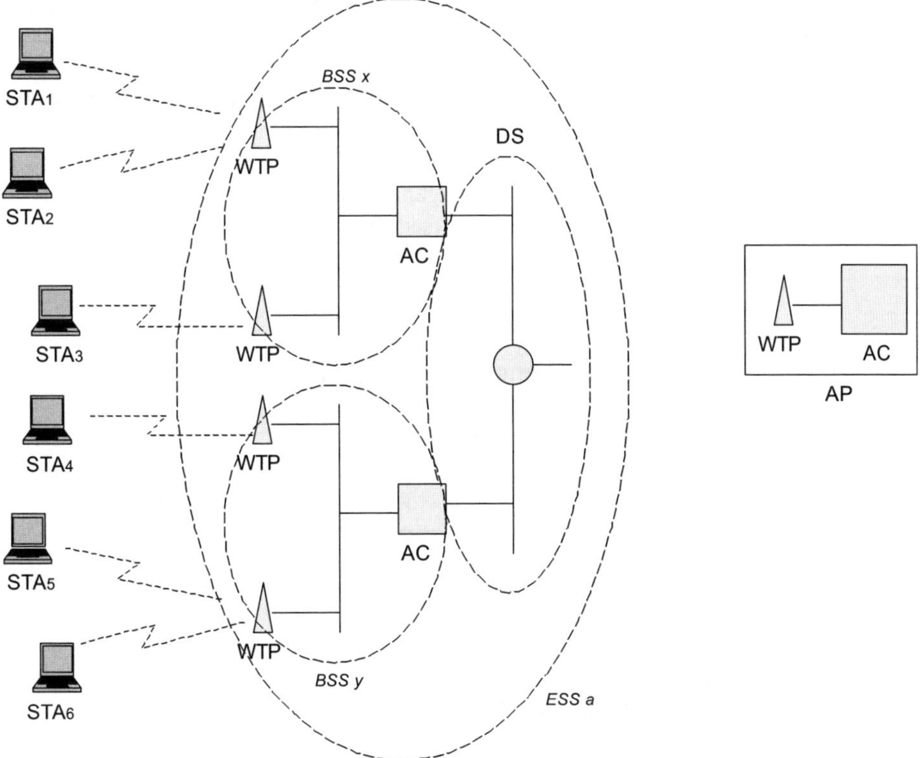

Figure 7.1 CAPWAP concepts and terminology

machinery. CAPWAP functions are defined to centralize authentication and policy enforcement, and are focused on radio management, configuration and monitoring, WTP or AC configuration or firmware upgrade, for instance. This distinction is essential to understand the difference between the level of 'control' functions: some control functions are exclusively related to the 802.11 transmission layer, whereas other management and provisioning functions of WLAN equipment are CAPWAP control functions.

From the logical separation of WTP and AC network functions, three distinct types of architecture have been identified in vendor solutions analyzed within the RFC 4118 survey [2]:

- *Autonomous WLAN architecture.* This architecture is formed by one or a set of stand-alone APs that embed both WTP and AC functions. This kind of AP is often used in home environments as well as in companies, usually manageable through a web interface or SNMP. This type of architecture does not provide any resource sharing, as each access point is completely autonomous.
- *Centralized WLAN architecture.* This architecture is hierarchical in the sense that multiple WTPs are connected to a centralized access controller that controls and manages WTPs, but also may be in the data path to achieve layer-2 or layer-3 treatment over the traffic coming from and to WTPs. In this kind of architecture, network administrators can manage WTPs by accessing to the centralized access controller. The centralized WLAN architecture itself has three variants:
  - split MAC architecture: some time-sensitive 802.11 management frames are processed in the WTP, whereas other management frames are transmitted and treated by the AC (see Figure 7.2);
  - remote MAC architecture: no 802.11 management frames are processed by the WTP, all frames are processed at the AC level. The WTP only ensures 802.11 PHY functions (see Figure 7.3);
  - local MAC architecture: most 802.11 management frames are handled by the WTP, the whole MAC function is local to the WTP (see Figure 7.4).

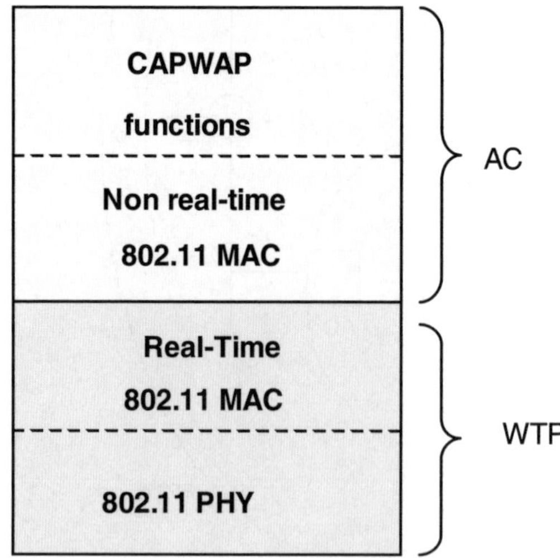

Figure 7.2 Function repartition with split MAC architecture

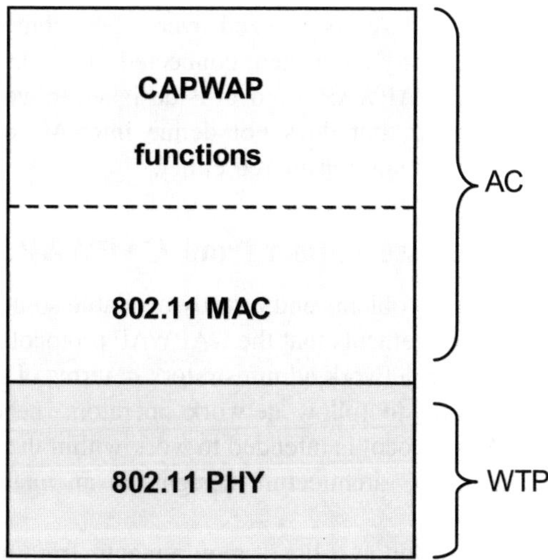

Figure 7.3 Function repartition with remote MAC architecture

- *Distributed WLAN architecture.* In this architecture, all wireless nodes are potentially part of the network infrastructure. Nodes are interconnected to each other through a 802.11 link, or wired link, and thus form a distributed and meshed network of wireless nodes.

It is interesting to note that CAPWAP does not intend to take into account the remote MAC subarchitecture described above for the CAPWAP protocol definition: the WTP is acting as a simple data transfer pass-through, and it is foreseen that CAPWAP will not provide any help to manage this kind of equipment. Therefore, the CAPWAP framework stays essentially focused on the centralized WLAN architecture in its split MAC and local MAC designs.

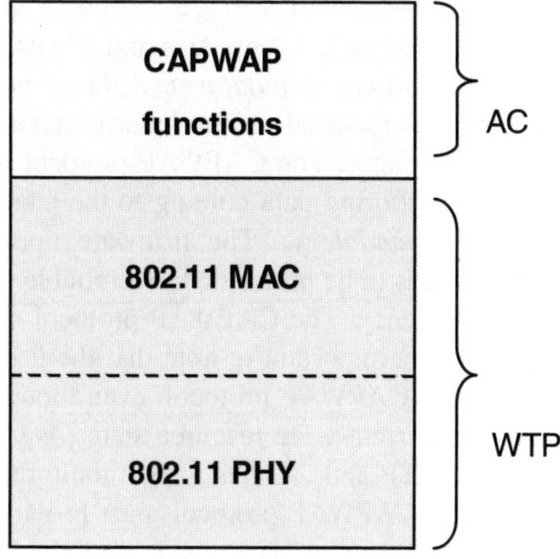

Figure 7.4 Function repartition with local MAC architecture

Connectivity between WTP and AC is realized over a 'switching segment' and can be achieved in different modes: directly connected, connected over a layer-2 (switch) element or connected over a router. The CAPWAP protocol is designed to work between WTPs and ACs over the switching segment, but does not define inter-AC communication. Direct communication between WTPs is not considered either.

7.3 Objectives: What do we Expect from CAPWAP?

Based upon the list of identified problems and already available solutions and architectures, CAPWAP proposed a set of requirements that the CAPWAP protocol has to meet in order to provide satisfactory solutions for network administrators in terms of security, operation and architecture, and more generally to follow network operators' needs, as documented in RFC 4564 [3]. The CAPWAP protocol is intended to work within the CAPWAP framework, i.e. in split MAC and local MAC architecture designs, in an interoperable and scalable manner.

The CAPWAP protocol has to follow a list of requirements, from mandatory to desirable, in the different aforementioned categories, namely general, security, operation and architecture:

- Logical groups (*architecture, mandatory*). A logical group can be defined by the concurrent usage of a physical WTP that can be used for different SSIDs. By using a single physical infrastructure, the network administrator is able to create multiple logical networks (virtual APs). The CAPWAP protocol must be compatible with this usage, by enabling management of separated logical groups over a single physical network.
- Support for traffic separation (*operations, mandatory*). As the CAPWAP protocol addresses communications between WTPs and ACs, data and control traffic have to be separated. This is important since it is usual to deploy logical groups, and therefore traffic separation is required for security reasons. Traffic separation also gives the opportunity to prioritize data and control traffic differently in the case of congestion, for instance.
- Wireless terminal transparency (*operations, mandatory*). Using the CAPWAP protocol must not have any compatibility issues with wireless terminals (stations), considering that CAPWAP communications only occur between ACs and WTPs.
- Configuration consistency (*operations, mandatory*). A large number of WTPs can be managed by a single AC, and it is required for the latter to have a consistent view of the current WTP configurations and states. The CAPWAP protocol must propose a mechanism to provide regular WTP monitoring data coming to the attached AC.
- Firmware trigger (*operations, mandatory*). The firmware update is usually a crucial administrative operation that needs to be performed in a reliable and fast manner, in order to ensure a plain WLAN consistency. The CAPWAP protocol must provide a trigger to start a WTP firmware update. It is important to note that the firmware transfer operation itself is not a requirement of the CAPWAP protocol, even though it can be supported.
- Monitoring and exchange of a system-wide resource state (*operations, mandatory*). The wireless environment of both WTP and AC has to be monitored as well, and has to be available for these devices. The CAPWAP protocol must provide a mechanism to allow bidirectional exchange of monitoring information regarding the wireless environment (such as the wireless medium status and the switched segment status).

- Resource control (*operations, mandatory*). The quality of service provided to the end-user is directly related to the parameters applied at both the wireless medium and the switched segment. By nature, QoS parameters are different for these two types of environment. The CAPWAP protocol must be able to provide AC and WTP QoS parameters corresponding to each medium constraint (e.g. providing 802.11e QoS parameters for the WLAN segment), and with appropriate adaptation in order to ensure coordination of QoS policies on both segments. Other IEEE 802 network resource parameters (still under definition) can also be applied to enhance user experience: the CAPWAP protocol must be able to adapt itself to these new parameters.
- Protocol security (*security, mandatory*). Owing to the sensitive nature of the WTP/AC exchanges, it is required for the CAPWAP protocol to provide a mechanism that allows mutual authentication between entities (in order to prevent a rogue WTP from connecting to the AC, compromising the whole WLAN infrastructure), as well as mechanisms to secure exchanges occurring between these entities. CAPWAP protocol must provide message integrity, authenticity and confidentiality.
- System-wide security (*security, mandatory*). Security threats are likely to come from the wireless segment, in addition to the switched segment mentioned above. The CAPWAP protocol has to be designed in a way that the centralized architecture cannot be compromised by attacks coming from malicious external (wireless) terminals.
- IEEE 802.11i considerations (*operations, mandatory*). Authentication and data encryption are obviously widely used in wireless environments. Usually, the stand-alone AP ensures the whole IEEE 802.11i operations, being both authenticator and encryption point. Since the centralized WLAN architecture splits the AP into two distinct parts (WTP and AC), applying 802.11i security is definitely a challenge. Depending on the subarchitecture used (local MAC or split MAC), the same security mechanisms can be applied in different equipment. The authenticator function is usually provided by the AC, but the encryption can be performed either by the AC (local MAC) or by the WTP (split MAC). When the WTP is in charge of encryption, the AC must provide appropriate keys to the WTP. In general, the CAPWAP protocol must be able to determine the exact role of each device in 802.11i security mechanisms, and provide appropriate transfer methods to provide keys whenever the WTP is the encryption point.
- Interoperability (*architecture, mandatory*). An access controller can be compatible with both split MAC and local MAC subarchitectures. Therefore, the CAPWAP protocol must provide mechanisms that enable the AC to determine actual WTP capabilities, in order to run different kinds of hardware on a single WLAN infrastructure. The required capability exchange mechanism must accommodate different modes of the split MAC case: MAC functions are split into two different pieces of equipment, but precise MAC functions have to be determined in order for WTP and AC to interoperate properly. The definition and denomination of each MAC function is provided by the IEEE 802.11 AP Functionality Ad-Hoc Committee.
- NAT traversal (*general, mandatory*). It is possible that, in some circumstances, the access controller and WTP can reach each other through a NAT gateway. In this case, the CAPWAP protocol must still operate as long as the NAT configuration is known from the parties.
- Multiple authentication mechanisms (*architecture, desirable*). The use of multiple logical groups may require different kinds of authentication procedure: one logical group may

adopt 802.11i, and another may choose to authenticate with PANA, or any other authentication protocol. The CAPWAP protocol should provide enough flexibility to configure logical groups with different authentication mechanisms.
- Support for future wireless technologies (*architecture, desirable*). Even though 802.11 access technology is widely available in public areas, it is also very likely that other technologies will be largely deployed in the future, trying to overcome WiFi drawbacks. One can think of WiMAX (IEEE 802.16), but there should not be any limitations in the access technology used for CAPWAP to work properly.
- Support for new IEEE requirements (*architecture, desirable*). IEEE 802.11 committees are still very active, and enhancements to the standards are regularly proposed. The CAPWAP protocol should remain compatible with these enhancements, considering that minor adaptation may be required to the protocol to integrate it.
- Interconnection (*architecture, desirable*). CAPWAP is an IP protocol, and therefore is intended to work over IPv4 and IPv6. This is the most efficient way to ensure interconnection compatibility between ACs and WTPs. In general, CAPWAP must not be dependent on the transport method chosen by the network administrator.
- Access control (*operations, desirable*). Additional information may be required by the AC to determine the exact nature of wireless terminal access or WTP access. For instance, the initial attachment of a WLAN terminal is different from a secondary attachment required in mobility situations.

The CAPWAP protocol also has to answer to more general requirements, such as vendor independence or vendor flexibility, where CAPWAP must not stick to any vendor architecture solution implementation but should be flexible enough to provide ways for vendors to provide specific enhancements to a basic set of features. A specific operator need is also taken into consideration, especially in what concerns the AP fast handoff: in mobility scenarios, the CAPWAP protocol must not affect the delay of fast handoff procedures. However CAPWAP may support optimizations that help to provide fast handoff.

7.4 CAPWAP Candidate Protocols

Many protocols have been put forward as candidates for the CAPWAP protocol. All these protocols have been evaluated in RFC 4565 [4] through a complete list of objectives which are mostly described above:

- CAPWAP tunneling protocol (CTP) [5];
- secure light access point protocol (SLAPP);
- wireless LAN control protocol (WICOP);
- lightweight access point protocol (LWAPP) [6].

The LWAPP protocol was chosen to be the CAPWAP protocol, as it complied with all requirements. Other candidates were not that far behind LWAPP in this evaluation: for example, SLAPP offered only partial compliance to three objectives, but CTP and WICOP failed to comply with security requirements.

Only the CAPWAP protocol (ex-LWAPP) will be described in the following section, as other candidate protocols will likely disappear in the future. Since CAPWAP modified large pieces of LWAPP, the latter will follow its own standardization path from now on.

7.5 The CAPWAP Protocol

Although the CAPWAP protocol (based on LWAPP) is not 100 % finished yet, the most important parts are present, and have already been discussed for years. Some implementations are already available, and major actors of the WLAN world have stated their intent to adopt this protocol in the future. There is no doubt that the CAPWAP protocol has a very promising future as a WLAN management protocol.

The LWAPP protocol was first designed in the context of seamless mobility at the IETF, in the SEAMOBY working group which was disbanded in fall 2004. Most of the SEAMOBY discussions have shifted to the MOBOPTS research group of the IRTF, but the provisioning part of the SEAMOBY activities found its second breathe within CAPWAP. LWAPP was adapted to CAPWAP objectives, specifically to support the local MAC architecture, since only split MAC was supported at first. For IETF, having the LWAPP protocol was a real opportunity to go ahead at a fast pace, meeting its requirements almost perfectly.

CAPWAP [7] is an extensible protocol that defines a protocol machinery and a set of messages intended to provide answers to CAPWAP objectives. Some CAPWAP messages are seen as generic, since they can be used in every environment making use of WTP and AC concepts. These messages are not technology specific, and are defined to orchestrate management exchanges between WTP and AC. CAPWAP also leaves space for specific messages that are more tied to the WLAN technology used: these messages define the 'binding' of CAPWAP to a specific WLAN technology. These bindings are not defined within the CAPWAP protocol specification, but rather in companion documents: the binding for 802.11 is defined in [8], and other bindings that will be needed in the future for other technologies such as 802.16 should be specified in the future. In order to do so, CAPWAP has been defined to make possible protocol extensions by extending the predefined set of messages and the pertaining content.

Even though the CAPWAP protocol is based upon LWAPP, some adaptations were required completely to fulfill CAPWAP objectives. As LWAPP continues its own life and evolution in the standardization process, CAPWAP imposes the use of DTLS (RFC 4347 [9]) for securing the communications between WTP and AC (inherited from the SLAPP security framework), and suppresses the possibility offered by LWAPP to convey frames from one entity to the other using its proprietary L2 encapsulation. DTLS encryption is only mandatory for control packets, and is optional for data frames. DTLS session establishment between WTP and AC is tightly bound to the CAPWAP state machine, and replaces the proprietary solution developed for LWAPP.

A CAPWAP key characteristic is to convey 802.11 frames from the WTP to the AC, using UDP/IPv4 or UDP-Lite/IPv6 as transport technology, and thereby complying with the connectivity modes shown in Figure 7.5. In split MAC mode, all 802.11 data and CAPWAP management frames are transmitted to the AC, whereas the local MAC mode makes it possible to tunnel management frames to the AC, and to bridge data frames locally or tunnel them in 802.3 frames to the AC.

CAPWAP data frames and control packets can easily be distinguished from each other since they make use of different UDP ports. Each CAPWAP packet embeds a specific

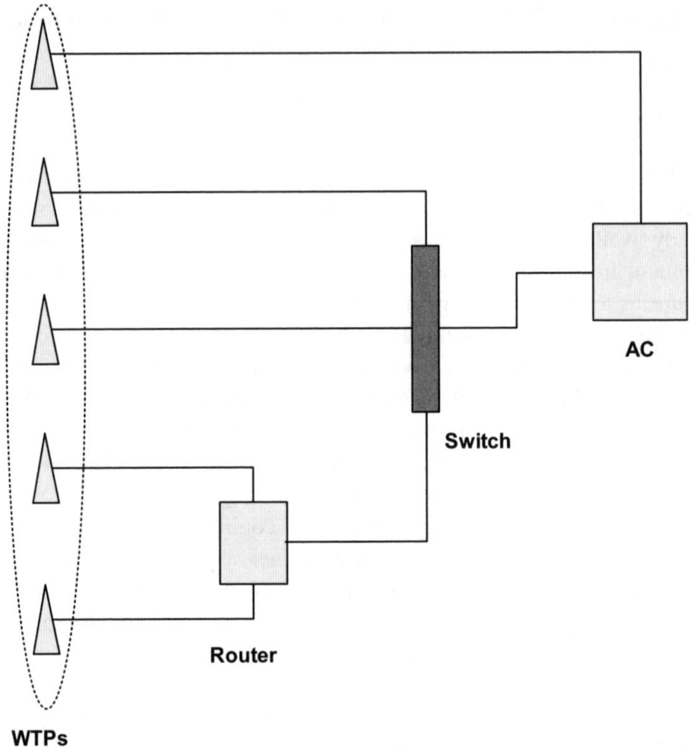

Figure 7.5 Possible connections between WTPs and an AC

CAPWAP header which is followed by a payload, constituting either a control header (followed by specific control messages) or the 802 data frame coming from the wireless terminal. All CAPWAP control packets are themselves encapsulated within a DTLS packet, except for discovery packets since DTLS session establishment is not performed yet at this stage. In order to optimize detection between DTLS-encrypted and unsecured CAPWAP packets, a common preamble has been defined, indicating the type of CAPWAP packet (DTLS or not). A view of the different types of possible CAPWAP packet is shown in Figure 7.6.

In spite of its use of UDP, the CAPWAP protocol has been designed to provide a reliable transportation for control messages. Firstly, each message conveying important provisioning information has a corresponding and explicit acknowledgment message. Secondly, a 'keep-alive' mechanism sends messages regularly between WTP and AC to ensure connectivity aliveness. Finally, CAPWAP defines several timers that must be used in order to coordinate WTP to AC exchanges, detect connectivity failure and ensure a proper backup procedure. It is important to note that CAPWAP messages are not ordered, except for fragmentation purposes: user data traffic ordering should be performed by upper layers (using the TCP/IP layer, for instance), and CAPWAP control messages have no specific need for that. A sequencing mechanism is defined in order to match each response message to the corresponding request. This way, multiple requests can be sent concurrently, but the protocol

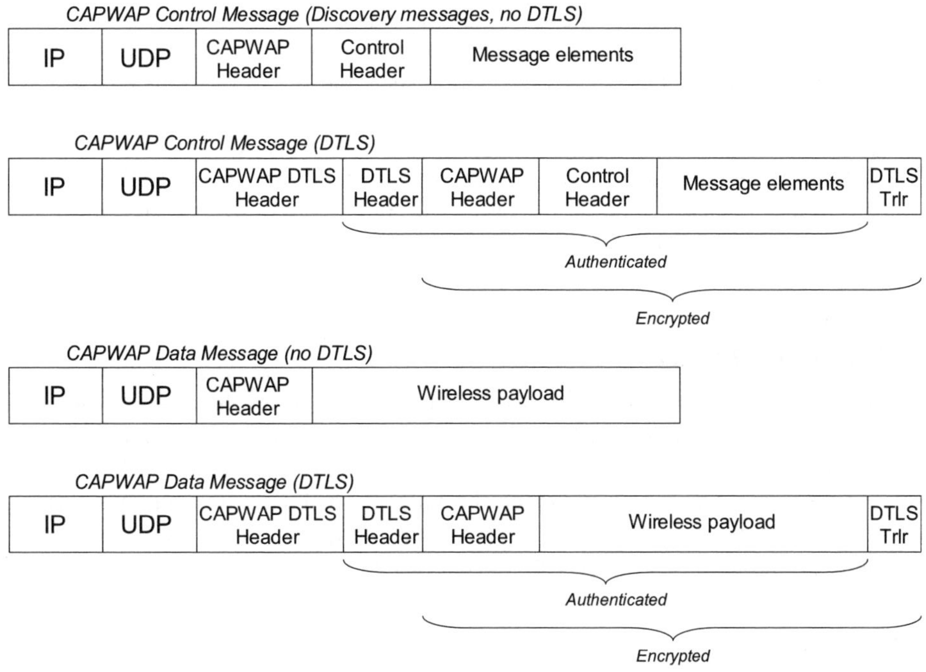

Figure 7.6 CAPWAP packets

is not defined to use this sequence number as a hint to order control messages. When massive management data have to be sent between WTP and AC (e.g. firmware download, debug info, etc.), each control message used for this is specifically acknowledged before the next to come. While this way to operate can be seen as an inefficient way to transfer a large amount of data, it has the advantage of simplifying the protocol machinery by avoiding sophisticated features only used for exceptional events.

CAPWAP control messages are themselves composed of 'message elements', which are equivalent to attributes in other protocols (RADIUS, Diameter) and are transmitted using the TLV format. Control elements are defined following the required information that has to be sent between WTP and AC, in adapted control messages.

CAPWAP operations are separated in distinct phases that have to be performed successively:

- *Discovery phase.* During this phase, wireless termination points run a discovery mechanism in order to gain knowledge of reachable access controllers with which it will be possible to establish a DTLS session.
- *DTLS session establishment phase.* The WTP and the AC start an exchange to establish a DTLS session that will be used to authenticate the messages and to encrypt them. After this phase, communication between the WTP and the AC is encrypted and authenticated.
- *Joining phase.* Once the DTLS session is established between the WTP and the AC, the WTP initiates the joining phase to check whether the image version present on the WTP is up-to-date.

- *Configuration phase.* During the configuration phase, the WTP will send its global configuration parameters, as well as specific parameters related to its WLAN technology binding. Depending on the WTP configuration state, the AC sends to the attached WTP the proper configuration to run.
- *Running phase.* Once in running phase, the WTP is operational. Other configuration steps can be needed on a dynamic basis, depending on the AC decision. The WTP can also initiate from this phase a special discovery and joining phase when a predefined – and preferred – AC is available.

7.6 CAPWAP Future

All concepts developed within CAPWAP will likely be used in the future, even though technical solutions are not yet ready. The CAPWAP protocol still requires some work to be properly defined at this time, and some consistency issues in the protocol definition still remain. However, its large review among the Internet community (especially on security issues), its wide support among WLAN vendors and the availability of partial implementation within products on sale are very promising factors for the future.

References

[1] O'Hara, B. *et al.*, 'Configuration and Provisioning for Wireless Access Points (CAPWAP) Problem Statement', RFC 3990, February 2005.
[2] Yang, L. *et al.*, 'Architecture Taxonomy for Control and Provisioning of Wireless Access Points (CAPWAP)', RFC 4118, June 2005.
[3] Govindan, S. *et al.*, 'Objectives for Control and Provisioning of Wireless Access Points', RFC 4564, July 2006.
[4] Loher, D. *et al.*, 'Evaluation of Candidate Control and Provisioning of Wireless Access Points (CAPWAP) Protocols', RFC 4565, July 2006.
[5] Singh, I. *et al.*, 'CAPWAP Tunneling Protocol (CTP)', draft-singh-capwap-ctp-02.txt, June 2005.
[6] Calhoun, P. *et al.*, 'Lightweight Access Point Protocol', draft-ohara-capwap-lwapp-04.txt, March 2007.
[7] Calhoun, P. *et al.*, 'CAPWAP Protocol Specification', draft-ietf-capwap-protocol-specification-07, June 2007.
[8] Calhoun, P. *et al.*, 'CAPWAP Protocol Binding for IEEE 802.11', draft-ietf-capwap-protocol-binding-ieee80211-04, June 2007.
[9] Rescorta, E. and Modadugu, N., 'Datagram Transport Layer Security', RFC 4347, April 2006.

Part II

Application Examples of Service Automation and Dynamic Resource Provisioning Techniques

8

Dynamic Enforcement of QoS Policies

8.1 Introduction

Undoubtedly, quality of service has been the most important fuel for investigating dynamic policy provisioning and enforcement schemes. The common open policy service (COPS) (RFC 2748 [1], RFC 3084 [2]) protocol has been primarily designed for resource admission control policy enforcement, while QoS rapidly became a major concern not only for service providers but also for customers who rightly expect that the service to which they have subscribed will be delivered with the relevant level of quality.

8.1.1 What is Quality of Service, Anyway?

Quality of service is a very generic term that is sometimes misused, especially when applied to IP/MPLS environments. In this document, the design and the enforcement of a QoS policy within IP/MPLS environments basically rely upon the following dimensions:

- A *forwarding* dimension, which consists in making a given router behave differently, depending on the kind of traffic it has to process (and forward, in particular). This gives rise to the need for traffic identification and classification, and possibly the activation of traffic conditioning, policing, scheduling and even discarding mechanisms. The forwarding dimension of a QoS policy is a notion that is **local** to a router (i.e. presumably independent of the (hopefully expected) behaviors of the other routers of the domain). The DiffServ (Differentiated Services, as described in RFC 2475 [3]) architecture is generally seen as the cornerstone of the forwarding dimension, introducing the notion of classes of service as well as behavior aggregates (BAs).
- A *routing* dimension, which consists in enforcing a traffic engineering policy at the scale of a DiffServ domain. Traffic engineering is a set of capabilities that allow the (hopefully dynamic) computation and selection of paths that will be used to convey

Service Automation and Dynamic Provisioning Techniques in IP/MPLS Environments C. Jacquenet, G. Bourdon and M. Boucadair

different kinds of traffic, depending on the (QoS) characteristics of such paths, which are supposed to comply with the QoS requirements that are related to the deployment of some value-added service offerings, and/or which may have been expressed and negotiated with customers. The traffic engineering capabilities of MPLS are a practical example of such a dimension.

- A *monitoring* dimension, which basically consists in qualifying the efficiency of a QoS policy on the basis of the use and the measurement of well-defined indicators. Such indicators include (but may not be limited to) the one-way transit delay, the interpacket delay variation (sometimes called the jitter), the packet loss rate or any combination of such indicators (RFC 2679 [4], RFC 2680 [5], RFC 2681 [6], RFC 3393 [7]).

8.1.1.1 DiffServ-based Forwarding Schemes

DiffServ-based forwarding relies upon the activation of a set of elementary capabilities, as defined in RFC 3290 [8]. Such activation reflects the actual enforcement of a QoS policy at the scale of a router, since the DiffServ architecture relies upon per-hop behaviors (PHBs) which indicate to the router how to process the incoming traffic according to some specific parameters that may possibly include (but are not limited to) the DiffServ code point (DSCP) marking.

The latter indication is usually associated with the notion of 'class of service' or 'behavior aggregate', whereas IP datagrams are gathered into these macroflows which are processed according to the PHBs that have been configured on the routers. Standardization has currently defined the 'assured forwarding' (AF) (RFC 2597 [9]) and the 'expedited forwarding' (EF) (RFC 3246 [10]) PHBs, in addition to the best effort PHB.

Each of these PHBs may or may not rely upon a set of elementary functions that include:

- *traffic conditioning* capabilities, based upon algorithms such as token buckets, and which often consist in indicating to the router what to do with the incoming traffic, depending on its envelope (whether or not there are enough tokens left in the bucket to process the traffic will yield different kinds of action taken by the router – simply forward in-profile traffic while out-of-profile traffic will be dropped, or remark out-of-profile traffic, etc.);
- *scheduling* capabilities, based upon queuing algorithms such as weighted fair queuing (WFQ) or class-based queuing (CBQ), and which often consist in indicating to the router what kind of outgoing traffic will be forwarded first, depending on the congestion occurrences as perceived by the router;
- *discarding* capabilities, based upon probabilistic algorithms such as random early detection (RED), and which consist in indicating to the router what kind(s) of traffic(s) should be discarded first, considering the average queue length and the congestion occurrences as perceived by the routers.

All of the aforementioned capabilities imply the need for traffic classification and recognition, and hence the use of criteria such as DSCP marking, but also the {source address, destination address} pair, the TCP/UDP protocol identifiers, etc.

8.1.1.2 About Traffic Engineering

Traffic engineering can be defined as a set of path computation techniques that help service providers in selecting specific routes whose characteristics will comply with requirements (e.g. QoS requirements defined in a service level specification contractually negotiated and invoked between a customer and a service provider) and constraints (e.g. the enforcement of the network planning policy that has been designed by the operator).

This means that a routing policy is implicitly a traffic engineering policy, at least for the traffic that needs to be forwarded to a given set of destination prefixes by using paths that are computed by the BGP protocol, owing to the manipulation of attributes that indicate, for example, the use of a specific exit point or a neighboring AS to reach the destinations mentioned in the BGP UPDATE messages. This is also true for IGP-based routing policies, whereas the use of OSPF-based shortcut ABR capabilities can dramatically enhance the convergence times within a domain.

Nevertheless, it is generally admitted that traffic engineering policies are designed with a flow granularity, where a flow is a set of IP datagrams that share at least one common characteristic, such as the destination address field. The use of MPLS as the switching technique is often seen as the cornerstone for the deployment of traffic engineering capabilities within a domain, based upon the use of a constraint-based shortest path first (CSPF) algorithm for the dynamic computation and selection of traffic-engineered paths that will be entitled to convey different kinds of traffic, depending on the QoS requirements associated with such flows.

Such requirements may include not only a set of indicators such as the one-way transit delay or the packet loss rate but also restoration capabilities in case of a link or a node failure. This section focuses on the use of MPLS-inferred traffic engineering capabilities (MPLS TE), but the reader should keep in mind that other (or complementary) options to MPLS-TE exist {the aforementioned routing policies, the use of multitopology capabilities (RFC 4915 [11], for example) given an efficient forwarding scheme, the use of the IPv6 flow label, etc.} and should deserve the appropriate investigation, depending on the requirements and/or the constraints of the service provider.

8.1.1.3 About Monitoring

Monitoring is the privileged means to qualify how efficient a QoS policy enforcement is, and how compliant it is with the QoS requirements that may have been contractually negotiated with customers. Monitoring relies upon a set of QoS indicators or metrics that include (but may not be limited to):

- the *interpacket delay variation* (sometimes called the 'jitter'), as defined in RFC 3393 [7], which is one of the key indicators reflecting the level of quality associated with the deployment of multimedia services, like VoIP and TV broadcasting;
- the *one-way packet loss rate*, as defined in RFC 2680 [5];
- the *one-way transit delay*, as defined in RFC 2679 [4], which is one of the indicators often used to reflect the quality of VPN service offerings, for example;
- the *round-trip delay* metric, as defined in RFC 2681 [6].

Service level specification (SLS) templates [12] can be instantiated with additional indicators, such as the overall *network availability* (NA), sometimes defined as NA $= 1 - \Sigma d_i t_i / DT$, where d_i is the bandwidth that has been affected during the t_i epoch, and D is the total amount of bandwidth assigned to the provisioning of a given service that is managed during the contractual T period (e.g. 24/24, 7/7).

The efficiency of the aforementioned metrics is obviously conditioned by the availability of reliable measurement methods, which are also defined and sometimes negotiated between customers and service providers. Such measurement techniques can rely upon probe technologies, which can be used either in passive mode (live traffic is captured by the probe which provides statistical information based upon such captures) or in active mode, where the probe sends test traffic on the network to qualify its performance according to one or a set of the aforementioned metrics.

8.1.2 The Need for Service Level Specifications

The deployment of value-added IP service offerings over the Internet has yielded a tremendous effort for the definition, the specification and possibly the standardization of the notion of quality of service (QoS), which generally encompasses a wide set of elementary parameters, such as the maximum transit delay, the interpacket delay variation or the packet loss rate.

Because the subscription to an IP service offering implies the definition of a contractual agreement between the customer and the corresponding IP service provider (ISP), the level of quality that will be associated with the deployment of such service will be based upon a set of the aforementioned parameters upon which both parties will have to agree.

The differentiated services specification effort has yielded the identification of a set of elementary functions and concepts, the respective interactions of which can be depicted according to a layered approach, as per Figure 8.1.

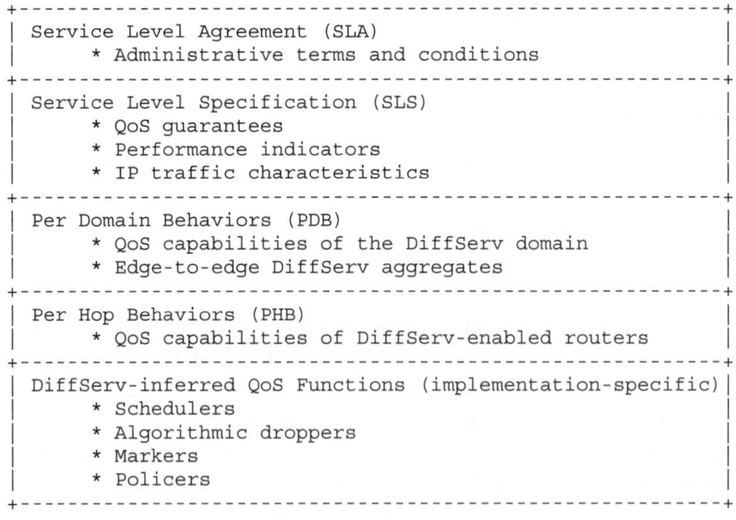

```
+------------------------------------------------------------+
| Service Level Agreement (SLA)                              |
|       * Administrative terms and conditions                |
+------------------------------------------------------------+
| Service Level Specification (SLS)                          |
|       * QoS guarantees                                     |
|       * Performance indicators                             |
|       * IP traffic characteristics                         |
+------------------------------------------------------------+
| Per Domain Behaviors (PDB)                                 |
|       * QoS capabilities of the DiffServ domain            |
|       * Edge-to-edge DiffServ aggregates                   |
+------------------------------------------------------------+
| Per Hop Behaviors (PHB)                                    |
|       * QoS capabilities of DiffServ-enabled routers       |
+------------------------------------------------------------+
| DiffServ-inferred QoS Functions (implementation-specific)  |
|       * Schedulers                                         |
|       * Algorithmic droppers                               |
|       * Markers                                            |
|       * Policers                                           |
+------------------------------------------------------------+
```

Figure 8.1 A layered model of DiffServ

In Figure 8.1, each layer displays its own QoS capabilities. According to the definition of a per-domain behavior (PDB) (RFC 3662 [13]), the specification of such PDBs should include the reference to the (lower-layer) PHB(s) upon which the PDB 'layer' relies.

8.2 An Example

The enforcement of a DiffServ-based policy is primarily justified by the identification of potential traffic bottlenecks that may jeopardize the level of quality associated with a given (set of) service offerings. As far as the core region of the IP/MPLS backbones is concerned, operators have often decided to enforce an overdimensioning policy, which is designed to absorb traffic peaks with presumably no impact on the overall throughput.

With the use of the Internet as a privileged underlying infrastructure to deploy value-added yet QoS-demanding service offerings, and also because of cost considerations that may affect the enforcement of the aforementioned overdimensioning policy, it is generally admitted that a DiffServ-based policy might be worthwhile in both the aggregation and access regions of the IP/MPLS backbone, where traffic congestion is likely to happen with the development of services such as TV broadcasting.

Table 8.1 provides examples of classes of service (CoS) that help in classifying the traffic according to the DSCP marking of the IP datagrams.

Table 8.1 Examples of traffic classes with associated DSCP marking

Traffic classes	DSCP (decimal)	DiffServ PHB	Binary	PF/EXP
D3 – in profile	10	AF_{11}	001010	1
D3 –out of profile	12	AF_{12}	001100	1
D2 – in profile	16	CS_2	010000	2
D2 – in profile	18	AF_{21}	010010	2
D2 – out of profile	20	AF_{22}	010100	2
D1 – in profile	26	AF_{31}	011010	3
D1 – out of profile	28	AF_{32}	011100	3
Real-time video	34	AF_{41}	100010	4
Real-time voice	46	EF	101110	5

In addition, a set of QoS-based services that elaborate on the QoS configuration parameters associated with the PHBs and other class selectors are illustrated in Table 8.2.

8.3 Enforcing QoS Policies in Heterogeneous Environments

8.3.1 SLS-inferred QoS Policy Enforcement Schemes

One of the key challenges that have to be addressed when considering the dynamic enforcement of QoS policies in heterogeneous environments is that these policies must dynamically align the resource allocation schemes to optimize the resource usage but also to address the requirements as they have been expressed by customers by means of negotiated SLS templates. The architecture that has been investigated and validated by the TEQUILA (Traffic Engineering for Quality of Service in the Internet at Large Scale) project [14] is a

Table 8.2 Examples of IP (QoS-based) service types

IP service type	CoS	SLS			DSCP	CoS implementation on PE
		RTD	LOSS	JITTER		
SILVER	D2 – 100 % of IP BW				18	N/A
GOLD3	D1 – 60 % of IP BW	√	√		26 (D1 – InP) 28(D1 – OOP)	LLQ: CB BW 60 % + WRED
	D2 – 30 % of IP BW	√			18 (D2 – InP) 20 (D2 – OOP)	LLQ: CB BW 30 % + WRED
	D3 – 10 % of IP BW				10 (D3 – InP) 12 (D3 – OOP)	LLQ: CB BW 9 % + WRED
GOLD2	D1 – 66% of IP BW	√	√		26 (D1 – InP) 28 (D1 – OOP)	LLQ: CB BW 60 % + WRED
	D2 – 33% of IP BW	√			18 (D2 – InP) 20 (D2 – OOP)	LLQ: CB BW 30 % + WRED
GOLD1	D2 – 100% of IP BW	√			18 (D2 – InP) 20 (D2 – OOP)	LLQ: CB BW 99 % + WRED
PLATINUM3 (IP BW $\geq$ 128 kbit/s)	RT – maximum 75 % of IP BW	√	√	√	46	LLQ: PQ
	D1 – 60 % of remaining data IP BW	√	√		26 (D1 – InP) 28 (D1 – OOP)	• LLQ: CB BW x % + WRED [x = 60 % * (remaining data IP BW/IP BW)] • FRF12 (for access speed $\leq$ 768 kbit/s)
	D2 – 30 % of remaining data IP BW	√			18 (D2 – InP) 20 (D2 – OOP)	• LLQ: CB BW y % + WRED [y = 30 % * (remaining data IP BW/IP BW)] • FRF12 (for access speed $\leq$ 768 kbit/s)
	D3 – 10 % of remaining data IP BW				10 (D3 – InP) 12 (D3 – OOP)	• LLQ: CB BW z % + WRED [z = 10 % * (remaining data IP BW/IP BW)] • FRF12 (for access speed $\leq$ 768 kbit/s)
PLATINUM2) (IP BW $\geq$ 128 kbit/s	RTD – maximum 75 % of IP BW	√	√	√	46	LLQ: PQ
	D1 – 66 % of remaining data IP BW	√	√		26 (D1 – InP) 28 (D1 – OOP)	• LLQ: CB BW x % + WRED [x = 66 % * (remaining data IP BW/IP BW)] • FRF12 (for access speed $\leq$ 768 kbit/s)

	D2 – 33 % of remaining data IP BW	√			18 (D2 – InP) 20 (D2 – OOP)	• LLQ: CB BW y % + WRED [y = 33 % * (remaining data IP BW/IP BW)] • FRF12 (for access speed ≤ 768 kbit/s)
PLATINUM1 (IP BW ≥ 128 kbit/s)	RT – maximum 75 % of IP BW	√	√	√	46	• LLQ: PQ • cRTP (for IP BW = 128 kbit/s)
	D2 – 100 % of remaining data IP BW	√			18 (D2 – InP) 20 (D2 – OOP)	• LLQ: CB BW y % + WRED [y = 99 % * (remaining data IP BW/IP BW)] • FRF12 (for access speed ≤ 768 kbit/s)
PLATINUM5 (IP BW <128 kbit/s)	RT – maximum 75 % of IP BW	√	√	√	46	LLQ: PQ
	D1 – 100 % of remaining data IP BW	√	√		26 (D1 – InP) 28 (D1 – OOP)	• LLQ: CB BW x % + WRED [x = 99 % * (remaining data IP BW/IP BW)] • FRF12
PLATINUM4 (IP BW <128 kbit/s)	RT – maximum 75 % of IP BW	√	√	√	46	LLQ: PQ
	D2 – 100 % of remaining data IP BW	√			18 (D2 – InP) 20 (D2 – OOP)	• LLQ: CB BW y % + WRED [y = 99 % * (remaining data IP BW/IP BW)] • FRF12
FLEXIBLE	RT – maximum 75 % of IP BW	√	√	√	46	LLQ: PQ
	RT-VI – w % (maximum 75 %) of remaining IP BW	√	√	√	34	LLQ: CB BW w % • FRF12 (for IP BW ≤ 768 kbit/s)
	D1 – x % of remaining IP BW	√	√		26 (D1 – InP) 28 (D1 – OOP)	• LLQ: CB BW x % + WRED • FRF12 (for IP BW ≤ 768 kbit/s)
	D2 – y % of remaining IP BW	√			18 (D2 – InP) 20 (D2 – OOP)	• LLQ: CB BW y % + WRED • FRF12 (for IP BW ≤ 768 kbit/s)
	D3 – z % of remaining IP BW				10 (D3 – InP) 12 (D3 – OOP)	• LLQ: CB BW z % + WRED • FRF12 (for IP BW ≤ 768 kbit/s)

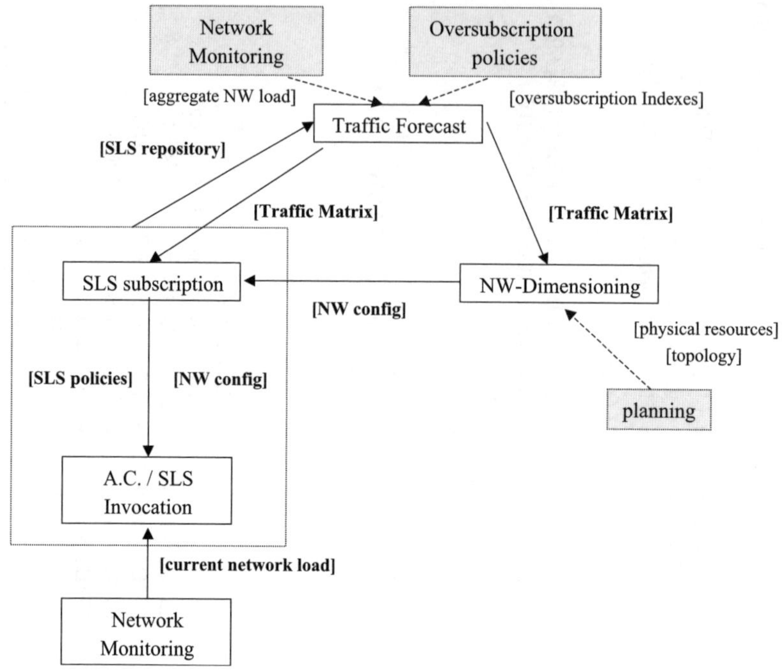

Figure 8.2 SLS management, traffic forecast and network dimensioning of the TEQUILA architecture

possible response to such challenges. This architecture relies upon a set of functional blocks that not only take care about the processing of incoming SLS templates but also deal with network planning and traffic forecast considerations, as outlined in Figure 8.2.

In accordance with Figure 8.2:

- SLS subscription is responsible for the permission/rejection of new/modified long-term SLS subscriptions. This is based on administrative policies and *available spare resources.*
- SLS subscription computes the available spare resources for new SLS subscription requests. Roughly speaking, one has:

$$\textit{Long-term spare resources} = \textit{network configuration} - \textit{long-term traffic load}$$

- Network configuration parameters include (but are not necessarily limited to):
  - per QoS class and per ingress/egress pair;
  - source/destination (ingress/egress) route set (in the case of multipath routing, for example);
  - route capacity (in terms of bandwidth usage, traffic identification, etc.);
  - QoS parameters such as the maximum transfer delay from ingress to egress.

In fact, the network configuration parameters have a similar structure to the traffic forecast (TF) matrix. However, the network configuration gives the resources that are actually allocated, while the TF matrix is a prediction of the load.

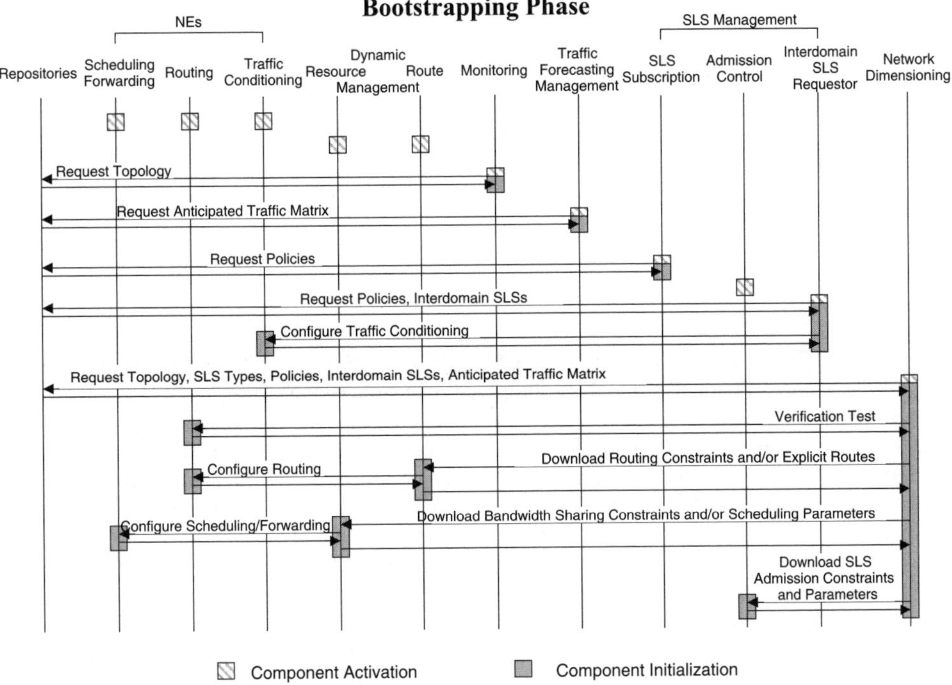

Figure 8.3 Possible message sequence chart (bootstrap phase) of a QoS policy enforcement scheme (NEs stands for network elements)

Figure 8.3 provides an example of a sequence chart depicting message flows during the bootstrap phase associated with a QoS policy enforcement scheme.

The SLS subscription process itself distinguishes three phases:

1. Retrieve the necessary information that will be processed by the SLS subscription block, enabling the permission or refusal of (new or changed) SLS subscription requests. This information will be stored and maintained in the QoS policy repository, which will be fed by the network dimensioning component, among other sources of information.
2. The actual SLS subscription process.
3. The SLS invocation process, where actions are triggered by the permission of a new (or changed) SLS subscription. Information is transmitted to admission control for handling SLS invocations. Information is also transmitted to the traffic forecast module.

8.3.2 Policy Rules for Configuring DiffServ Elements

RFC 3317 [15] provides an example of a DiffServ policy information base, which is directly derived from RFC 3290 [8], which specifies the model of a DiffServ router. Functional elements of such routers are described in RFC 3317 [15] through a set of provisioning classes (PRCs) that include: Data Path, Classifier, Classifier Element, Meter, Token Bucket Parameter, DSCP Mark Action, Algorithmic Dropper, Random Dropper.

The proposed approach enables the network administrator flexibly to define generic policy types that can be applied to any kind of DiffServ device (a router, a bridge, a gateway, etc.), and that can then be instantiated (PRI) with device-specific parameters.

When DiffServ elements enforce a QoS policy, the policy actions are those provided by the (COPS) management interface of the data path capabilities that compose a given PHB [16].

References

[1] Boyle, J., Cohen, R., Durham, D., Herzog, S., Raja R. and Sastry A., 'The COPS (Common Open Policy Service) Protocol', RFC 2748, Proposed Standard, January 2000.

[2] Ho Chan, K., Durham, D., Gai, S., Herzog, S., McLoghrie, K., Reichmeyer, F., Seligson, J., Smith, A. and Yavatkar, R., 'COPS Usage for Policy Provisioning (COPS-PR)', RFC 3084, March 2001.

[3] Blake, S. *et al.*, 'An Architecture for Differentiated Services', RFC 2475, December 1998.

[4] Almes, G. *et al.*, 'A One-way Delay Metric for IPPM', RFC 2679, September 1999.

[5] Almes, G. *et al.*, 'A One-way Packet Loss Metric for IPPM', RFC 2680, September 1999.

[6] Almes, G. *et al.*, 'A Round-trip Delay Metric for IPPM', RFC 2681, September 1999.

[7] Demichelis, C. and Chimento, P., 'IP Packet Delay Variation Metric for IP Performance Metrics (IPPM)', RFC 3393, November 2002.

[8] Bernet, Y. *et al.*, 'An Informal Management Model for Diffserv Routers', RFC 3290, May 2002.

[9] Heinanen, J. *et al.*, 'Assured Forwarding PHB Group', RFC 2597, June 1999.

[10] Davie, B. *et al.*, 'An Expedited Forwarding PHB (Per-Hop Behavior)', RFC 3246, March 2002.

[11] Psenak, P. *et al.*, 'Multi-Topology (MT) Routing in OSPF', RFC 4915, June 2007.

[12] Goderis, D. *et al.*, 'Attributes of a Service Level Specification (SLS) Template', draft-tequila-sls-03.txt, Work in Progress, October 2003.

[13] Bless, R. *et al.*, 'A Lower Effect Per-Domain Behavior (PDB) for Differentiated Services', RFC 3662, December 2003.

[14] Goderis, D. *et al.*, 'D1.1: Functional Architecture and Top Level Design', September 2000. See also http://www.ist-tequila.org.

[15] Chan, K. *et al.*, 'Differentiated Services Quality of Service Policy Information Base', RFC 3317, March 2003.

[16] Lymberopoulos, L., Lupu, E. and Sloman, M., 'An Adapative Policy Based Management Framework for Differentiated Services Networks', in *Proc. 3rd IEEE Workshop on Policies for Distributed Systems and Networks*, Washington, DC, June 2002, pp. 147–158.

9

Dynamic Enforcement of IP Traffic Engineering Policies

9.1 Introduction

The deployment of value-added IP services (such as quality-of-service-based IP virtual private networks) over the Internet has become one of the greatest challenges for service providers, as well as a complex technical issue, from a (dynamic) resource provisioning perspective.

From this perspective, the COPS protocol (RFC 2748 [1]) and its usage for the support of policy provisioning (RFC 3084 [2]) have been designed to help service providers by introducing a high level of automation for the dynamic production of a wide range of services, including dynamic capabilities for the enforcement of service-specific policies.

Such policies include routing and traffic engineering policies, and their aim is to appropriately provision, allocate/deallocate and use the switching and the transmission resources of an IP network (i.e. the routers and the links that connect these routers respectively) according to a set of constraints such as quality of service (QoS) requirements (e.g. rate, one-way delay, interpacket delay variation, etc.) that have been possibly negotiated between the customers and the service providers, as well as routing metrics, which can reflect the network conditions.

Within the context of this chapter, the actual enforcement of IP routing and traffic engineering policies is primarily based upon the activation of both intra- and interdomain routing protocols that will be activated in the network adequately to compute, select, install, maintain and possibly withdraw routes that will comply with the aforementioned QoS requirements and/or specific routing constraints, depending on the type of traffic that will be conveyed along these routes.

It is therefore necessary to provide the route selection processes with the information that will depict the routing policies that are to be enforced within a domain and, whenever appropriate, the aforementioned constraints and metrics, given that the dynamic routing

Service Automation and Dynamic Provisioning Techniques in IP/MPLS Environments C. Jacquenet, G. Bourdon and M. Boucadair

protocols actually support traffic engineering capabilities for the calculation and the selection of such routes.

These capabilities have been specified in RFC 3630 [3] and RFC 3784 [4] for the open shortest path first (OSPF) (RFC 2328 [5]) and the intermediate system to intermediate system (IS–IS) (ISO 8473 [6]) interior routing protocols respectively, while there is an equivalent specification effort for the border gateway protocol, version 4 (BGP-4), as described in Refs [7] and [8], for example.

To provide the routers that will participate in the dynamic enforcement of an IP routing and/or traffic engineering policy with the appropriate configuration information (such as the values of metrics), one option is to use the COPS protocol for policy provisioning purposes. To do so, a new COPS client-type is required, called the 'IP traffic engineering' (IP TE) client-type.

9.2 Terminology Considerations

The enforcement of an IP routing/TE policy is based upon the processing of configuration information that reflects the characteristics of the metric values of these policies e.g. the values of the BGP attributes, the QoS requirements and/or constraints, etc.

This information is called the 'QoS-related' information within the context of this chapter.

Then, this QoS-related information must be taken into account by the routing processes that will participate in the calculation, the selection, the installation and the maintenance of the routes that will comply with the aforementioned requirements. The algorithms invoked by the routing processes include the cost metrics (whose corresponding values can possibly be inferred by a DiffServ code point (DSCP) value (RFC 2474 [9]) that has been assigned by the network administrators.

This metric-related information is called the 'IP TE'-related information within the context of this chapter.

Thus, there is a distinction between QoS-related information and IP TE-related information, where:

- QoS-related information is negotiated between customers and service providers;
- IP TE-related configuration information is dynamically provided to routers, and is exchanged between routers so that they can compute, select, install and maintain the (traffic-engineered) routes accordingly.

From this perspective, QoS-related information provides information about the traffic to be forwarded in the network (such as source address, destination address, protocol identification, DSCP marking, etc.), whereas IP TE-related information provides information for the routing processes that will indicate to the routers of the network how to forward the aforementioned traffic, i.e. compute and select the routes that will convey such traffic.

Given these basic assumptions, a COPS-based IP TE client-type has the following characteristics:

- The IP TE client-type is supported by the policy enforcement point (PEP) capability which allows a router to enforce a collection of policies owing to a COPS communication that has been established between the PEP and the PDP.

- The actual enforcement of an IP routing/TE policy is based upon the TE-related configuration information that will be exchanged between the PDP and the PEP, and that will be used by the router for selecting, installing, maintaining and possibly withdrawing IP TE routes.

9.3 Reference Model

The use of the COPS protocol for dynamically enforcing an IP routing/TE policy yields the generic model depicted in Figure 9.1.

As depicted in Figure 9.1, the routers embed the following components:

- A PEP capability, which supports the IP TE client-type. The support of the IP TE client-type is notified by the PEP to the PDP, and is unique for the area covered by the IP routing/traffic engineering policy, so that the PEP can treat all the COPS client-types it supports as non-overlapping and independent namespaces.
- A local policy decision point (LPDP), which can be assimilated to the routing processes that have been activated in the router. The LPDP will therefore contribute to the computation and the selection of the IP routes.
- Several instances of routing information bases (RIBs), according to the different (unicast and multicast) routing processes that have been activated – one can easily assume the activation of at least one interior gateway protocol ((IGP), like OSPF) and BGP-4.
- Conceptually, one forwarding information base (FIB), which will store the routes that have been selected by the routing processes, but within this section we do not make any

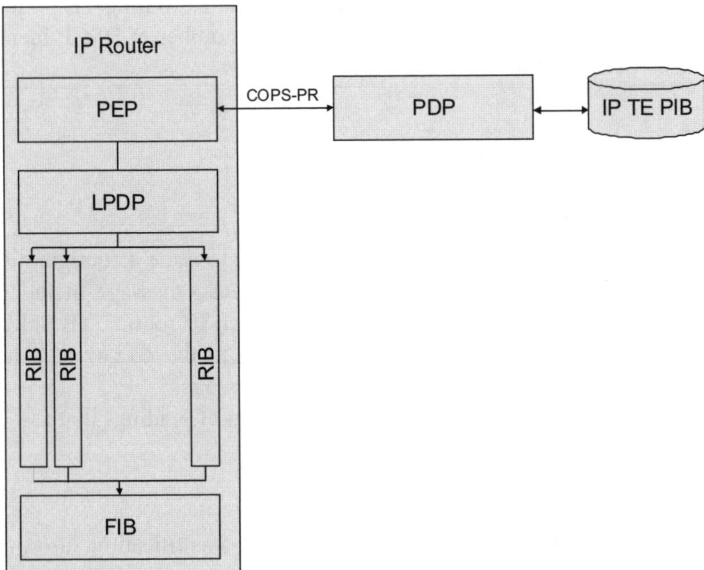

Figure 9.1 Reference model of an IP routing/traffic engineering policy enforcement scheme

assumption about the number of FIBs that can be supported by a router [e.g. within the context of an IP virtual private network (VPN) service offering].

As suggested in Ref. [10], the enforcement of an IP routing/traffic engineering policy is based upon the use of a policy server (the PDP in Figure 9.1) that sends IP TE-related information to the PEP capability embedded in the IP router.

The IP TE-related information is stored and maintained in an IP TE policy information base [11], which will be accessed by the PDP to retrieve and update the IP TE-related information whenever necessary.

The IP TE-related information is conveyed between the PDP and the PEP owing to the establishment of a COPS-PR connection between these two entities. The COPS-PR protocol assumes a named data structure (the PIB) so as to identify the type and purpose of the policy information that is sent by the PDP to the PEP for the provisioning of a given policy.

The data structure of the PIB refers to the IP routing/TE policy which is described in the PIB as a collection of provisioning classes (PRCs). Furthermore, these classes contain attributes that actually describe the IP TE-related policy provisioning data that will be sent by the PDP to the PEP. Some of these attributes consist of the link and traffic engineering metrics that will be manipulated by the routing processes being activated in the routers to compute the IP routes.

The IP TE classes are instantiated as multiple provisioning instance (PRI) instances, each of which are identified by a provisioning instance identifier (PRID). A given PRI specifies the data content carried in the IP TE client-specific objects. An IP TE PRI typically contains a value for each attribute that has been defined for the IP TE PRC.

Currently, the IP TE PIB has identified a per-DSCP IP TE PRC instantiation scheme, because the DSCP value conveyed in each IP datagram that will be processed by the routers is one of the key criteria to make forwarding decisions within the context of a QoS-based routing scheme. Such a routing scheme aims to reflect the IP routing/TE policies that have been defined by a service provider, assuming a restricted number of DSCP-identified classes of service that will service the customers' requirements.

9.4 COPS Message Content

9.4.1 Request Messages (REQ)

The COPS REQ message is sent by the IP TE client-type to issue a configuration request to the PDP, as specified in the COPS context object. The `REQ` message includes the current configuration information related to the enforcement of an IP routing/TE policy.

Such configuration information is encoded according to the `ClientSI` format that is defined for the named `ClientSI` object of the `REQ` message.

The configuration information is encoded as a collection of bindings that associate a PRID object and encoded provisioning instance data (EPD).

Such information may consist of:

- The identification information of the router, e.g. the identification information that is conveyed in OSPF link state advertisement (LSA) type 1 messages. The use of a loopback interface IP address is highly recommended for the instantiation of the corresponding EPD.

- The link metric values that have been currently assigned to each (physical/logical) interface of the router, as described in RFC 2328 [5], for example. Such values may vary with an associated DSCP value, i.e. the link metric assigned to an interface is a function of the DSCP value encoded in each IP datagram that this router may have to forward.
- The traffic engineering metric values that specify the link metric values for traffic engineering purposes, as defined in RFC 3630 [3], for example. These values may be different from the above-mentioned link metric values, and they may also vary according to DSCP values.

9.4.2 Decision Messages (DEC)

The **DEC** messages are used by the PDP to send IP TE policy provisioning data to the IP TE client-type. **DEC** messages are sent in response to a **REQ** message received from the PEP, or they can be unsolicited, e.g. subsequent **DEC** messages can be sent at any time to supply the PEP with additional or updated IP TE policy configuration information without the solicited message flag set in the COPS message header, since such messages correspond to unsolicited decisions.

DEC messages typically consist of *install* and/or *remove* decisions, and, when there is no Decision Flags set, the **DEC** message includes the named decision data (provisioning) object.

Apart from the aforementioned identification information, and according to the kind of (PRID, EPD) bindings that may be processed by the PEP, **DEC** messages may refer to the following decision examples:

- Assign new link/traffic engineering metric values each time a new interface is installed/ created on the router. These new values will obviously yield the generation of LSA messages in the case of the activation of the OSPF protocol, and/or the generation of BGP-4 UPDATE messages {e.g. in the case of a new instantiation of the MULTI_EXIT_DISC (MED) attribute (RFC 4271 [12])}. This will in turn yield the computation of (new) IP routes that may be installed in the router's FIB.
- Modify previously assigned metric values, owing to a remove/install decision procedure (obviously, this may yield a modification of the router's FIB as well).
- Remove assigned metric values, e.g. the corresponding interfaces may not be taken into consideration by the routing algorithms anymore (or during a specific period of time, e.g. for maintenance purposes).

9.4.3 Report Messages (RPT)

The *report* message allows the PEP to notify the PDP with a particular set of IP routing/TE policy provisioning instances that have been successfully or unsuccessfully installed/ removed.

When the PEP receives a **DEC** message from the PDP, it sends back a **RPT** message to the PDP. The **RPT** message will contain one of the following report-types:

- *Failure*. Notification of errors that occurred during the processing of the (PRID, EPD) bindings contained in the **DEC** message. Such a notification procedure can include a failure report in assigning an updated value of a given metric, for example.
- *Success*. Notification of the successful assignment of metric values and/or successful installation of IP routes in the router's FIB. From this perspective there may be routes that will be installed in the router's FIB without any explicit decision sent by the PDP to the PEP with respect to the calculation/installation of the aforementioned route. This typically reflects a normal dynamic routing procedure, whenever route advertisement messages are received by the router, including messages related to a topology change. In any case (i.e. whatever the effect that yielded the installation of a route in the router's FIB), an **RPT** message is sent by the PEP to the PDP to notify such an event, so that the IP TE PIB will be updated by the PDP accordingly.
- *Accounting*. The accounting **RPT** message will carry statistical information related to the traffic that will transit through the router. This statistical information may be used by the PDP possibly to modify the metric values that have been assigned when thresholds have been crossed; for example, if the **RPT** message reports that **x** % of the available rate associated with a given interface has been reached, then the PDP may send an unsolicited **DEC** message in return, so that potential bottlenecks can be avoided.

9.5 COPS-PR Usage of the IP TE Client-Type

After having opened a COPS connection with the PDP, the PEP sends a REQ message to the PDP that will contain a 'client handle'. The 'client handle' is used to identify a specific request state associated with the IP TE client-type supported by the PEP. The REQ message will contain a 'configuration request' context object.

This **REQ** message will also carry the named client specific information [including the (default) configuration information]. Default configuration information includes the information available during the bootstrap procedures of the routers.

The routes that have been installed in the router's FIB may be conveyed in specific (PRID, EPD) bindings in the **REQ** message as well.

Upon receipt of the **REQ** message, the PDP will send back a **DEC** message towards the PEP. This **DEC** message will carry an IP TE named decision data object that will convey all the appropriate installation/removal of (PRID, EPD). One of the basic goals of this named decision object consists in making the routers enforce a given IP routing/TE policy.

Upon receipt of a **DEC** message, the IP TE-capable PEP will (try to) apply the corresponding decisions by making the network device (and its associated implementation-specific command line interface, if necessary) and install the named IP TE policy data (e.g. assign a metric value to a recently installed interface).

Then, the PEP will notify the PDP about the actual enforcement of the named IP TE policy decision data by sending the appropriate **RPT** message back to the PDP. Depending on the report-type that will be carried in the **RPT** message, the contents of the message may include:

- Successfully/unsuccessfully assigned new/updated metric values.
- Successfully installed routes from the router's FIB. Note that the notion of 'unsuccessfully installed routes' is meaningless.
- Successfully/unsuccessfully withdrawn routes from the router's FIB. Route withdrawal is subject not only to the normal IGP and BGP-4 procedures (thus yielding the generation of the corresponding advertisement messages) but also to named IP TE policy decision data (carried in a specific **DEC** message), such as those data related to the lifetime of a service.

The **RPT** message may also carry the 'accounting' report-type.

9.6 Scalability Considerations

The use of the COPS-PR protocol for the dynamic enforcement of an IP traffic engineering policy raises some scalability issues as far as the volume of configuration information that will be exchanged not only between the routers themselves (because of the OSPF machinery, for example) but also between the PEP components embedded in the routers and the PDP with which they communicate is concerned.

While the concern strictly related to the design of a routing policy is outside the scope of this book, the dynamic provisioning of metric values as well as the reports related to the actual enforcement of decisions taken by the PDP deserves some elaboration.

9.6.1 A Tentative Metric Taxonomy

The metrics that will be taken into account by the shortest path first (SPF) algorithms for IP TE route calculation can be classified into two basic categories:

- Metrics assigned on a long-term basis, which basically consist of the 'usual' cost metrics, like those defined in RFC 2328 [5]. These metrics are those that are assigned on a (logical) interface basis, and they aim to reflect the link quality to which the corresponding interface is attached.
- Metrics assigned on a (very) short-term basis, which may consist of the following information:
  - the available bandwidth [e.g. based upon the information provided by simple network management protocol (SNMP) (RFC 1157 [13]) counters like **ifInOctets** and **ifOutOctets**];
  - the amount of bandwidth that can be reserved;
  - the amount of reserved bandwidth.

While 'long-term' metric values should not change frequently by definition, the 'short-term' metric values may vary like the ongoing usage of the resources of the network.

Therefore, the performance of short-term metric value processing should remain comparable with SPF computation, since newly assigned values yield the spontaneous generation of link state update (LSU) messages. Thus, the traffic generated by the IP traffic engineering provisioning data should be minimized according to precomputation engineering recommendations like those described in RFC 2676 [14].

9.6.2 Reporting the Enforcement of an IP Traffic Engineering Policy

Likewise, the actual enforcement of policy decisions implies the activation of a reporting mechanism, as described in the COPS-PR specification.

From this perspective, within this section we assume that the corresponding reports sent by the PEP components of the routers to the PDP should include the 'traffic-engineered' routes that have been computed by the routers, at least for network planning purposes: the service subscription requests will be negotiated according to the knowledge of the network resources that are actually available, and this information includes the routes that could very well service the aforementioned requirements, without any extra computation.

Therefore, the volume of traffic generated by the notification reports of the installed routes should remain comparable with the volume of traffic generated by the route announcement procedures of the IP routing protocol machineries (like OSPF), and it is assumed that the volume of the corresponding COPS-PR traffic is also highly dependent on the precomputation engineering recommendations that have been mentioned earlier.

In other words, scalability issues that may be encountered by network operators when considering the dynamic enforcement of an IP traffic engineering policy are the results of an inefficient design, not because of the activation of the COPS-PR protocol as a means to convey the corresponding IP TE provisioning data.

9.7 IP TE PIB Overview

The dynamic enforcement of an IP traffic engineering policy relies on the activation of intra- and interdomain routing protocols that will have the ability to take into account traffic-engineering-related information for the computation and the selection of routes that will comply as much as possible with the QoS requirements that have been contractually defined between customers and service providers.

This traffic-engineering-related information is basically composed of metric values that will aim to reflect an IP TE policy, as well as the result of the enforcement of such a policy, so that customers and providers can check anytime that the IP service is provisioned with the appropriate (and contractual) levels of quality (which can be expressed in terms of service availability, for example).

Therefore, the IP TE PIB mainly aims to:

- store and maintain the configuration information that will be used by the routers to compute and select the routes that will comply with a collection of QoS requirements, such as the one-way maximum transit delay, or the maximum interpacket delay variation;
- store and maintain the information related to the traffic-engineered routes that have been installed in the routers' forwarding information bases, so that the service providers have the permanent knowledge of the network's resource availability

From this perspective, the IP TE PIB is organized into the following provisioning classes:

- The *forwarding* classes (**`ipTeFwClasses`**). The information contained in these classes is meant to provide a detailed description of the traffic-engineered routes. Only one table is defined, the IP TE route table, which describes the information related to TE routes that have been installed by the routers in their FIBs.

- The *metrics* classes (**ipTeMetricsClasses**). The information stored in the tables of this class is meant to provide a description of the metric values that will be taken into account by intra- and interdomain routing protocols for the computation and the selection of traffic-engineered routes. So far, two groups have been identified: the first group is based upon the traffic engineering extensions of intradomain routing protocols; the second group is related to QoS-related information that can be conveyed in BGP-4 messages.
- The *statistics* classes (**ipTeStatsClasses**). The information contained in these classes is meant to provide statistics on the enforcement of the TE policies.

The detailed description of the IP TE PIB can be found in Appendix 3.

9.8 COPS Usage for IP TE Accounting Purposes

Traffic engineering is one of the possible means for solving congestion problems and permitting efficient use of the network resources. Several tools have been proposed to achieve this goal. Nevertheless, only a few solutions introduce a high level of automation for the allocation of resources and the configuration operations. This section provides some insights into a candidate solution making use of COPS and its capacity to manipulate and retrieve accounting data related to IP TE usage. This solution is documented in Boucadair [15].

The design of an IP Traffic Engineering (IP TE) policy implies the manipulation of a large amount of configuration information that includes routing considerations, traffic forecast, available resources, etc. These parameters are provisioned as configuration information to the network devices (such as routers) by means of a COPS-based communication scheme, owing to the use of a specific client-type as described in Section 9.1.2. However, there remains the choice of the appropriate parameters to meet network constraints as well as quality of service (QoS) requirements, and also to observe the impact of such a choice/decision on the stability of the network. From this standpoint, several methods can be adopted: use either statistical data based on mathematical models or data resulting from measurements. The advantage of the second method is that it allows for real-time statistics.

To implement the dynamic mode in a COPS environment, the actual enforcement of a traffic engineering policy requires a feedback mechanism to qualify not only the efficiency of such enforcement but also the impact of future decisions made by the policy decision point (PDP) and installed by the PDP at the policy enforcement point (PEP)-embedded devices. Figure 9.2 illustrates that the actual enforcement of an IP TE policy is conditioned by the manipulation of information such as traffic forecast (according to customers' requests, for example [16]) and traffic load calculation.

This section introduces a set of IP TE accounting usage policy rule classes (PRCs) that will be monitored, recorded and/or reported by the PEP [15]. Those PRCs complement the PRC classes that have been defined in the framework of COPS-PR PIB for policy usage (RFC 3483 [17]).

Within the context of this book and for illustration purposes, the data recorded, monitored and/or reported by the PEP are the results of the activation of dynamic routing processes [e. g. open shortest path first (OSPF) and border gateway protocol version 4 (BGP-4)].

The IP TE report classes are instantiated as multiple provisioning instances (PRIs), each of which is identified by a provisioning instance identifier (PRID). These classes contain attributes that actually describe the accounting IP TE-related information collected in the network. The PIB defined within the context of IP traffic engineering for accounting

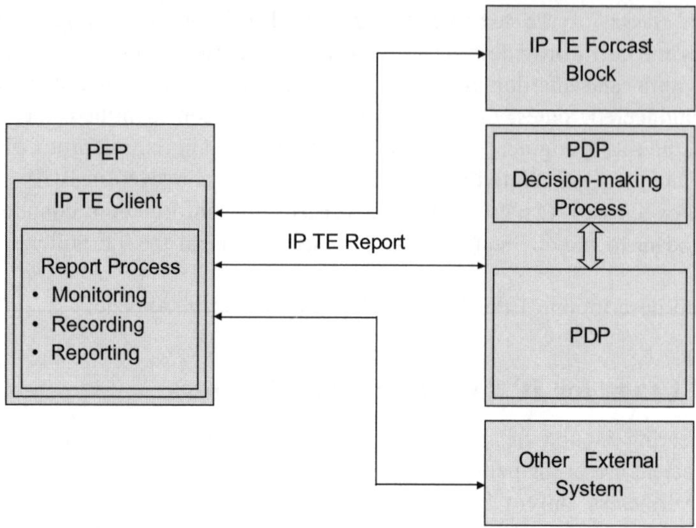

Figure 9.2 IP TE reporting mechanism

purposes has the goal of completing the whole COPS TE reporting machinery. This PIB contains the following tables:

- *ospfTeRouterUsageTable*. This class defines the usage attributes to be reported, which are related to the router identified by the *Router-Id*.
- *ospfTeUsageTable*. This class defines the usage attributes to use for OSPF TE purposes.
- *isisTeUsageTable*. This class defines the usage attributes to use for IS-IS TE purposes.
- *bgpTeTable*. This table contains a set of accounting information related to the activation of the BGP process, enabling exchange of QoS information.
- *ospfTeThresholdTable*. This class defines the threshold attributes corresponding to OSPF TE usage attributes specified in **ospfTeUsageTable**.
- *isisTeThresholdTable*. This class defines the threshold attributes corresponding to IS-IS TE usage attributes specified in **isisTeUsageTable**.
- *bgpTeThresholdTable*. This class defines the threshold attributes corresponding to BGP usage attributes specified in **`bgpTeUsageTable`**.

The detailed description of the IP TE accounting PIB can be found in Appendix 4.

References

[1] Boyle, J., Cohen, R., Durham, D., Herzog, S., Raja R. and Sastry A., 'The COPS (Common Open Policy Service) Protocol', RFC 2748, Proposed Standard, January 2000.

[2] Ho Chan, K., Durham, D., Gai, S., Herzog, S., McLoghrie, K., Reichmeyer, F., Seligson, J., Smith, A. and Yavatkar, R., 'COPS Usage for Policy Provisioning (COPS-PR)', RFC 3084, March 2001.

[3] Katz, D., Yeung, D. and Kompella, K., 'Traffic Engineering Extensions to OSPF', RFC 3630, Work in Progress, September 2003.

[4] Smit, H. and Li, T., 'IS-IS Extensions for Traffic Engineering', RFC 3784, June 2004.

[5] Moy, J., 'OSPF Version 2', RFC 2328, April 1998.

[6] ISO/IEC 10589, 'Intermediate System to Intermediate System, Intra-Domain Routing Exchange Protocol for use in Conjunction with the Protocol for Providing the Connectionless-mode Network Service (ISO 8473)', June 1992.

[7] Jacquenet, C., 'Providing Quality of Service Indication by the BGP-4 Protocol: the QOS_NLRI Attribute', draft-jacquenet-qos-nrli-04.txt, Work in Progress, March 2002.

[8] Boucadair, M., 'QoS-Enhanced Border Gateway Protocol', Internet-Draft, draft-boucadair-qos-bgp-spec, Work in progress.

[9] Nichols, K., Blake, S., Baker, F. and Black, D., 'Definition of the Differentiated Services Field (DS Field) in the IPv4 and IPv6 Headers', RFC 2474, December 1998.

[10] Apostopoulos, G., Guerin, R., Kamat, S. and Tripathi, S.K., '*Server Based QOS Routing*', Proceedings of the 1999 GLOBCOMM Conference.

[11] Boucadair, M. and Jacquenet, C., 'An IP Forwarding Policy Information Base', draft-jacquenet-ip-fwd-pib-00.txt, Work in Progress, January 2003.

[12] Rekhter, Y. and Li, T., 'A Border Gateway Protocol 4 (BGP-4)', RFC 4271, January 2006.

[13] Case, J. *et al.*, 'A Simple Network Management Protocol', RFC 1157, May 1990.

[14] Guerin, R. *et al.*, 'QoS Routing Mechanisms and OSPF Extensions', RFC 2676, August 1999.

[15] Boucadair, M., 'An IP Traffic Engineering PIB for Accounting purposes', draft-boucadair-ipte-acct-pib-02.txt, June 2003.

[16] Goderis, D. *et al.*, 'Attributes of a Service Level Specification (SLS) Template', draft-tequila-sls-03.txt, Work in Progress, October 2003.

[17] Rawlins, D. *et al.*, 'Framework for Policy Usage Feedback for Common Open Policy Service with Policy Provisioning (COPS-PR)', RFC 3483, March 2003.

10

Automated Production of BGP/MPLS-based VPN Networks

10.1 Introduction

An IP virtual private network (IP VPN) can be defined as a collection of switching and transmission resources that will be used by a dedicated set of authorized users to exchange information over a public IP infrastructure, like the Internet.

IP VPN networks may use different and complex technologies, thus giving rise to the need for a high level of automation to dynamically provision such networks. To do so, the network resources that will be involved in the forwarding of the traffic for a given IP VPN will have to process quite a large amount of configuration information, which includes (but is not necessarily limited to):

- topology information (e.g. location of the sites that will be interconnected via the IP VPN);
- addressing information (e.g. identification of the IP networks and hosts that will access the IP VPN facility);
- routing information (e.g. definition of a routing policy within the IP VPN, and how the Internet can be accessed through the IP VPN);
- security information (e.g. establishment and activation of filters);
- quality of service (QoS) information related to the service offering (e.g. the QoS parameters that will be conveyed in a particular service level specification (SLS) template [1] and that will be (dynamically) negotiated and invoked between the customer and the service provider, like the bandwidth, the one-way delay and the interpacket delay variation (RFC 2678 [2]).

Service Automation and Dynamic Provisioning Techniques in IP/MPLS Environments C. Jacquenet, G. Bourdon and M. Boucadair

The end result of IP VPN configuration tasks is to align the network elements to provide consistent treatment of the corresponding IP VPN traffic. The network elements will require a combination of capabilities depending largely on their location in the topology and the technology being used.

In addition, the IP VPN policy information model can help in defining a standard interface to VPN facilities supported by an IP network. This interface is useful for dynamic and customizable definition of provided VPN services based upon customer needs.

Therefore, the motivation for an IP VPN policy information model basically consists in providing a common understanding of how the corresponding IP VPN service is to be deployed over the network according to instances of the above information, from the IP VPN service level to the network element level involved in the design, the deployment and the operation of an IP VPN service offering.

BGP/MPLS-based VPN services (RFC 2917 [3], RFC 4364 [4]) may be delivered between premises of the same company, or between different companies. BGP/MPLS-based VPN services are deployed and operated by the combined activation of a set of elementary capabilities, which can be classified according to the following taxonomy:

- *Topological* considerations. These capabilities correspond to the information needed for the deployment of BGP/MPLS-based VPN topologies. This information includes, but is not limited to, the identification of the endpoints that will be interconnected via the VPN, the VPN-specific forwarding and routing policies to be enforced by the participating network devices and the topology of VPN membership.
- *Quality of service* considerations. These capabilities correspond to the information that characterizes the level of quality provided with the VPN service offering. QoS parameters include, but are not limited to, VPN traffic classification and marking capabilities, traffic conditioning and scheduling capabilities, as well as VPN traffic engineering capabilities.
- *Security* considerations. Any BGP/MPLS-based VPN that is deployed and operated across multiple domains [or autonomous systems (ASs)] (RFC 1772 [5]) may also encourage the need for identification, authentication and, potentially, VPN traffic encryption capabilities. This includes the possible identification and authentication of the resources that participate in the establishment and operation of a VPN, as well as the ability to check the integrity of VPN route announcements exchanged between ASs.
- *Management* considerations. It is assumed that the operation of QoS-based BGP/MPLS-based VPN services is part of the management tasks performed by service providers within their own ASs. Additional operational tasks are, however, needed in order to enable the management of VPN services across multiple ASs.
- *Measurement and monitoring* considerations. the ability to measure and monitor service delivery is of paramount importance, especially when such services span multiple ASs.

10.2 Approach

The IP VPN information model aims to provide a common understanding of how the corresponding IP VPN service is to be deployed over the network. This objective is achieved by identifying the various kinds of configuration information that need to be provided for defining, deploying and operating an IP VPN. Figure 10.1 provides a graphical view of

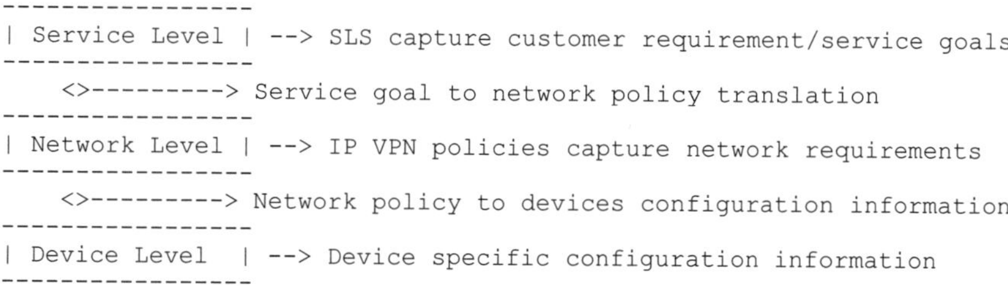

Figure 10.1 Dynamic VPN provisioning model, from SLS negotiation to VPN device configuration

where the information model fits in, with respect to the service goals and the device configuration.

The dynamic provisioning of the appropriate configuration information to the devices involved in the deployment and the operation of the IP VPN has significant advantages. It will introduce a high level of automation into the actual provisioning of the IP VPN service offering. It will provide some guarantees as far as the consistency of such configuration information is concerned.

The IP VPN information model includes policy components and topology components. The policies make references to the physical topology components. The provisioning system creates a virtual topology to meet the requirements captured in the policies.

The information required for the provisioning of an IP VPN service offering are captured in the form of rules. The rules reflect the customer requirements at the service level, translated into network requirements. The device configuration information is generated using these rules.

The device configuration information results in the creation of a virtual topology over the physical topology. The physical topology identifies policy targets for IP VPN deployment. The virtual topology is used by the provisioning system to track the current status of the network resource allocation owing to the previous IP VPN-related configuration.

This section provides an overview of the IP VPN policy information model. Subsequent sections will elaborate on the components of the IP VPN policy information model identified in this section.

The topology information model seizes the network status from a dual standpoint: the physical and the virtual. Physical topology classes represent the physical structure of the network that supports the IP VPN service offering. The IP VPN policy information model uses them in order to identify policy targets for the IP VPN deployment. The end result of such deployment is the creation of a virtual topology. The latter is captured by the virtual topology classes.

This model assumes that the IP VPN is provisioned over a provider network as depicted in RFC 2764 [6], and according to the 'customer premises equipment (CPE)-based'/'Network-based' taxonomy.

This is summarized by the reference models in Figure 10.2.

The IP VPN policy information model supports both network-based and CPE-based types of IP VPN network. In order to have a single model for both types, the following

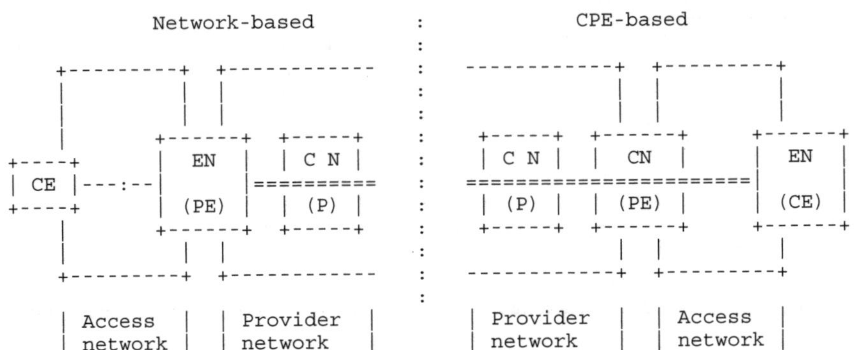

Figure 10.2 Reference models for IP VPN

generalization has been adopted: as far as IP VPNs are concerned, devices can be divided into IP VPN-aware nodes and IP VPN-unaware nodes. The former are grouped as 'edge nodes', while the latter are grouped as 'core nodes', irrespective of where devices physically reside (be they located at a provider network border or at customer premises).

10.3 Use of Policies to Define Rules

There are different ways of defining rules. The rule definition approach described in this chapter is based upon policies defined in RFC 3060 [7]. The core classes address common rule definition requirements, such as prioritization, and reuse of rule building blocks, such as conditions and actions.

The core classes have been extended to address the requirements that are specific to the IP VPN domain. The storage and distribution recommendations in RFC 2753 [8] can be applied to the storage and distribution of IP VPN policies. The corresponding lightweight directory access protocol (LDAP) (RFC 4510 [9]) implementations could be built on the basis of the 'policy core LDAP schema' (RFC 3703 [10]) and the 'Policy QoS information model' (RFC 3644 [11]) implementations.

The IP VPN policy information model also references QoS (RFC 2475 [12]), IP-secure (IPsec) (RFC 2401 [13], RFC 4301 [14]) and MPLS (RFC 3031 [15]) classes where appropriate. Some of the work in this area is directed at device configuration. The IP VPN policy information model, however, aims to capture the network requirements for deploying IP VPN networks, whereas the generation of the device configuration information is delegated to policy servers.

10.4 Instantiation of IP VPN Information Model Classes

The IP VPN provisioning system can instantiate the required classes to capture the network requirements for an IP VPN. The provisioning system needs to take into account the customer requirements and the physical topology to instantiate the classes with the appropriate values, as depicted in Figure 10.3.

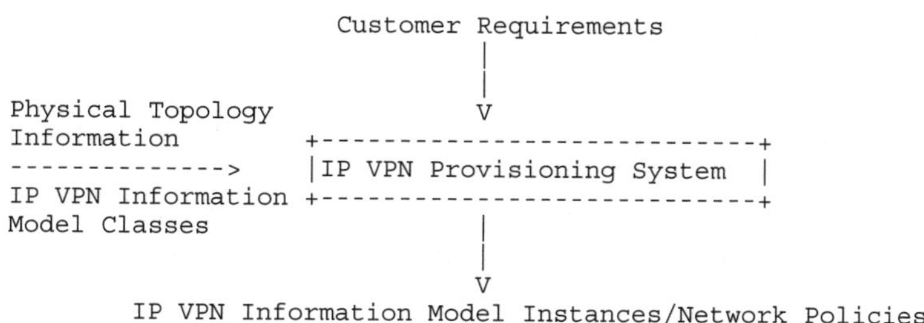

Figure 10.3 Instantiation of IP VPN information model classes

10.5 Policy Components of an IP VPN Information Model

The IP VPN information model consists of a set of IP VPN configuration action classes that are combined together with the rule and condition classes defined in RFC 3060 [7] and RFC 3460 [16] in order to obtain the IP VPN provisioning rules (Figure 10.4).

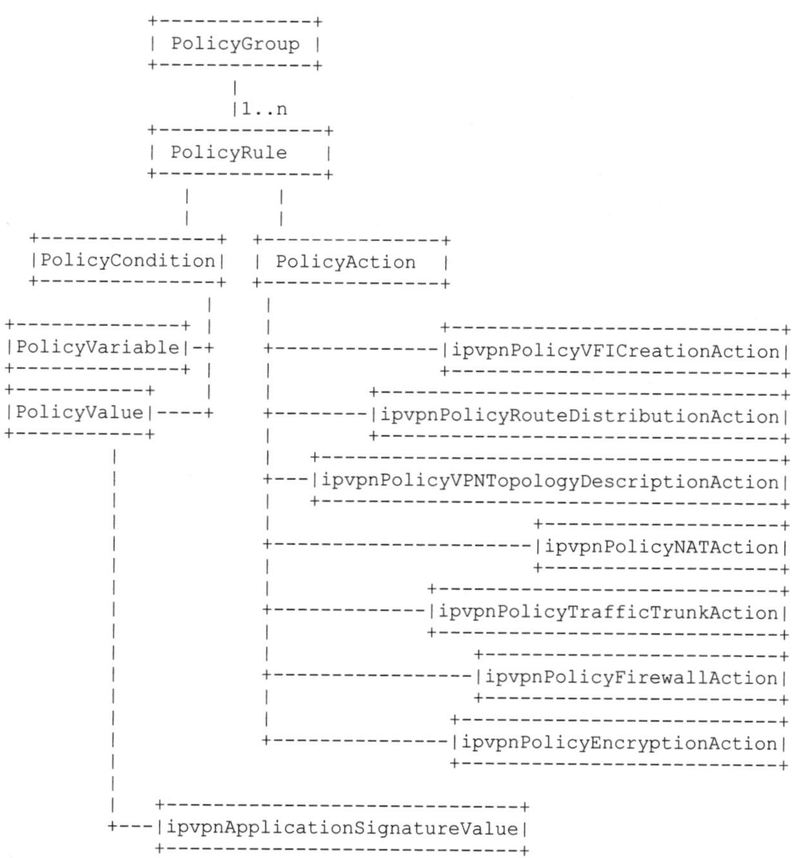

Figure 10.4 Policy components of an IP VPN information model

The important classes to be highlighted in Figure 10.4 are:

- The *ipvpnPolicyVFICreationAction* specifies the virtual forwarding instance (VFI) to be created on the edge nodes if the chosen IP VPN implementation is compliant with RFC 4364 [4].
- The *ipvpnPolicyRouteDistributionAction* specifies the connectivity between the edge nodes in the IP VPN and enables the IP VPN to be implemented as specified in RFC 4364 [4].
- The *ipvpnPolicyVPNTopologyDescriptionAction* provides a description of the IP VPN topology according to the connectivity requirements of the IP VPN service and enables CE-based IP VPNs to be implemented with IPsec, as described in RFC 4111 [17].
- The *ipvpnPolicyNATAction* enables the network address translation (NAT) (RFC 2663 [18]) requirements of an IP VPN to be captured.
- The *ipvpnPolicyTrafficTrunkAction* aggregates the requirements for the traffic trunks that can be used to transport the IP VPN traffic over the provider network.
- The *ipvpnPolicyFirewallAction* enables the firewall requirements of an IP VPN to be captured.
- The *ipvpnPolicyEncryptionAction* enables the encryption requirements of an IP VPN to be captured.
- The *ipvpnApplicationSignatureValue* specifies the layer-4 to layer-7 characteristics of an IP VPN packet. This class enables the policies to capture the application layer requirements of the customer with regards to treatment for specific VPN traffic.

The policy components make references to physical topology components which are defined as part of the complete set of topology components which are classified into physical topology components and virtual topology components, as further elaborated in Sections 10.5.1 and 10.5.2.

10.5.1 Physical Components of an IP VPN Information Model

The physical topology components are used to capture the physical topology of the network, as shown in Figure 10.5.

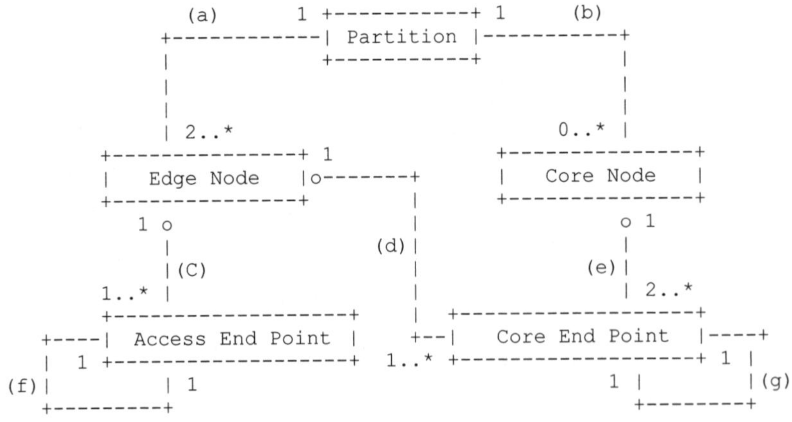

Figure 10.5 Overview of physical topology classes and relationships

In Figure 10.5 the relationships are labeled as follows:

(a) EdgeNodeInPartition;
(b) CoreNodeInPartition;
(c) AccessEndPointInEdgeNode;
(d) CoreEndPointInEdgeNode;
(e) CoreEndPointInCoreNode;
(f) AccessLink;
(g) CoreLink.

Network nodes are classified as core nodes (P or PE) and edge nodes (PE or CE). Edge nodes provide IP VPN connectivity to customers by means of one or more AccessEndPoints. The set of AccessEndPoints represents the set of interfaces towards IP VPN customers; interfaces can be either virtual or physical.

Note that the term 'interface' does not refer to physical adapters. Edge nodes are also associated with a second set of interfaces, called core endpoints, which represent the attachment points to the core network (note that 'core' is defined with respect to the IP VPN service).

On the other hand, core nodes are associated with core interfaces only (see the aggregation labeled as (e) in Figure 10.5). Core interfaces are represented by instances of the core endpoint class. Core interfaces are interconnected by core links, which represent the transmission resources that interconnect routers.

These physical topology classes are referenced by the different policy action classes defined in this information model. These classes are described in further detail in subsequent sections under topology class definitions.

10.5.2 Virtual Components of an IP VPN Information Model

The configuration generated as a result of the enforcement of IP VPN policies will result in a virtual topology, which can be modeled using the classes and relationships described in this section.

Figure 10.6 depicts the class diagram of virtual topology entities.

In Figure 10.6 the relationships are labeled as follows:

(h) EdgeProviderNodeInIPVPN;
(i) VirtualEndPointInEdgeProviderNode;
(l) VFIInEdgeProviderNode;
(m) VirtualLink;
(n) AccessEndPointInVFI;
(o) VirtualEndPointInVFI.

An IP VPN is identified as a set of edge nodes (ENs) that participate in the interconnection of IP VPN sites. As far as the IP VPN service is concerned, the role of an EN is to forward IP VPN traffic from access links to the correct paths, and vice versa. A virtual forwarding instance can be defined to accomplish this task, if the chosen implementation of IP VPN is

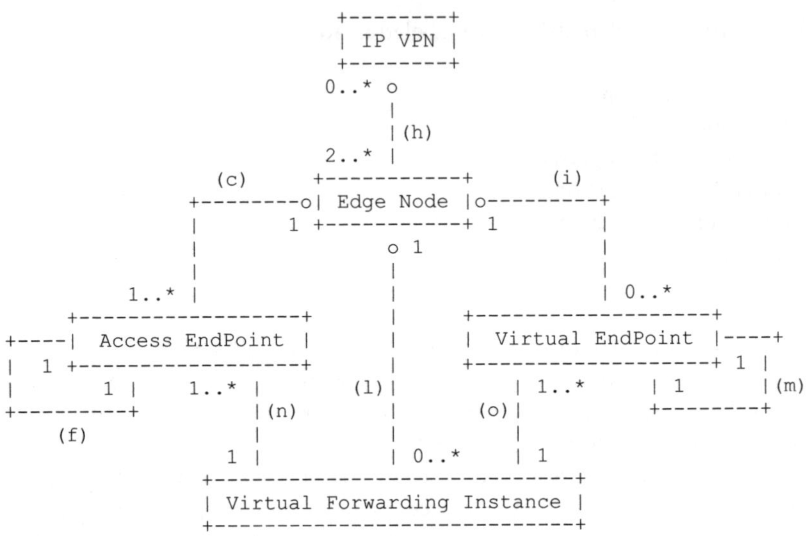

Figure 10.6 Overview of virtual topology classes and relationships

compliant with RFC 4364 [4]. Hence, VFI instances can be activated on access and virtual endpoints.

10.5.3 Inheritance Hierarchy

The inheritance hierarchy shows the various classes used to define the IP VPN policy information model. This information model is policy driven, so we start with the classes derived from the policy base class.

10.5.3.1 Inheritance Hierarchy for Policy Classes

The tree structure depicted in Figure 10.7 reflects the inheritance relationships between the generic policy classes as defined in RFC 3060 [7], RFC 3460 [16] and RFC 3644 [11] and those policy classes that are specific to an IP VPN service offering.

The new classes introduced by the IP VPN information model are as follows:

- The *ipvpnPolicyVFICreationAction* specifies the VFI to be created on the edge nodes if the chosen IP VPN implementation is compliant with RFC 4364 [4].
- The *ipvpnPolicyRouteDistributionAction* specifies the connectivity between the edge nodes in the IP VPN and enables the IP VPN to be implemented as specified in RFC 4364 [4].
- The *ipvpnPolicyVPNTopologyDescriptionAction* provides a description of the IP VPN topology according to the connectivity requirements of the IP VPN service and enables CE-based IP VPNs to be implemented with IPsec (RFC 4111 [17]).
- The *ipvpnPolicyNATAction* enables the NAT requirements of an IP VPN to be captured.

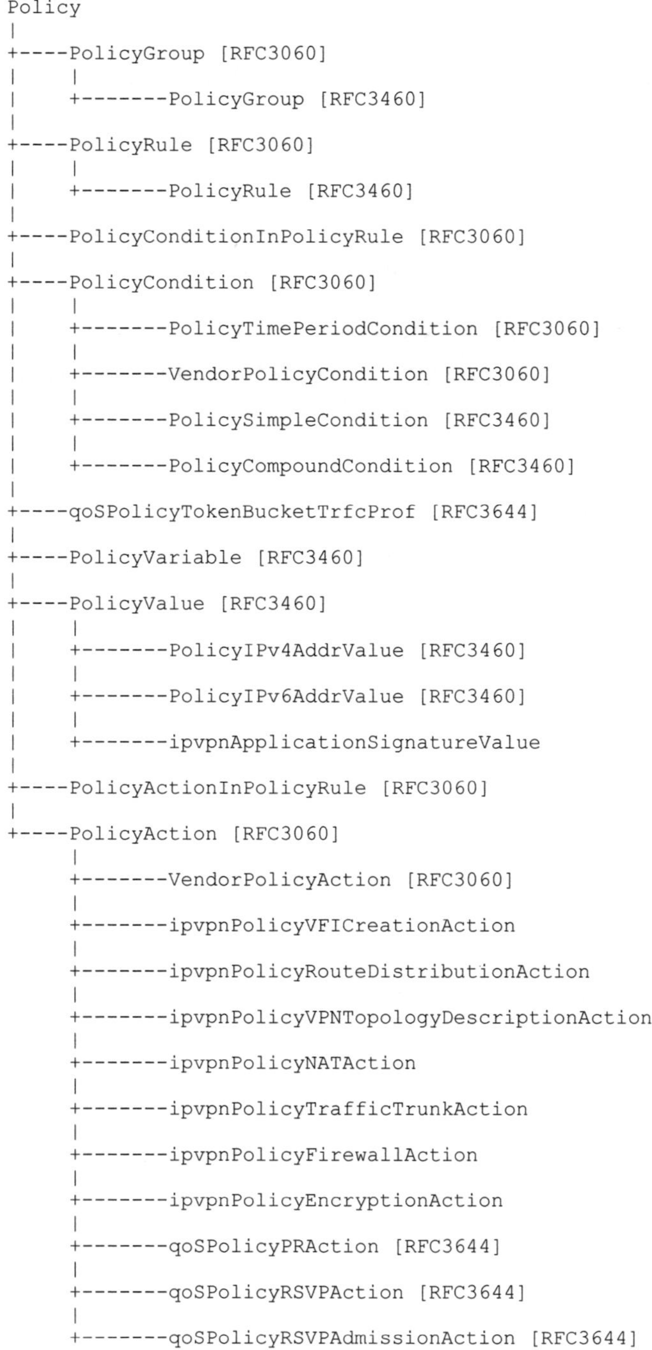
Policy
+----PolicyGroup [RFC3060]
+-------PolicyGroup [RFC3460]
+----PolicyRule [RFC3060]
+-------PolicyRule [RFC3460]
+----PolicyConditionInPolicyRule [RFC3060]
+----PolicyCondition [RFC3060]
+-------PolicyTimePeriodCondition [RFC3060]
+-------VendorPolicyCondition [RFC3060]
+-------PolicySimpleCondition [RFC3460]
+-------PolicyCompoundCondition [RFC3460]
+----qoSPolicyTokenBucketTrfcProf [RFC3644]
+----PolicyVariable [RFC3460]
+----PolicyValue [RFC3460]
+-------PolicyIPv4AddrValue [RFC3460]
+-------PolicyIPv6AddrValue [RFC3460]
+-------ipvpnApplicationSignatureValue
+----PolicyActionInPolicyRule [RFC3060]
+----PolicyAction [RFC3060]
+-------VendorPolicyAction [RFC3060]
+-------ipvpnPolicyVFICreationAction
+-------ipvpnPolicyRouteDistributionAction
+-------ipvpnPolicyVPNTopologyDescriptionAction
+-------ipvpnPolicyNATAction
+-------ipvpnPolicyTrafficTrunkAction
+-------ipvpnPolicyFirewallAction
+-------ipvpnPolicyEncryptionAction
+-------qoSPolicyPRAction [RFC3644]
+-------qoSPolicyRSVPAction [RFC3644]
+-------qoSPolicyRSVPAdmissionAction [RFC3644]

Figure 10.7 Inheritance hierarchy for policy components

- The *ipvpnPolicyTrafficTrunkAction* aggregates the requirements for the traffic trunks that can be used to transport the IP VPN traffic over the provider network.
- The *ipvpnPolicyFirewallAction* enables the firewall requirements of an IP VPN to be captured.
- The *ipvpnPolicyEncryptionAction* enables the encryption requirements of an IP VPN to be captured.
- The *ipvpnApplicationSignatureValue* specifies the layer-4 to layer-7 characteristics of the packet. This class enables the policies to capture the application layer requirements of the customer with regards to treatment for specific IP traffic.

10.5.3.2 Inheritance Tree for Policy Classes

Classes related to the topology model are shown in Figure 10.8. They are derived from the classes mentioned on the DMTF website [19].

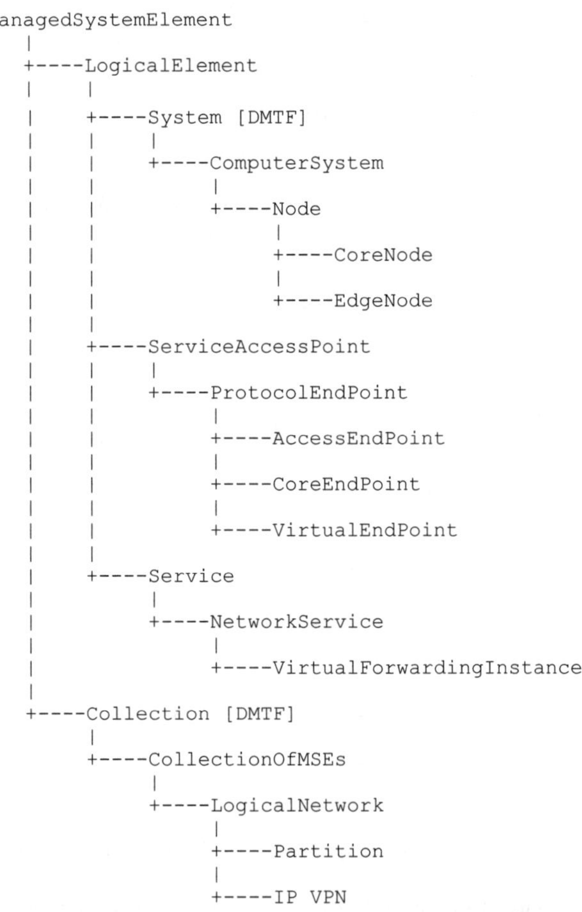

Figure 10.8 Class inheritance for topology components

The inheritance hierarchy (Figure 10.9) shows the various classes used to define relationships between topology classes.

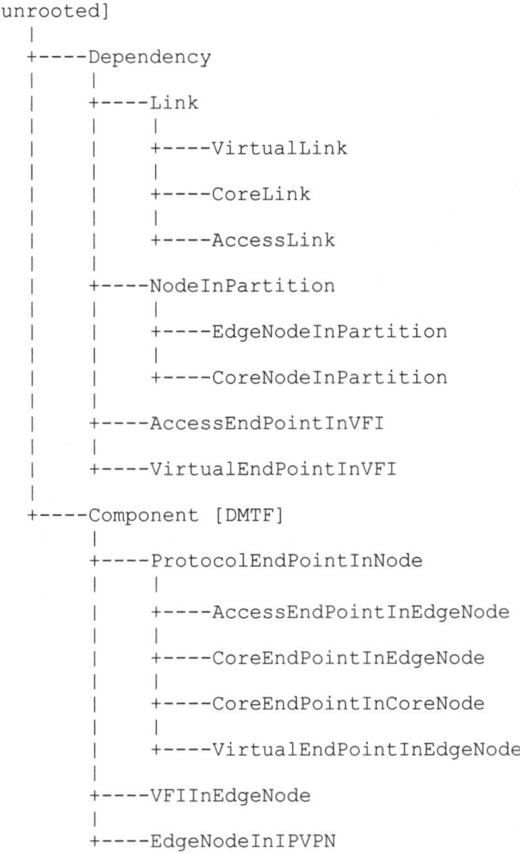

Figure 10.9 Association inheritance for topology components

A detailed description of the IP VPN-specific classes that have been introduced in Section 10.5 can be found in Appendix 5.

10.6 Dynamic Production of IP VPN Services

The physical topology reflects the physical layout of the devices and their interfaces. They are referenced by the policy action classes defined in the IP VPN information model. The role of the policy server in the policy management framework is detailed in RFC 2753 [8]. The policy servers use this information to generate the configuration information that will be processed by the IP VPN participating devices.

The topology of an IP VPN is an implicit result of the (device) configuration information, i.e. the topology is displayed/described once the devices have been configured accordingly, in terms of architecture, QoS, security and management, as per a 'global' IP VPN deployment policy.

The network devices involved in the forwarding of the IP VPN traffic as well as the virtual links generated by the configuration represent the IP VPN topology.

10.7 Context of a Multidomain Environment

10.7.1 A Bit of Terminology

Some of the terminology used in this chapter is taken from RFC 4026 [20] and RFC 4031 [21]. 'VPN' in the context of this document refers specifically to BGP/MPLS-based VPN services. In order to clarify the requirements listed in this document, it is necessary further to define and introduce new terminology specific to multiprovider VPN services.

10.7.1.1 Agent

For the purposes of this chapter, an agent is a VPN service provider (VSP) who is responsible for the management of multiparty business processes, negotiations and fulfillment that allow the multiprovider VPN to function. The agent manages this responsibility by either operationally complying with or coordinating policies across all parties involved in delivering their customer's end-to-end service. Policy compliance or multiparty policy coordination is achieved either in a distributed or in a centralized manner:

- For *distributed* policy enforcement, cooperating VSPs agree upon the enforcement of consistent policies for VPN service provisioning purposes. In this case, end-to-end policy enforcement is distributed across multiple VSPs, each of which is a stakeholder in the supply and enforcement of a fixed set of policies within the shared multi-AS environment. A VPN customer defines its VPN service requirements with an agent who then maps these requirements to the set of policies that may have been predefined by the VSP, and possibly negotiated with the customer, then adapted to the customer's specific requirements.
- For *centralized* policy enforcement, the agent coordinates, per customer, a set of policies related to the management of the customer's VPN service. In contrast to a shared multi-AS environment, where policy enforcement is distributed across multiple VSPs, this agent will coordinate policies and integrate VPN services per customer, thereby creating a customer-specific environment that is dedicated to the agent's customers only. The agent may independently agree and manage service level specification (SLS) with each partner VSP and offer an aggregated end-to-end SLS to their customer's.

10.7.1.2 Multidomain Environment (MDE)

Two or more autonomous systems that may or may not be owned by separate administrative authorities, and which are used to interconnect service endpoints (sites) of one or more VPNs.

10.7.1.3 VPN Peering Location (VPL)

A VPL is a physical location where VPN services delivered by one or more VSPs are interconnected. Examples of VPLs include VSP central offices, a collocation facility or any building common to one or more VSPs. A VPL could be operated by a single VSP, a consortium of VSPs or a neutral third party.

10.7.1.4 VPN Service Provider (VSP)

A VSP is an operator who participates in the delivery of a single domain or multidomain (MDE-wide) VPN service. In delivering the VPN service, the VSP may own a subset or all of the participating network elements. Examples of VSPs include network service providers (NSPs), systems integrators (SIs), network integrator (NIs), mobile operators (MOs) and virtual network operators (VNOs).

10.7.2 Reference Model

Figure 10.10 shows a generic reference model for a multidomain environment. It shows the relationships that exist between the various parties involved in the establishment of VPN services that span multiple VSP-administered networks.

The MDE consists of sites interconnected via ASs. ASs are then interconnected via VPLs, within the context of delivering VPN service offerings. There is potentially a one-to-many relationship between VSPs and ASs, similarly there is potentially a one-to-many relationship between VPL operators and VPLs. Note that VSPs may be remotely interconnected, that is, VSPs do not necessarily need to be directly connected to each other through a given VPL.

With reference to Figure 10.10, the following sections detail examples of the coordination of the various parties in the enforcement of a set of policies. These policies relate to the provisioning of an MDE-wide VPN service and include (but are not limited to) addressing, routing, QoS and security.

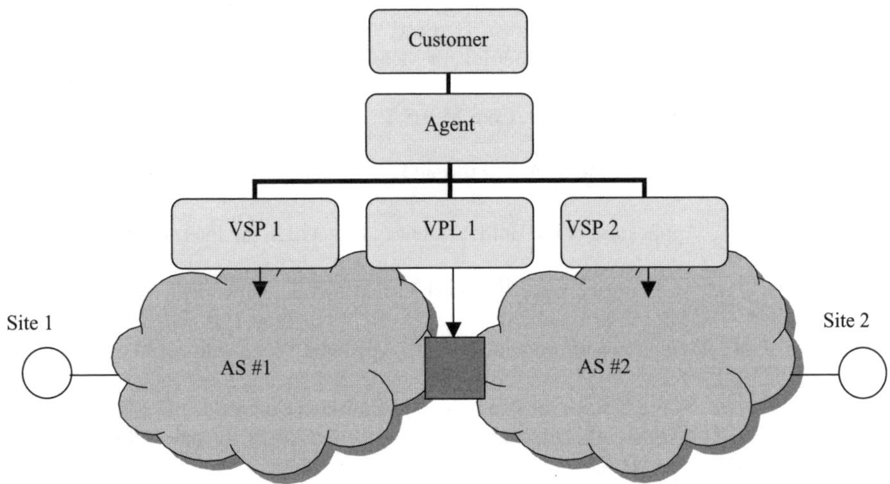

Figure 10.10 The MDE reference model

10.7.2.1 Distributed Policy Enforcement Scheme

The customer's agent is a single VSP (VSP1 in this example). VSP1 manages the relationship with the customer, which includes the specification, instantiation, possible negotiation and invocation of the customer's SLS.

Cooperating VSPs, VSP1 and VSP2, have preagreed an enforced set of policies, including the management of VPL1. The customer's requirements are mapped by VSP1 to the preagreed policies of VSP1 and VSP2.

10.7.2.2 Centralized Policy Enforcement Scheme

The customer's agent is a systems integrator (SI). The SI does not own the network infrastructure (the VPN networking elements), but manages the VPLs, as well as the processes involved in connecting the VPLs with each relevant VSP-owned AS.

VSP1 and VSP2 independently provide customer-specific VPN services and associated SLS instantiated templates to the SI who is then responsible for integrating each VSP's VPN service components and negotiating/invoking an end-to-end SLS with/for their customer.

10.8 Possible Extensions of the VPN Model

The IP VPN policy information model describes the IP VPN basic features – namely connectivity, security and QoS. The IP VPN policy information model can be extended to support new requirements generated as a result of new functions for the deployment of value-added IP VPN services, like the integration of IP multicast transmission schemes within the IP VPN.

References

[1] Goderis, D., T'Joens, Y., Jacquenet, C., Memenios, G., Pavlou, G., Egan, R., Griffin, D., Georgatsos, P. and Georgiadis L., 'Attributes of a Service Level Specification (SLS) Template', draft-tequila-sls-03.txt, Work in Progress, October 2003.

[2] Mahdavi, J. and Paxson, V., 'IPPM Metrics for Measuring Connectivity', RFC 2678, September 1999.

[3] Muthukrishnan, K. and Malis, A., 'A Core MPLS IP VPN Architecture', RFC 2917, September 2000.

[4] Rosen, E. and Rekhter, Y., 'BGP/MPLS IP Virtual Private Networks (VPNs)', RFC 4364, February 2006.

[5] Rekhter, Y. *et al.*, 'Application of the Border Gateway Protocol in the Internet', RFC 1772, March 1995.

[6] Gleeson, B. *et al.*, 'A Framework for IP Based Virtual Private Networks', RFC 2764, February 2000.

[7] Moore, B. *et al.*, 'Policy Core Information Model – Version 1 Specification', RFC 3060, February 2001.

[8] Yavatkar, R. *et al.*, 'A Framework for Policy-based Admission Control', RFC 2753, January 2000.

[9] Zeilenga, K. *et al.*, 'Lightweight Directory Access Protocol (LDAP): Technical Specification Road Map', RFC 4510, June 2006.

[10] Strassner, J. *et al.*, 'Policy Core LDAP Schema', RFC 3703, February 2004.

[11] Snir, Y. *et al.*, 'Policy Quality of Service (QoS) Information Model', RFC 3644, November 2003.
[12] Blake, S. *et al.*, 'An Architecture for Differentiated Services', RFC 2475, December 1998.
[13] Kent, S. and Atkinson, R. 'Security Architecture for the Internet Protocol', RFC 2401, November 1998.
[14] Kent, S. and Seo, K., 'Security Architecture for the Internet Protocol', RFC 4301, December 2005.
[15] Rosen, E. *et al.*, 'Multiprotocol Label Switching Architecture', RFC 3031, January 2001.
[16] Moore, B. *et al.*, 'Policy Core Information Model Extensions', RFC 3460, January 2003.
[17] Fang, L. *et al.*, 'Security Framework for Provider-Provisioned Virtual Private Networks (PPVPNs)', RFC 4111, July 2005.
[18] Srisuresh, P. and Holdredge, M., 'IP Network Address Translator (NAT) Terminology and Considerations', RFC 2663, August 1999.
[19] Distributed Management Task Force, Inc., 'DMTF Technologies: CIM Standards CIM Schema: Version 2.5', available via links on the following DMTF web page: http://www.dmtf.org/spec/cim_schema_v25.html.
[20] Andersson, L. and Madsen, T., 'Provider-Provisioned Virtual Private Network (VPN) Terminology', RFC 4026, March 2005.
[21] Carugi, M. *et al.*, 'Service Requirements for Layer 3 Provider Provisioned Virtual Private Networks (PPVPNs)', RFC 4031, April 2005.

11

Dynamic Enforcement of Security Policies in IP/MPLS Environments

Enforcing security policies in an IP/MPLS network usually involves the AAA chain. However, it is worthwhile describing a little bit further what we define here as a security policy. The most basic definition of a security policy is to define the ability of an end-user to get IP connectivity from the network. This binary approach can be extended to the notion of network resource; for instance, granting or not granting access to a VPN, to the Internet or to multicast flows. In this section, we will focus on security policies attached to this definition, even though network resources can take much more sophisticated aspects.

Through two different application examples extracted from real network architecture scenarios, this section intends to provide the reader with an overview of various ways dynamically to enforce security policies in networks:

- Wi-Fi access control through a web-based captive portal;
- company network access control using 802.1X.

11.1 Enforcing Security Policies for Web-based Access Control

Web-based access control is something often used for Wi-Fi access or line access in nomadic situations (hotels, for instance). The scenario is simple: the end-user connects his/her terminal to a public wire or activates his/her Wi-Fi connection. Putting aside the layer-2 connectivity (especially for Wi-Fi access), the terminal gets an IP address through a public DHCP server. After opening his/her browser, the user requests an URL that is automatically redirected to a captive portal, asking for a login and password, or another credential (room number for billing in hotels, magic number printed on scratch cards, etc.). If the security test

Service Automation and Dynamic Provisioning Techniques in IP/MPLS Environments C. Jacquenet, G. Bourdon and M. Boucadair

is passed, the user can access the network service (Internet access, VPN, etc.), depending on his/her authorization profile.

A lot of solutions exist in the market to put in place such scenarios. From very simple solutions that are free to be used by individuals or SMEs who are interested in offering a prepaid Internet access (like Chillispot) to the most complex and scalable solutions sized for operators or major players with a consequent customer base [like Cisco subscriber access and management (SAM)]. In this example, we will focus on the industrial solution proposed by Cisco – the SAM.

The SAM is actually a package of two elements, the SSG service selection gateway (SSG) and the subscriber edge services manager (SESM):

- The SSG is the gateway through which all IP traffic from and to the subscriber has to go. This equipment is dedicated to subscriber connection aggregation, and to enforcement of policies. The SSG will redirect user's traffic to the SESM as long as the requested service is not authorized and activated. The SSG is in charge of enforcing policies according to the subscriber and service profiles managed by the AAA server and the SESM. It acts as proxy-RADIUS for all authentication/authorization messages exchanged between the AAA server and the SESM. The SSG is typically an edge router from the Cisco 7200, 7400 or 7600 series, for instance, at least running IOS 12.2(2)B.
- The SESM is the portal with which the subscriber interacts for authentication and service selection. Actually, the SESM is not only an HTTP server capable of interacting with the SSG, it is also a set of applications that are able to manage customer and service profiles, depending on the degree of complexity in which it is used. The SESM communicates with the SSG and the AAA server with the RADIUS protocol. For simplicity, here we will consider the SESM in a simple configuration. The SESM can be hosted in a single machine running Linux, Windows or Solaris.

Figure 11.1 shows how SESM and SSG are integrated in a web-based access control architecture for Wi-Fi.

The SESM and the SSG use RADIUS to communicate to the AAA server. This AAA server holds user profiles and service profiles: user profiles contain basic elements (login, password, etc.) but also the list of services allowed for this user. The service profile is a description of the services, with the technical parameters that are required to achieve each service. The services can be grouped into 'service groups', to ease service management for user subscriptions to bundles of services: a user profile can mention a set of service groups.

Firstly, here is the sequence of events occurring between the subscriber and the network in order to get the service:

1. The subscriber turns on his/her terminal and activates Wi-Fi.
2. The user's terminal attaches itself to the access point.
3. The user's terminal requests an IP address through DHCP.
4. The user opens his/her web browser, and the traffic is redirected to the SESM at the first URL request.
5. The browser displays the captive portal HTTPS page pushed by the SESM, requiring a login and a password to connect.

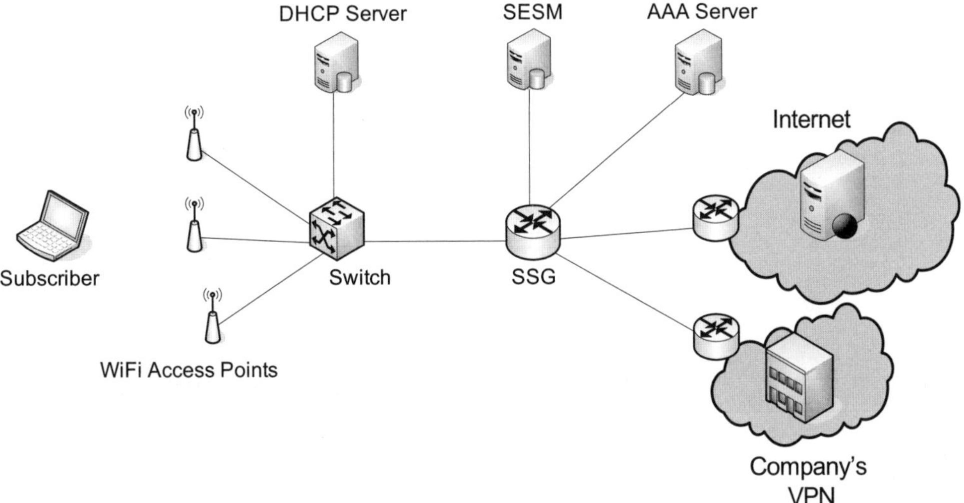

Figure 11.1 SSG and SESM in a web-based access control architecture for Wi-Fi

6. The user enters his/her login and password through HTTPS.
7. The browser displays an SESM page with the services available for the user to select.
8. The user selects the service on this page.
9. The browser displays an SESM page indicating that the service is active.
10. The user can now use the network service.

From this simple event sequence, multiple actions and messages are exchanged between the terminal, the SESM, the SSG and the AAA server. We consider here that layer-2 association as well as DHCP sequence has already occurred.

Two authorization phases are performed before the user can use his/her service:

- The user authentication and authorization phase, where the network authenticates the user and identifies the services to which he/she has subscribed.
- The service activation and authorization phase, where the user selects the services s/he wants to use, and where the network authorizes the selected service and enforces the associated policies.

The user authentication phase corresponds to steps 4 to 6 in the event user sequence listed above. The RADIUS messages that are exchanged in this scenario are shown in Figure 11.2:

- Phase *a* (layer-2 attachment and DHCP exchanges are already done). The user sends a request to the network to reach a URL through HTTP. The request is intercepted by the SSG, which has a default policy rule to redirect HTTP traffic to the SESM when the IP address is not authorized for available services. The HTTP request is redirected to the

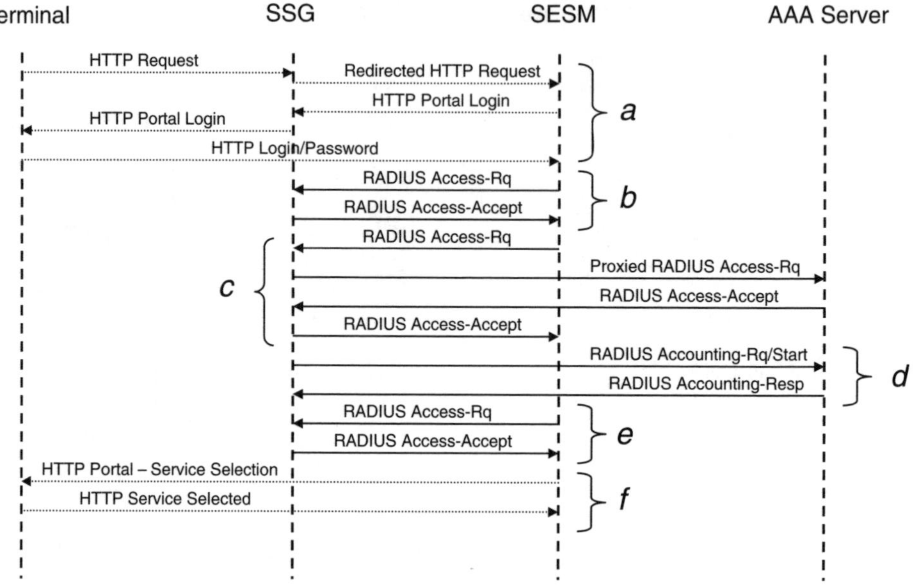

Figure 11.2 Protocol exchanges between the SSG, the SESM and the AAA server, from access control to service selection

SESM, which sends the user its portal login page to ask the user for authentication (login/password). The user sends his/her login/password to the SESM using the HTTPS protocol. The SESM has the user's credentials and begins the user authentication and authorization phase.

- Phase *b*. The SESM sends a request to the SSG to get the user's status, i.e. if a service is already activated or not (and which one), based on the IP address of the subscriber. The SSG answers to the SESM indicated that the user is not logged on.
- Phase *c*. The SESM sends a RADIUS *Access-Request* with the login/password entered by the subscriber. The request is sent to the SSG which proxies the request to the AAA server. The SESM can use either PAP or CHAP for authentication. The AAA server successfully authenticates the user and sends back, embedded in an *Access-Accept* message, the user profile with the list of authorized services. This message is proxied to the SESM by the SSG.
- Phase *d*. As soon as the SSG proxies the connection acceptation message to the SESM, it sends an accounting message to the AAA server notifying that the connection is now up. The AAA server acknowledges this message, but the time in which this acknowledgment is received does not influence the sequence of the following phases.
- Phase *e*. Once again, the SESM queries the SSG to get the status of the end-user session, in order to consolidate its own state, and by ensuring the SSG has created the end-user context for the next exchanges.
- Phase *f*. After the synchronization phase that occurred in phase *e*, the SESM is able to push a personalized web page to the end-user, notifying him/her of the successful authentication phase, and proposing a set of services that are available. The end-user has to select one service among the list proposed, and the choice is sent to the SESM.

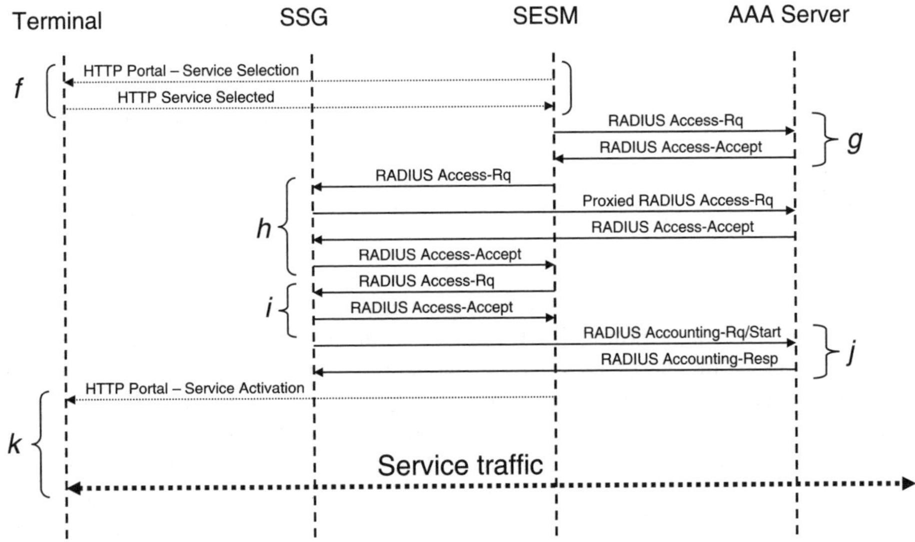

Figure 11.3 Protocol exchanges between the SSG, the SESM and the AAA server for service selection to policy enforcement

Then comes the service authorization phase, which really enforces service policies into the SSG. The protocol sequence that follows is shown in Figure 11.3, corresponding to steps 6 to 10 in the event user sequence listed above:

- Phase *g*. Once the service has been selected by the end-user, the SESM queries the AAA server to get service details. This is the only case where the SESM is in direct contact with the AAA server without being proxied by the SSG. This phase is optional since the service description can still be in SESM memory from the previous choice of another user.
- Phase *h*. Following answers and possibly following guidelines imposed by locally defined policies, the SESM requests from the AAA server the authorization for the end-user to get the selected service. The AAA requests are exchanged through the SSG. The service description embedded in the reply message (*Access-Accept*) is then used by the SSG to enforce the policy to be applied on the user's traffic.
- Phase *i*. After receipt of the *Access-Accept* message from the AAA server (through the SSG), the SESM queries the SSG to make sure the policy is enforced and the service is available for the end-user. The SSG answers with a 'logged on' message including the active service applied.
- Phase *j*. The SSG starts the accounting for the service as soon as it is activated. The *Accounting-Request* message *Start* is thus sent to the AAA server which acknowledges it.
- Phase *k*. The SESM pushes a web page to the end-user, indicating that the selected service is now available to be used. The traffic sent and received by the end-user goes through the SSG as long as the policies are enforced in the SSG.

The way Cisco uses RADIUS in this example is quite interesting. The AAA protocol is not just used to convey authentication data, accounting data and service policies to enforce on the SSG. RADIUS is also used here between the SESM and the SSG as a query/response protocol to get the status of the end-user. In steps *b*, *e* and *i*, the SESM queries the SSG to get the status of the end-user, in order to synchronize HTTP pages to be presented to him/her. This is a very specific way of using this protocol, and Cisco uses specific AVPs with proprietary codes for this purpose. These queries are considered as part of the whole solution, and no compatibility with other vendors is required, nor compatibility between operators since the SESM and SSG are managed by the same administrative entity.

Note that two accounting exchanges take place for service activation. Firstly, there is a connection accounting that is triggered as soon as the subscriber is authenticated. Secondly, a service accounting is activated as soon as the service is enforced in the SSG. The SSG is in charge of sending accounting information to the AAA server because the SSG is aware of service usage (traffic exchanged), not the SESM.

Once the subscriber wants to end his/her session, s/he has to notify the SESM that s/he wants to log out; otherwise, the service will still be active. There are two possible ways to avoid keeping a service session up while the terminal is no longer connected: either configuring a maximum session timeout after which the subscriber is forced to reconnect to the service through SESM, or a maximum idle timeout specifying the maximum delay of inactivity after which the subscriber is automatically logged out by the system. These parameters are sent with the user profile during phase *c*. This mechanism is particularly useful for this case where there is no notion of session, apart from the DHCP lease time allowed during connection establishment to the network. In the case of network failure, this helps to free unused resources, but it is also helpful if the service is billed on the basis of the time spent, either prepaid or post-paid. In our example shown in Figure 11.4, we consider that the subscriber manually logs out from the service by hitting a disconnect button available on the SESM portal web page

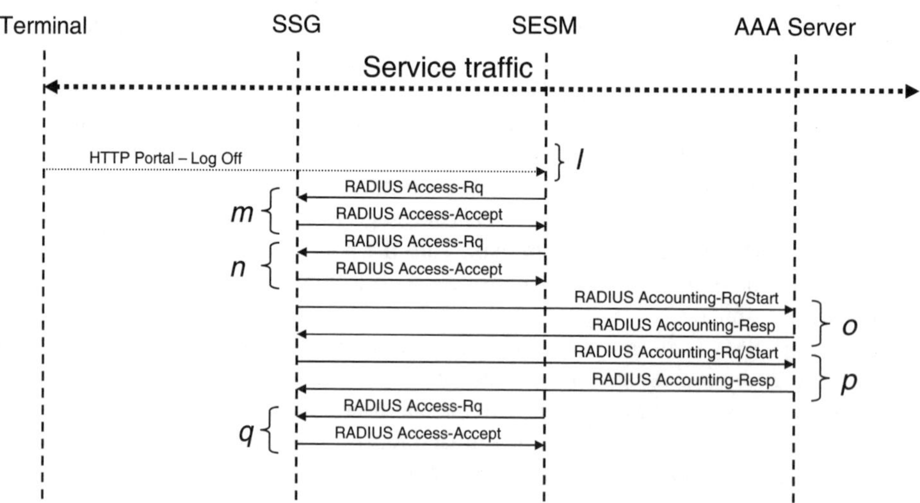

Figure 11.4 Protocol exchanges between the SSG, the SESM and the AAA server when the service is stopped

- Phase *l*. The subscriber goes to the SESM web page portal to log out. Logging out also means that the service activated in the context of his/her connection will be terminated.
- Phase *m*. The SESM queries the SSG to get information about the user's session. The SSG replies back by indicating that the user is connected to the system, with a mention of the service activated.
- Phase *n*. The SESM sends a request to disconnect the subscriber from the SSG. It is interesting to note that this request uses *Access-Request/Access-Accept* messages instead of *Disconnect-Request/Disconnect-Response* messages defined in RFC 3576 (see Chapter 3) for this purpose. Once again, this choice is due to multiple factors: SESM/SSG is a Cisco proprietary solution without any requirement to interoperate, and the IETF standard was not yet ready when the system was conceived.
- Phase *o*. The SSG stops the connection of the subscriber, and restores the default policies for the subscriber's IP address. The SSG first notifies the AAA server that the service activated for this user is terminated.
- Phase *p*. The SSG then sends the accounting message notifying the AAA server that the connection is terminated for this user. We must understand here that 'connection' refers to the relationship maintained between the terminal and the SESM. It must not be understood as the connectivity that is established with the access nodes (APs): at this stage, the subscriber still has his/her Wi-Fi connectivity as well as his/her IP address.
- Phase *q*. The SESM queries the SSG to consolidate the states on both pieces of equipment, to make sure that the context pertaining to the subscriber has been released on the SSG as well.

Once these steps are over, the default security policy is enforced back for the subscriber IP address, meaning that all HTTP traffic issued by the subscriber will be redirected to the SESM.

To enforce policies on the SSG, it is interesting to focus on some of the attributes sent by the AAA server when transmitting the user profile in step *c* as well as the service profile in steps *g* and *h*. Along with standard attributes that are used for traditional authentication, five Cisco-specific sub-AVPs are used, which, for some of them, are split into subactions:

- Cisco-AVPair (AVP 26, vendor ID 9, subattribute ID 1). This AVP is used to convey general parameters that are used in different Cisco solutions. Cisco defined its own syntax to discriminate between the different actions. The most interesting parameters are the access control lists (ACLs), downstream or upstream, that can be associated with the subscriber. It is possible to use this sub-AVP to define ACLs related to a user profile (if there is a specific policy for this user), or related to a service profile (when all subscribers attached to a service must have the same set of ACLs applied).
- Cisco-SSG-Account-Info (AVP 26, vendor ID 9, subattribute ID 250). This AVP can be used for five different usages specifically dedicated to describe the user profile or the service group profile. We will focus a little bit more on this AVP later.
- Cisco-SSG-Service-Info (AVP 26, vendor ID 9, subattribute ID 251). This AVP can be used for 14 different usages specifically dedicated to describe the service. This does not mean that the service will only be described by this AVP, however.

- Cisco-SSG-Command-Code (AVP 26, vendor ID 9, subattribute ID 252). This AVP is used for specific exchanges between the SESM and SSG for specific queries and actions (user status, subscriber log out).
- Cisco-SSG-Control-Info (AVP 26, vendor ID 9, subattribute ID 253). This AVP is used to help in designating and defining next hop entries.

Taking as an example of how the SSG is able to create a user context by interpreting RADIUS attributes sent by the AAA server, we will focus on the Cisco-SSG-Account-Info attribute used to define the user's profile. Five meanings are possible for this AVP by following a specific syntax of the type: <**command**>**value**. The <**command**> is a letter that gives the meaning of the attribute:

- A for Auto Service indicates the service name the SSG has to enforce by default.
- I for Service Group Description gives a description of a service group in plain text, to be displayed to the subscriber by the SESM, for instance.
- H or U indicates the default URL to which the subscriber wishes to be connected.
- G for Service Group indicates the name of a service group that is attached to the subscriber.
- N indicates a Service Name that is attached to the subscriber.

Let's take the example of the profile of user 1, who is eligible to Internet access and to the Company's VPN access (ACME). If we consider that user 1 wants to be connected by default (without service selection phase) to the Internet, the user's profile held in the AAA server (Free RADIUS format) will look like this:

```
User1 Password = ''userpwd'' ,
  Session-Timeout = 3600,
Account-Info = ''Ninternet'' ,
Account-Info = ''Nacme-vpn'' ,
Account-Info = ''Ainternet''
```

In the same manner, the services 'internet' and 'acme-vpn' will be described by using attributes embedded in Cisco-SSG-Service-Info, following a close syntax but with other possible actions.

In conclusion, it is important to mention that this kind of architecture for a Wi-Fi access has some drawbacks, especially in terms of security. Firstly, the communications are not encrypted by this system. This means that the user will have to activate a specific VPN client on his/her terminal if s/he wants to secure his/her traffic to a specific endpoint (such as a company's VPN gateway). Everyone within the 802.11 radio range will be able to inspect the user's traffic. The second security drawback of this solution is the risk of the connectivity being stolen by a malicious user: it is still possible to 'kick out' a terminal from its Wi-Fi connection and to take its place by spoofing its MAC address. It is important to note that, at this time, 802.11i [1], Wi-Fi protected access (WPA) or WPA2 remains the most secure way to connect to a wireless access point. However, these technologies are not user friendly (it is often necessary to install certificates on the terminal and to set up complex security infrastructures), which explains why web-based security access is so popular nowadays. The bottom line is always to keep in mind the degree of security required considering the usage of the network infrastructure by the end-users. For a wireless ISP it is clear that security considerations have to be taken into account, not from a subscriber perspective, but from a business perspective, by protecting itself from attacks and service theft. For a company that wants to provide access to its VPN through Wi-Fi, the security requirements are much greater and must take into account user integrity as well. The example provided here is definitely not suitable in this case.

11.2 Enforcing Security Policies in Companies with 802.1X

One usual drawback encountered by companies is to secure their LAN access. In order to do this, IEEE 802.1X has been defined to secure the LAN by only allowing terminals that are able to present known and verifiable credentials to the network. In the context of IP/MPLS networks, one might wonder why 802.1X, exclusively defined for layer-2 connectivity, is helpful. Actually, several items of layer-2 equipment available on the market are also capable of layer-3/4 treatments, so it is just moving one step forward to take the opportunity to use the AAA exchanges performed during the 802.1X authentication phase to convey additional policies to be enforced at the IP level. Of course, this is not the way the IEEE standardized access control, but it is an efficient way to enforce customized policies transparently for the end-user.

To illustrate this, we will take the example of the ACME company which wants to offer two levels of security in its internal network: the first security level is accessible to everyone who is employed by the company, whereas the second level is reserved for the high management of the company who have access to sensitive applications and databases (see Figure 11.5). The advantage of treating the security aspect on the IP level rather than the layer-2 level is to dissociate the right to access to resources from the VLAN aspect (not always possible when the company has multiple sites interconnected through an L3 VPN). It is also possible to apply security rules in equipment placed on the path used to access the servers needing protection, based on the IP address of the terminal requesting access. In the latter case, the solution would consist in assigning a specific IP address to terminals that are allowed to access resources, by placing firewall rules to stop packets coming from non-authorized IP addresses. This way to proceed can become burdensome, as it requires managing several pools of addresses, with the risk of a security breach when the access lists and the IP pools are not consistent. Moreover, the process of address assignment, usually performed through DHCP, relies on the MAC address: this address can easily be forged; DHCP user databases have to be refreshed as soon as a user changes his/her terminal hardware. Finally, this method relies on the assumption that MAC addresses are unique worldwide, which has long been proven to be false because of the casual process followed by some Ethernet card vendors.

The solution proposed here makes use of the filtering capabilities of L2/L3 switches in order to block IP packets generated by terminals that are not allowed access to some parts of the network. This way to proceed is the most efficient way to enforce network IP security policies directly bound to the credentials presented by the terminal. It is not linked to the hardware of the terminal, it is not linked to the port to which the terminal is connected and it

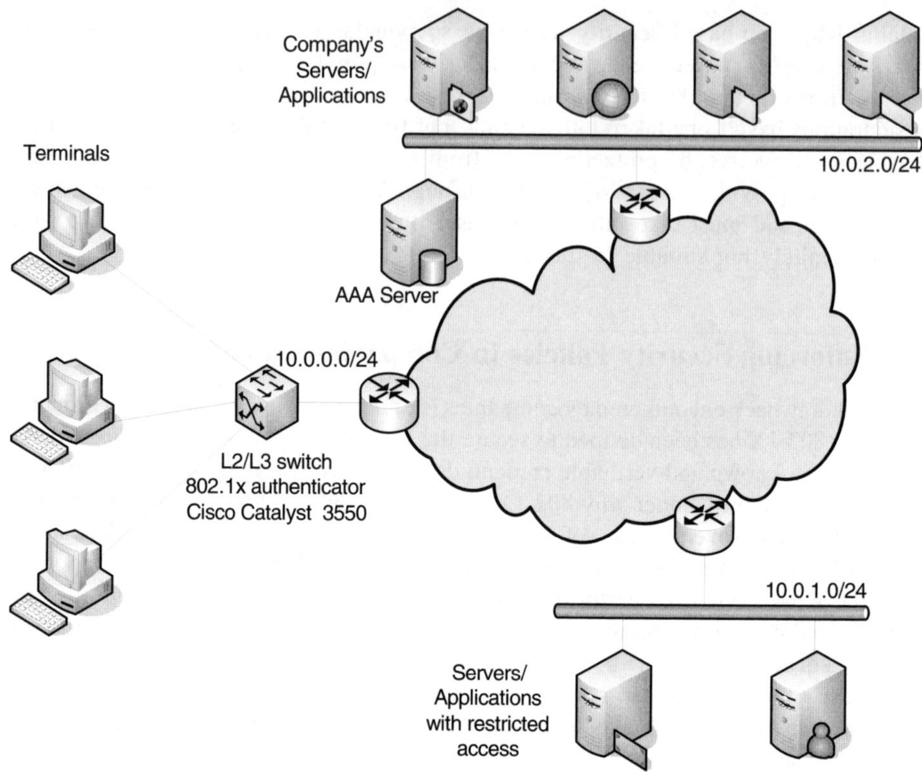

Figure 11.5 Network architecture for dynamic security enforcement

is not linked to the IP address obtained by the terminal from the DHCP server. The elements of the solution are as follows:

- terminals embedding an 802.1X module, with address assignment either done dynamically through DHCP or statically configured;
- an L2/L3 switch connected to the company's network backbone, activating the 802.1X authenticator role with remote authentication through RADIUS;
- an AAA server that stores the individual security policy to be enforced, running RADIUS and EAP methods enforced by the company.

The choice of the L2/L3 switch brand can be crucial here: it is possible to use a vendor-specific solution (which usually is the easiest way to proceed, even though it is a drawback for network evolution), or to configure each switch in the network with the policy to be enforced, the user profile making reference to this profile. It is a matter of fact that the latter is not the preferred solution, as it requires reconfiguring all switches if there is a change in IP address filters. Even with an automated update of managed equipment through SNMP MIBs, the risk is higher than transferring IP policies dynamically from a centralized point.

In this example, we will make use of Cisco Catalyst 3550 equipment, which is capable of assuming L2 connectivity, an 802.1X authenticator and an ACL for layer-3 filtering.

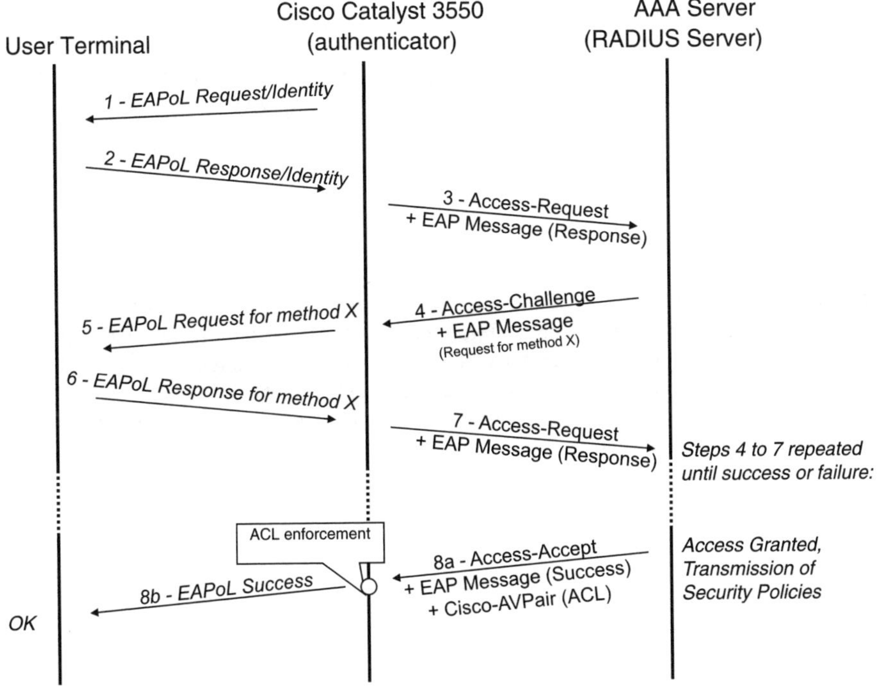

Figure 11.6 EAP exchanges for dynamic provisioning of individual security policy

This equipment is also capable of applying individual filters using the two methods mentioned above: either by using the RADIUS *Filter-Id* attribute (see Section 3.2.5), or by using the vendor-specific AVP Cisco-AVPair attribute embedding the proper extended access lists (either inbound or outbound), as used in the previous example (Section 11.1).

The general architecture of our company's network is shown in Figure 11.5.

Figure 11.6 (applying the diagram presented in Section 3.2.1.2) shows the message exchange that takes place between the terminal and the switch (using EAPoL, as specified in 802.1X [2]), and between the switch and the AAA server holding the security policy to be applied for this user. This user is considered as a regular employee and therefore is not allowed to access subnet 10.0.1.0/24. The corresponding ACL can be statically configured in the switch with the following command line:

```
Switch (config) # access-list 101 deny ip any any 10.0.1.0 0.0.0.255
```

The sequence of messages is detailed in Section 3.2.1.2. The *Access-Accept* message sent by the RADIUS server in *8a* to the switch embeds a specific AVP, the Cisco-AVPair, which contains the filter to apply in order to block access to the 10.0.1.0/24 subnet, which is reserved to managers. The following line has to be added to the user profile (formatting the aforementioned ACL into the Cisco-AVPair attribute):

```
Cisco-AVPair = "ip:inacl#101=deny ip any any 10.0.1.0 0.0.0.255"
```

The example given here is suitable not only for wired Ethernet access but also when a WLAN access is provided to access the company's VPN. In such a situation, the authentication phase is more complex and requires using 802.11i, WPA or WPA2 (Wi-Fi protected access, adopted by the Wi-Fi Alliance and based on 802.1X work). Policy security provisioning would be realized the same way as described here (conveyed in the RADIUS *Access-Accept* message).

References

[1] IEEE 802.11i – Medium Access Control (MAC) Security Enhancements, Amendment 6 to IEEE Standard for Information technology – Telecommunications and Information Exchange Between Systems – Local and Metropolitan Area Networks – Specific Requirements – Part 11: Wireless Medium Access Control (MAC) and Physical Layer (PHY) Specifications.

[2] IEEE Standard 802.1X-2001. IEEE Standard for Local and Metropolitan Area Networks – Port-Based Network Access Control.

12

Future Challenges

12.1 Introduction

In today's Internet, configuration procedures are achieved by technical personnel who are required to have an ever-growing level of expertise because of the various technologies and features that need to be used, configured and activated to deploy a wide range of IP service offerings. This level of expertise has become mandatory as each equipment manufacturer has developed its own interfaces and configuration schemes. In addition, as IP services may rely upon the activation of a set of sophisticated yet complex features, the time needed to adequately provision such services is also increasing.

As a consequence, the specification and the use of standardized protocol (for conveying configuration information) and interfaces should dramatically help in facilitating, if not automating, the configuration process and the operational production of a wide range of IP services.

12.1.1 Current Issues with Configuration Procedures

This section aims to list issues that should be carefully studied when dealing with configuration tasks. The items below should be taken into account when designing a protocol for configuration purposes.

12.1.1.1 Protocol Diversity

The production of a whole set of IP yet complex services relies upon the activation of a set of capabilities in the participating devices. In particular, a large set of protocols need to be configured, such as routing protocols, management protocols and security protocols, not to mention capabilities that relate to addressing scheme management, QoS policy enforcement, etc.

Service Automation and Dynamic Provisioning Techniques in IP/MPLS Environments C. Jacquenet, G. Bourdon and M. Boucadair

Such a diversity of features and protocols may increase the risk of inconsistencies. Therefore, the configuration information that is forwarded to the whole set of participating devices for producing a given service or a set of services should be consistent, whatever the number of features/services to be activated/deployed in the network.

12.1.1.2 Topology Discovery

Network operators should have means to dynamically discover the topology of their network. This topology information should be as elaborate as possible, including details like the links that connect network devices, including information about their capacity, such as the total bandwidth, the available bandwidth, the bandwidth that can be reserved, etc.

12.1.1.3 Scalability

As far as scalability is concerned, adequate indicators should be specified in order to qualify the ability of a given technical means to support a large number of configuration processes. The maintenance of these processes should not impact on the performance of a given system (a system is a set of elements that compose the key fundamentals of an architecture that aims to deliver configuration data).

Therefore, configuration operations should be qualified with performance indicators in order to check whether the architecture designed for configuration management is scalable in terms of:

- the volume of configuration data to be processed per unit of time and according to the number of capabilities and devices that need to be configured;
- the volume of information generated by any reporting mechanism that may be associated with a configuration process;
- the number of processes that are created in order to achieve specific configuration operations.

12.1.2 Towards Service-driven Configuration Policies

Current configuration practice basically focuses on elementary functions, i.e. configuration management for a given service offering breaks down into a set of elementary tasks. Thus, the consistency of configuration operations for producing IP services must be checked by any means appropriate, while current configuration methods can, at best, only check if provisioning decisions are correctly enforced by a single device.

A network device should be seen as a means to deploy a service and not just as a component of such a service. Thus, service configuration and production techniques should not focus on a set of devices taken one by one, but on the service itself, which will rely upon a set of features that need to be configured and activated in various regions of the network that supports that service.

Service providers could dedicate centralized entities that would be responsible for the provisioning and the management of participating devices. The main function of these centralized entities would be to make appropriate decisions and generate convenient configuration data that would be delivered to the participating devices. In addition, these centralized entities would make sure of the consistency of the decisions that are taken to produce the service, as per a dynamic configuration policy enforcement scheme.

Service-oriented configuration should rely upon the following requirements:

- The data models must be service driven.
- The configuration protocol(s) should reuse existing data and information models.
- The configuration protocol(s) should be flexible enough for further enhancement and addition of new functionalities that in the future could prove to be a must.
- The configuration protocol(s) should provide means to assess the consistency of the configuration tasks and to check the validity (or correctness) of the configuration of the service before its operation.

12.2 Towards the Standardization of Dynamic Service Subscription and Negotiation Techniques

The previous chapters of this book have discussed the various issues and solutions related to automated deployment and operation of services over IP/MPLS network infrastructures by means of dynamic policy enforcement techniques. In the past 2 years, investigation in the field of dynamic service subscription and negotiation techniques has dramatically progressed by forming a community of vendors and operators who have decided to initiate the standardization of the architectural framework where such techniques would naturally serve as a cornerstone.

Within a multiple-play service environment, this community, which has been named the IPsphere Forum (IPSF) [1] is currently addressing the following very basic questions from a standardization standpoint:

- What if customers had the ability to dynamically subscribe to a wide range of service offerings and even negotiate the level of quality associated with such services?
- What if service providers had the ability to automate the service production process, whether such a service can be deployed at the scale of the Internet or its access limited to the provider's own domain?

12.2.1 Basic Motivation

The IPSF aims to enhance the commercial framework for IP services to provide a win–win scenario for all stakeholders, where buyers of network-facilitated services can enjoy an expanding and evolving set of potential experiences, where sellers are able to offer an increasingly richer array of valued services and where collaborating service providers have a clear economic incentive for contributing their network, IT and associated resources.

12.2.2 Commercial Framework

The IPsphere framework [2] specifies mechanisms by which the next generation of IP services, and their associated market and revenue potential, can be realized. These mechanisms create a commercial framework by which resource improvements that add value to service delivery (e.g. dynamic provisioning techniques, automated service production procedures) can be offered.

The major high-level design goals of the IPsphere framework are as follows:

- First and foremost, all providers can present whatever contribution they care to make to a service mix at whatever price they find compelling, allowing customers to select the best deal overall.
- The investments of any provider in a service's underlying infrastructure must be protected.
- Business relationships are able to be flexibly established via a loosely coupled, network-technology-agnostic business layer, called the service structuring stratum (SSS) in the IPSF reference model.
- Services can be composed from a provider's own resource capabilities and – where desired or required – from the offerings of other providers, assuming agreements between such providers that can be contractually defined through the use of the aforementioned SSS layer capabilities.
- Providers need only publish the service–business relationships that they are willing to allow others to select from.
- Service publication can be tailored to give different views to various potential buyers/partners.

Any conceivable business model can be tried, and so the best model is likely to be available for selection over time, thereby creating an optimum market.

12.2.3 A Service-oriented Architecture

The IPsphere technical framework is based upon the service-oriented architecture (SOA) composition model. As shown in Figure 12.1, this approach provides for the connection of the functional units of an application, called 'services' (or 'elements' in the IPsphere

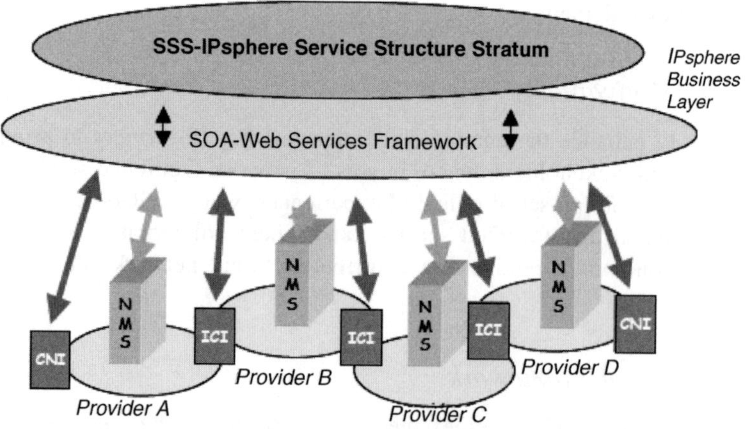

Figure 12.1 The IPSF model

vernacular), through well-defined interfaces [namely the customer-to-network Interface (CNI), which to some degree can be compared with user-to-network interfaces (UNIs) in other environments, and the intercarrier interface (ICI), which to some degree can be compared with network-to-network Interfaces (NNIs) in other environments].

The generic identification of network management systems (NMSs) in Figure 12.1 reflects the need for storing, maintaining and updating the various policies (routing, traffic engineering, security, etc.) that will be derived from the SSS-based negotiations, and that will be enforced by the network resources involved in the deployment and the management of the IPSF-inferred services.

SOA also emphasizes loose coupling between services. Loose coupling precludes undocumented interactions between services, for example through shared data, and it also supports the independent evolution of interfaces and IPsphere elements.

SOA also has the advantage that an element's interface is defined in a way that is independent of the hardware platform, the operating system, the hosting middleware and the programming language used to implement the element. This allows elements, built on a variety of systems, to interact with each other in a uniform and universal manner. In addition, the applications' interfaces and elements are expressed using business terms and concepts – they are not technology focused.

The benefit of a loosely coupled system is in its agility – the ability to accommodate changes in the structure and implementation of the internals of an element. By contrast, tight coupling means that the interfaces between the different components of an application are dependent on the form of implementation, making the system brittle when changes are made to components.

The agile nature of loosely coupled systems serves the need of a business to adapt rapidly to changes in policies, the business environment, product offerings, partnerships and regulatory requirements.

12.2.4 Publishing and Accessing Services

The IPsphere framework is made up of a federation of providers who share a service structuring stratum connection for business coordination. The interconnected networks use standard packet protocols for service signaling, but at key points in and between network jurisdictions there are 'IPsphere interface points' where business relationship management must be structured (Figure 12.2).

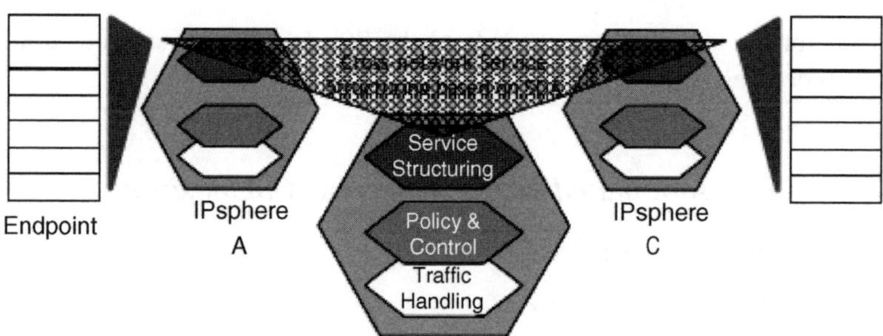

Figure 12.2 A pan-provider environment

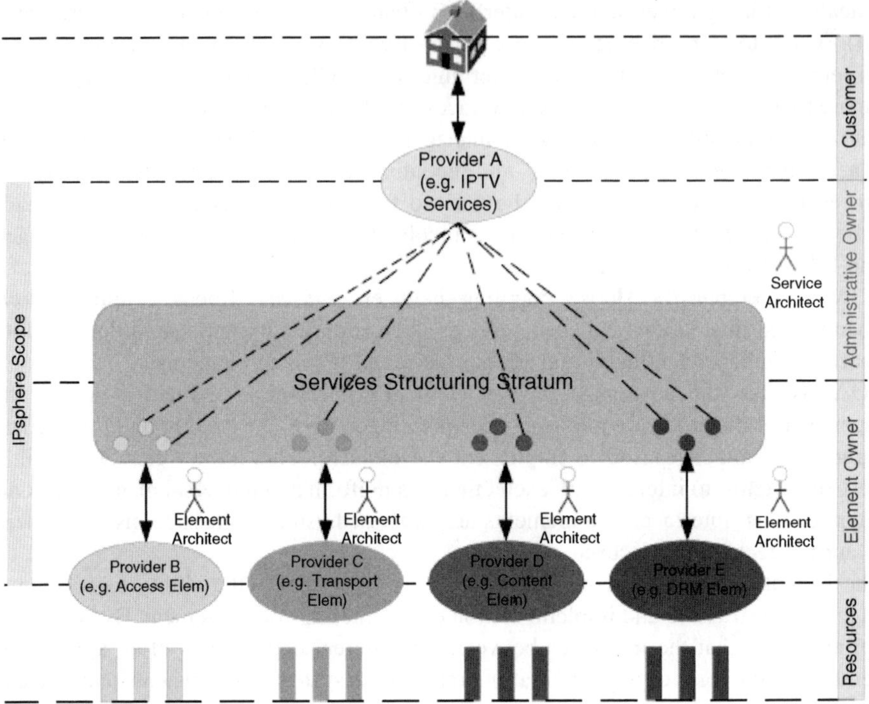

Figure 12.3 Role of the IPsphere's service structuring stratum in the dynamic service publication and composition processes

The policy and control stratum is to some degree equivalent to the control and management planes in classical representations, while the traffic handling stratum can be easily compared with the capabilities that are activated at the data plane.

A provider makes an explicit business decision to participate in a pan-provider service, and that decision is communicated by publishing a list of 'elements' representing service offers the provider is willing to make and the commercial conditions under which the offer is valid. These are combined with elements from other providers to create services.

Figure 12.3 further elaborates on the role of the SSS layer in the service publication and composition processes.

12.2.5 Example of Automated IP VPN Service Composition

12.2.5.1 VPN Service Subscription Phase

This is where the SSS layer steps in by:

- presenting the available services for the customer's selection;

- providing receiver access control capabilities that will be solicited for receiver identification and authentication purposes;
- providing means for electronic payment.

12.2.5.2 VPN Service Negotiation Phase

The delivery of pan-provider VPN services is conditioned by several contexts, which include (but are not necessarily limited to):

- *The access capabilities of each VPN site.* These may possibly influence traffic encoding schemes (yielding access-inferred adaptive capabilities that should be supported by the source, based upon dynamic notification capabilities of the SSS or the PCS layers) as well as the traffic forwarding policies (yielding DiffServ-based encoding schemes and dynamic per-hop behavior (PHB) enforcement through the use of the PCS layer).
- *The network conditions.* These may yield the design and the enforcement of VPN-specific traffic engineering policies, hence soliciting the PCS layer on the basis of the number of instantiated service level specification templates being processed by the SSS layer in a given period of time.
- *Time considerations.* The service to be accessed by the users might be available for a limited period of time only (working hours, for example). From this standpoint, the corresponding information should be explicitly described in the SLS to be processed by the SSS layer, possibly yielding the activation of specific resource provisioning cycles through the PCS layer.

12.2.5.3 VPN Service Operation

Once the subscription and negotiation phases are completed (including the completion of the transactional payment procedure, if any), the VPN service must be delivered to the customer. This means that all the relevant configuration information must have been provided to all the devices that are involved in the provisioning of the service up to the customer.

The corresponding configuration policy will be instantiated by means of PCS capabilities, based upon the SLS-formatted information that has been processed by the SSS layer. And it will be enforced by some policy enforcement point (PEP)-equivalent capabilities to be embedded in the aforementioned participating devices, including the support of monitoring capabilities that should help in assessing how efficient such a policy enforcement is, and how compliant the service is, as far as the customer's requirements (QoS, security, management, etc.) are concerned.

12.2.5.4 VPN Communication Flows

Figure 12.4 depicts the communication flows between a customer and a VPN service provider.

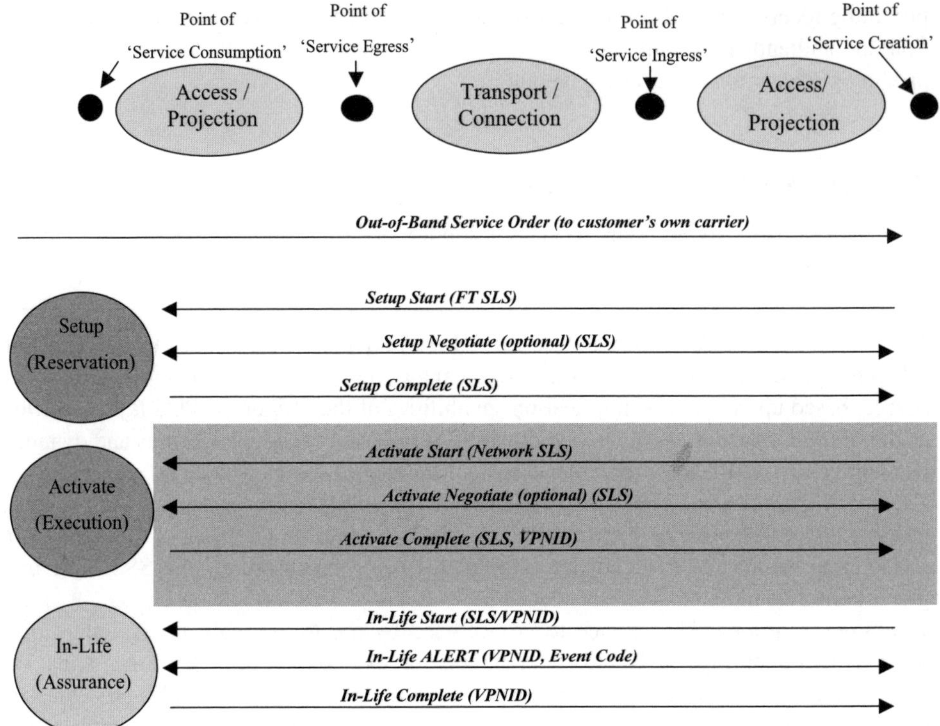

Figure 12.4 Communication flows of VPN service composition and deployment

12.3 Introducing Self-organizing Networks

12.3.1 What is a Self-organizing Network?

The deployment and the operation of a wide range of service offerings in highly volatile, dynamic environments (such as wireless mesh networks) is the next challenge that service providers will have to address. Such a challenge relates to the so-called ATAWAD (Any Time, Any Where, Any Device) paradigm, which basically suggests that any kind of service offering should be accessed whatever the access technology (fixed or mobile), whatever the terminal device technology (cell phone, PC, TV set, etc.) and whether the customer is in motion or not. In addition, access to the service may sometimes yield the establishment of environment-constrained intermittent communications, hence questioning the level of quality associated with the delivery of the service.

These environmental issues have encouraged investigation in the field of self-organizing networks (SONs) [3], where self-organization means that a functional structure appears and maintains itself spontaneously: a self-organized system is a system that evolves towards an organized form autonomously. Self-organization can be found in physics, biology or economics. As an example, a flock of seagulls will behave similarly at a collective level.

A self-organizing system can further be defined as a system that organizes itself by means of local interactions. This implicitly assumes that such a system is organized without any central control as far as the global behavior of the system is concerned. From this perspective, a SON network is deployed with no preconfigured support for interconnecting the SON devices.

SON networks are essentially highly dynamic: while 'classical' networks are designed according to a network planning policy that will affect the importance of the mesh and the dimensioning of both the transmission and the switching resources, SON devices organize themselves into a network architecture and deal with the networking processes (forwarding, routing) by themselves at runtime, intelligently and autonomously.

In a SON network, traffic may very well modify the behavior of the SON devices: a typical example is mobile ad hoc networks (MANETs) [4], where mobile devices (laptops, cell phones, etc.) communicate with each other over a wireless infrastructure in a distributed manner, so as to provide the relevant forwarding capabilities in the absence of a fixed infrastructure. Therefore, the forwarding policy enforced by MANET routers will evolve according to the motion of the users.

12.3.2 Characteristics of SON Networks and Devices

Generally speaking, networks consist of resources (processing, bandwidth and memory) and constraints (power and latency). Although SON networks do require power and memory (because of the support of autonomic capabilities), they optimize the resource/constraint ratio; for example, active caching uses processing and memory to reduce bandwidth usage and latency, yielding the deployment of cache servers in points of presence (POP) locations that are as close to the users as possible.

SON nodes collect the information about the network state (congestion occurrences, moving users, etc.) and will use this information better to address the following issues:

- dynamic location of contents, taking into account the location of the user as well as his/her access conditions (whether the user is in motion or not, how the access rate can dynamically influence the way the traffic will be encoded/delivered/engineered to the user as per his/her request, how a user in motion may trigger a relocation process of the information to be retrieved, etc.);
- dynamic publishing of a range of services that may be accessed by a user, based upon his/her profile, and/or his/her location (at home, at work, in motion, etc.) and/or his/her interests (e.g. dynamically informing the user about flight delays while he/she is approaching the airport), etc.;
- dynamic (re)configuration of network devices, depending on the traffic load conditions, the provisioning of additional content servers, etc.;
- dynamic notification of users about the efficiency of a quality of service (QoS) and/or security policy, possibly based upon the dynamic activation of monitoring capabilities in network devices that will facilitate the accounting and invoicing of the services subscribed to by the user, etc.

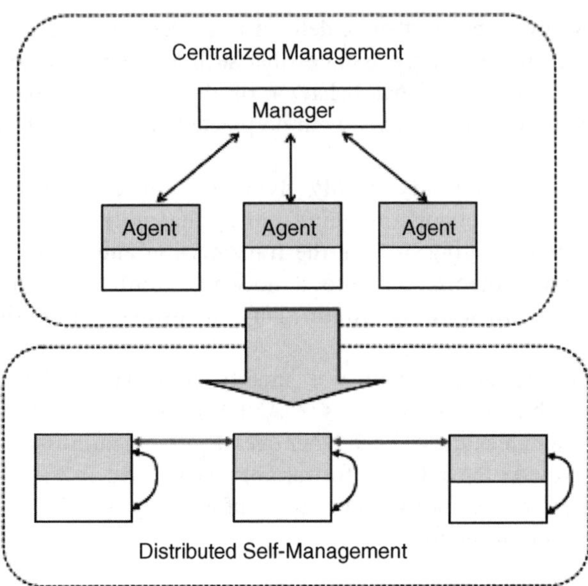

Figure 12.5 SON management architectures

12.3.3 On Self-management

SON nodes manage and control themselves and can be operated in a stand-alone fashion. Management actions can be coordinated for the sake of consistency, hence not excluding a mix of centralized and distributed management architectures, as depicted in Figure 12.5.

12.3.4 SON Algorithms and How to Use Them for Enhancing Dynamic Policy Enforcement Schemes

SON nodes organize themselves according to four phases:

- *Discovery.* Each SON node identifies its set of neighbors and defines the radius of data transmission so as to cope with scalability constraints.
- *Organization.* Each SON node computes, maintains and distributes its forwarding table according to the information exchanged during the discovery phase. Once this phase is completed, the network is said to be organized.
- *Maintenance.* SON nodes notify their neighbors about their presence, and about any change in the topology that may have affected their own forwarding decision process. SON nodes update their forwarding tables accordingly, including the connectivity graph which is derived from such tables.
- *Reorganization.* Whenever an event affects the information maintained by the SON nodes (link/node failures, for example).

Such advanced capabilities could be used for dynamic policy enforcement purposes in highly volatile, dynamic environments, where the quality of services is likely to be degraded because of the constraints of such environments (like still receiving a TV content with no perception of degraded quality when jumping on a bus that enters a tunnel).

In that case, SON techniques would be part of the policy enforcement design scheme, where:

- Local rules that achieve global properties are enforced, e.g. 'access-aware' multicast-enabled SON nodes to forward IPTV traffic with the required level of quality;
- implicit coordination is exploited, e.g. hello mechanisms are used for maintaining connectivity graph information beyond SON node discovery;
- long-lived state information is minimized, e.g. to accommodate zapping behaviors of IPTV customers (in motion);
- dynamically adaptive protocols are preferred, e.g. to notify cluster nodes about any topology change and enforce (updated) forwarding policies accordingly.

12.3.5 SON-inferred Business Opportunities

SON techniques are often seen as key drivers for addressing the aforementioned ATAWAD paradigm. Context-aware SON techniques can be seen as a means to develop business, e.g. by employing user location information as an advanced routing metric to accelerate the development and the production of user-centric services, such as interactive TV. The automated production of complex, QoS-demanding services in highly dynamic environments can also be facilitated by SON capabilities.

References

[1] http://www.ipsphereforum.org

[2] Alateras, J. *et al.*, 'IPsphere Framework Technical Specification (Release 1)', June 2007.

[3] Nakano, T. and Suda, T., 'Self-organizing Network Services with Evolutionary Adaptation', *IEEE Transactions on Neural Networks*, **16**(5), September 2005, 1269–1278,

[4] Hoebeke, H. *et al.*, 'An Overview of Mobile Ad-Hoc Networks: Applications and Challenges', Proceedings of the 43rd European Telecommunication Congress, Ghent, Belgium, November 2004, pp. 60–66.

Appendices

APPENDIX 1

XML Schema for NETCONF RPCs and Operations

```
<?xml version="1.0" encoding="UTF-8"?>
<xs:schema xmlns:xs="http://www.w3.org/2001/XMLSchema"
           xmlns="urn:ietf:params:xml:ns:netconf:base:1.0"
           targetNamespace="urn:ietf:params:xml:ns:netconf:
           base:1.0"
           elementFormDefault="qualified"
           attributeFormDefault="unqualified"
           xml:lang="en">
  <!--
    import standard XML definitions
  -->

  <xs:import namespace="http://www.w3.org/XML/1998/namespace"
             schemaLocation="http://www.w3.org/2001/xml.xsd">
  </xs:import>

  <!--
    message-id attribute
  -->

  <xs:simpleType name="messageIdType">
    <xs:restriction base="xs:string">
      <xs:maxLength value="4095"/>
    </xs:restriction>
  </xs:simpleType>
```

Service Automation and Dynamic Provisioning Techniques in IP/MPLS Environments C. Jacquenet, G. Bourdon and M. Boucadair

```
<!--
  Types used for session-id
-->

<xs:simpleType name="SessionId">
  <xs:restriction base="xs:unsignedInt">
    <xs:minInclusive value="1"/>
  </xs:restriction>
</xs:simpleType>
<xs:simpleType name="SessionIdOrZero">
  <xs:restriction base="xs:unsignedInt"/>
</xs:simpleType>

 <!--
   <rpc> element
   -->

 <xs:complexType name= "rpcType">
   <xs:sequence>
     <xs:element ref="rpcOperation"/>
   </xs:sequence>
   <xs:attribute name="message-id" type="messageIdType"
     use="required"/>

 <!--
   Arbitrary attributes can be supplied with element
   -->

   <xs:anyAttribute processContents="lax"/>
 </xs:complexType>
 <xs:element name="rpc" type="rpcType"/>

 <!--
   data types and elements used to construct rpc-errors
   -->

 <xs:simpleType name="ErrorType">
   <xs:restriction base="xs:string">
     <xs:enumeration value="transport"/>
     <xs:enumeration value="rpc"/>
     <xs:enumeration value="protocol"/>
     <xs:enumeration value="application"/>
   </xs:restriction>
 </xs:simpleType>
 <xs:simpleType name="ErrorTag">
```

```
      <xs:restriction base="xs:string">
        <xs:enumeration value="in-use"/>
        <xs:enumeration value="invalid-value"/>
        <xs:enumeration value="too-big"/>
        <xs:enumeration value="missing-attribute"/>
        <xs:enumeration value="bad-attribute"/>
        <xs:enumeration value="unknown-attribute"/>
        <xs:enumeration value="missing-element"/>
        <xs:enumeration value="bad-element"/>
        <xs:enumeration value="unknown-element"/>
        <xs:enumeration value="unknown-namespace"/>
        <xs:enumeration value="access-denied"/>
        <xs:enumeration value="lock-denied"/>
        <xs:enumeration value="resource-denied"/>
        <xs:enumeration value="rollback-failed"/>
        <xs:enumeration value="data-exists"/>
        <xs:enumeration value="data-missing"/>
        <xs:enumeration value="operation-not-supported"/>
        <xs:enumeration value="operation-failed"/>
        <xs:enumeration value="partial-operation"/>
    </xs:restriction>
  </xs:simpleType>
  <xs:simpleType name="ErrorSeverity">
    <xs:restriction base="xs:string">
      <xs:enumeration value="error"/>
      <xs:enumeration value="warning"/>
    </xs:restriction>
  </xs:simpleType>
  <xs:complexType name="errorInfoType">
    <xs:sequence>
      <xs:choice>
        <xs:element name="session-id" type="SessionId
          OrZero"/>
        <xs:sequence minOccurs="0" maxOccurs="unbounded">
          <xs:sequence>
            <xs:element name="bad-attribute" type="xs:QName"
              minOccurs="0" maxOccurs="1"/>
            <xs:element name="bad-element" type="xs:QName"
              minOccurs="0" maxOccurs="1"/>
            <xs:element name="ok-element" type="xs:QName"
              minOccurs="0" maxOccurs="1"/>
            <xs:element name="err-element" type="xs:QName"
              minOccurs="0" maxOccurs="1"/>
```

```
          <xs:element name="noop-element" type="xs:QName"
            minOccurs="0" maxOccurs="1" />
          <xs:element name="bad-namespace" type="xs:QName"
            minOccurs="0" maxOccurs="1" />
        </xs:sequence>
      </xs:sequence>
    </xs:choice>

<!-- Elements from any other namespace are also allowed
    to follow the NETCONF elements
-->

    <xs:any namespace="##other"
      minOccurs="0" maxOccurs="unbounded" />
  </xs:sequence>
</xs:complexType>
<xs:complexType name="rpcErrorType" >
  <xs:sequence>
    <xs:element name="error-type" type="ErrorType" />
    <xs:element name="error-tag" type="ErrorTag" />
    <xs:element name="error-severity" type="ErrorSever-
        ity" />
    <xs:element name="error-app-tag" type="xs:string"
                minOccurs="0" />
    <xs:element name="error-path" type="xs:string" min-
                Occurs="0" />
    <xs:element name="error-message" minOccurs="0" >
      <xs:complexType>
        <xs:simpleContent>
          <xs:extension base="xs:string" >
            <xs:attribute ref="xml:lang" use="optional" />
          </xs:extension>
        </xs:simpleContent>
      </xs:complexType>
    </xs:element>
    <xs:element name="error-info" type="errorInfoType"
      minOccurs="0" />
  </xs:sequence>
</xs:complexType>


<!--
  <rpc-reply> element
  -->
```

```
<xs:complexType name=''rpcReplyType'' >
  <xs:choice>
    <xs:element name=''ok'' />
    <xs:group ref=''rpcResponse'' />
  </xs:choice>
  <xs:attribute name=''message-id'' type=''messageIdType''
   use=''optional'' />

  <!- -
    Any attributes supplied with <rpc> element must be returned
    on <rpc-reply>
  - ->
  <xs:anyAttribute processContents=''lax'' />
</xs:complexType>
<xs:group name=''rpcResponse'' >
  <xs:sequence>
    <xs:element name=''rpc-error'' type=''rpcErrorType''
      minOccurs=''0'' maxOccurs=''unbounded'' />
    <xs:element name=''data'' type=''dataInlineType''
      minOcc urs=''0'' />
  </xs:sequence>
</xs:group>
<xs:element name=''rpc-reply'' type=''rpcReplyType'' />

<!- -
  Type for <test-option> parameter to <edit-config>
  - ->

<xs:simpleType name=''testOptionType'' >
  <xs:restriction base=''xs:string'' >
    <xs:enumeration value=''test-then-set'' />
    <xs:enumeration value=''set'' />
  </xs:restriction>
</xs:simpleType>

<!- -
  Type for <error-option> parameter to <edit-config>
  - ->

<xs:simpleType name=''errorOptionType'' >
  <xs:restriction base=''xs:string'' >
    <xs:annotation>
      <xs:documentation>
        Use of the rollback-on-error value requires
        the: rollback-on-error capability.
```

```
      </xs:documentation>
    </xs:annotation>
    <xs:enumeration value="stop-on-error" />
    <xs:enumeration value="continue-on-error" />
    <xs:enumeration value="rollback-on-error" />
  </xs:restriction>
</xs:simpleType>

<!--
  rpcOperationType: used as a base type for all
  NETCONF operations
  -->
<xs:complexType name="rpcOperationType" />
<xs:element name="rpcOperation"
      type="rpcOperationType" abstract="true" />

<!--
  Type for <config> element
  -->

<xs:complexType name="configInlineType" >
  <xs:complexContent>
    <xs:extension base="xs:anyType" />
  </xs:complexContent>
</xs:complexType>

<!--
  Type for <data> element
  -->

<xs:complexType name="dataInlineType" >
  <xs:complexContent>
    <xs:extension base="xs:anyType" />
  </xs:complexContent>
</xs:complexType>

<!--
  Type for <filter> element
  -->

<xs:simpleType name="FilterType" >
  <xs:restriction base="xs:string" >
    <xs:annotation>
      <xs:documentation>
        Use of the xpath value requires the '':xpath`` capability.
```

```
      </xs:documentation>
    </xs:annotation>
    <xs:enumeration value="subtree" />
    <xs:enumeration value="xpath" />
  </xs:restriction>
</xs:simpleType>
<xs:complexType name="filterInlineType" >
  <xs:complexContent>
    <xs:extension base="xs:anyType" >
      <xs:attribute name="type"
                    type="FilterType" default="subtree" />
      <!- - if type="xpath" , the xpath expression
      appears in the select element ->
      <xs:attribute name="select" />
    </xs:extension>
  </xs:complexContent>
</xs:complexType>

<!- -
  configuration datastore names
  - ->

<xs:annotation>
  <xs:documentation>
    The startup datastore can be used only if the ":startup"
    capability is advertized. The candidate datastore can
    be used only if the: candidate datastore is advertized.
  </xs:documentation>
</xs:annotation>
<xs:complexType name="configNameType" />
<xs:element name="config-name"
            type="configNameType" abstract="true" />
<xs:element name="startup" type="configNameType"
            substitutionGroup="config-name" />
<xs:element name="candidate" type="configNameType"
            substitutionGroup="config-name" />
<xs:element name="running" type="configNameType"
            substitutionGroup="config-name" />

<!- -
  operation attribute used in <edit-config>
  - ->

<xs:simpleType name="editOperationType" >
  <xs:restriction base="xs:string" >
```

```
      <xs:enumeration value="merge" />
      <xs:enumeration value="replace" />
      <xs:enumeration value="create" />
      <xs:enumeration value="delete" />
    </xs:restriction>
  </xs:simpleType>
  <xs:attribute name="operation"
                type="editOperationType" default="merge" />

  <!- -
    <default-operation> element
    - ->
  <xs:simpleType name="defaultOperationType" >
    <xs:restriction base="xs:string" >
      <xs:enumeration value="merge" />
      <xs:enumeration value="replace" />
      <xs:enumeration value="none" />
    </xs:restriction>
  </xs:simpleType>

  <!- -
    <url> element
    - ->

  <xs:complexType name="configURIType" >
    <xs:annotation>
      <xs:documentation>
        Use of the url element requires the ":url" capability.
      </xs:documentation>
    </xs:annotation>
    <xs:simpleContent>
      <xs:extension base="xs:anyURI" />
    </xs:simpleContent>
  </xs:complexType>

  <!- -
    Type for element (except <get-config>)
    - ->

  <xs:complexType name="rpcOperationSourceType" >
    <xs:choice>
      <xs:element name="config" type="configInlineType" />
      <xs:element ref="config-name" />
      <xs:element name="url" type="configURIType" />
    </xs:choice>
  </xs:complexType>
```

```
<!--
  Type for <source> element in <get-config>
  -->

<xs:complexType name="getConfigSourceType">
  <xs:choice>
    <xs:element ref="config-name"/>
    <xs:element name="url" type="configURIType"/>
  </xs:choice>
</xs:complexType>

<!--
  Type for <target> element
  -->

<xs:complexType name="rpcOperationTargetType">
  <xs:choice>
    <xs:element ref="config-name"/>
    <xs:element name="url" type="configURIType"/>
  </xs:choice>
</xs:complexType>

<!--
  <get-config> operation
  -->

<xs:complexType name="getConfigType">
  <xs:complexContent>
    <xs:extension base="rpcOperationType">
      <xs:sequence>
        <xs:element name="source"
                    type="getConfigSourceType"/>
        <xs:element name="filter"
                    type="filterInlineType" minOccurs="0"/>
      </xs:sequence>
    </xs:extension>
  </xs:complexContent>
</xs:complexType>
<xs:element name="get-config" type="getConfigType"
            substitutionGroup="rpcOperation"/>

<!--
  <edit-config> operation
  -->
```

```
<xs:complexType name="editConfigType" >
  <xs:complexContent>
    <xs:extension base="rpcOperationType" >
      <xs:sequence>
        <xs:annotation>
          <xs:documentation>
            Use of the test-option element requires the
            ":validate" capability. Use of the url element
            requires the ":url" capability.
      </xs:documentation>
    </xs:annotation>
    <xs:element name="target"
                type="rpcOperationTargetType" />
    <xs:element name="default-operation"
                type="defaultOperationType"
                minOccurs="0" />
    <xs:element name="test-option"
                type="testOptionType"
                minOccurs="0" />
    <xs:element name="error-option"
                type="errorOptionType"
                minOccurs="0" />
    <xs:choice>
      <xs:element name="config"
                  type="configInlineType" />
      <xs:element name="url"
                  type="configURIType" />
    </xs:choice>
   </xs:sequence>
  </xs:extension>
 </xs:complexContent>
</xs:complexType>
<xs:element name="edit-config" type="editConfigType"
            substitutionGroup="rpcOperation" />

 <!--
   <copy-config> operation
   -->

 <xs:complexType name="copyConfigType" >
   <xs:complexContent>
     <xs:extension base="rpcOperationType" >
       <xs:sequence>
         <xs:element name="target" type="rpcOperationTarget-
                     Type" />
```

```
      <xs:element name="source" type="rpcOperationSource-
                  Type" />
     </xs:sequence>
    </xs:extension>
  </xs:complexContent>
</xs:complexType>
<xs:element name="copy-config" type="copyConfigType"
          substitutionGroup="rpcOperation" />

<!--
  <delete-config> operation
  -->

<xs:complexType name="deleteConfigType" >
  <xs:complexContent>
    <xs:extension base="rpcOperationType" >
      <xs:sequence>
        <xs:element name="target" type="rpcOperationTarget-
                  Type" />
      </xs:sequence>
    </xs:extension>
  </xs:complexContent>
</xs:complexType>
<xs:element name="delete-config" type="deleteConfigType"
          substitutionGroup="rpcOperation" />

<!--
  <get> operation
  -->

<xs:complexType name="getType" >
  <xs:complexContent>
    <xs:extension base="rpcOperationType" >
      <xs:sequence>
        <xs:element name="filter"
                  type="filterInlineType" minOccurs="0" />
      </xs:sequence>
    </xs:extension>
  </xs:complexContent>
</xs:complexType>
<xs:element name="get" type="getType"
          substitutionGroup="rpcOperation" />
<!--
  <lock> operation
  -->
```

```
<xs:complexType name="lockType">
  <xs:complexContent>
    <xs:extension base="rpcOperationType">
      <xs:sequence>
        <xs:element name="target"
                    type="rpcOperationTargetType"/>
      </xs:sequence>
    </xs:extension>
  </xs:complexContent>
</xs:complexType>
<xs:element name="lock" type="lockType"
            substitutionGroup="rpcOperation"/>

<!--
  <unlock> operation
  -->

<xs:complexType name="unlockType">
  <xs:complexContent>
    <xs:extension base="rpcOperationType">
      <xs:sequence>
        <xs:element name="target" type="rpcOperationTarget-
                    Type"/>
      </xs:sequence>
    </xs:extension>
  </xs:complexContent>
</xs:complexType>
<xs:element name="unlock" type="unlockType"
            substitutionGroup="rpcOperation"/>

<!--
  <validate> operation
  -->

<xs:complexType name="validateType">
  <xs:annotation>
    <xs:documentation>
    The validate operation requires the ":validate" capability.
    </xs:documentation>
  </xs:annotation>
<xs:complexContent>
  <xs:extension base="rpcOperationType">
    <xs:sequence>
      <xs:element name="source" type="rpcOperationSource-
                  Type"/>
```

```
      </xs:sequence>
    </xs:extension>
  </xs:complexContent>
</xs:complexType>
<xs:element name="validate" type="validateType"
            substitutionGroup="rpcOperation" />

<!--
  <commit> operation
  -->

<xs:simpleType name="confirmTimeoutType" >
  <xs:restriction base="xs:unsignedInt" >
    <xs:minInclusive value="1" />
  </xs:restriction>
</xs:simpleType>
<xs:complexType name="commitType" >
  <xs:annotation>
    <xs:documentation>
     The commit operation requires the ":candidate" capability.
    </xs:documentation>
  </xs:annotation>
  <xs:complexContent>
    <xs:extension base="rpcOperationType" >
      <xs:sequence>
        <xs:annotation>
          <xs:documentation>
          Use of the confirmed and confirm-timeout elements
          requires the ":confirmed-commit" capability.
        </xs:documentation>
      </xs:annotation>
      <xs:element name="confirmed" minOccurs="0" />
      <xs:element name="confirm-timeout"
                  type="confirmTimeoutType"
                  minOccurs="0" />
      </xs:sequence>
    </xs:extension>
  </xs:complexContent>
</xs:complexType>
<xs:element name="commit" type="commitType"
            substitutionGroup="rpcOperation" />

<!--
  <discard-changes> operation
  -->
```

```
<xs:complexType name="discardChangesType">
  <xs:annotation>
    <xs:documentation>
      The discard-changes operation requires the
      ":candidate" capability.
    </xs:documentation>
  </xs:annotation>
  <xs:complexContent>
    <xs:extension base="rpcOperationType"/>
  </xs:complexContent>
</xs:complexType>
<xs:element name="discard-changes"
            type="discardChangesType"
            substitutionGroup="rpcOperation"/>

<!--
   <close-session> operation
  -->

<xs:complexType name="closeSessionType">
  <xs:complexContent>
    <xs:extension base="rpcOperationType"/>
  </xs:complexContent>
</xs:complexType>
<xs:element name="close-session" type="closeSessionType"
            substitutionGroup="rpcOperation"/>

<!--
   <kill-session> operation
  -->

<xs:complexType name="killSessionType">
  <xs:complexContent>
    <xs:extension base="rpcOperationType">
      <xs:sequence>
        <xs:element name="session-id"
                    type="SessionId" minOccurs="1"/>
      </xs:sequence>
    </xs:extension>
  </xs:complexContent>
</xs:complexType>
<xs:element name="kill-session" type="killSessionType"
            substitutionGroup="rpcOperation"/>
<!--
  <hello> element
  -->
```

```
    <xs:element name=''hello'' >
      <xs:complexType>
        <xs:sequence>
          <xs:element name=''capabilities'' >
            <xs:complexType>
              <xs:sequence>
                <xs:element name=''capability'' type=''xs:any
                   URI'' maxOccurs=''unbounded'' />
              </xs:sequence>
            </xs:complexType>
          </xs:element>
        <xs:element name=''session-id''
                    type=''SessionId'' minOccurs=''0'' />
      </xs:sequence>
    </xs:complexType>
  </xs:element>
</xs:schema>
```

APPENDIX 2

XML Schema for NETCONF Notifications

```
?xml version="1.0" encoding="UTF-8"?>
<xs:schema xmlns:xs="http://www.w3.org/2001/XMLSchema"
  xmlns="urn:ietf:params:netconf:capability:notification:1.0"
  xmlns:netconf="urn:ietf:params:xml:ns:netconf:base:1.0"
  targetNamespace=
     "urn:ietf:params:netconf:capability:notification:1.0"
  elementFormDefault="qualified"
  attributeFormDefault="unqualified"
   xml:lang="en">

 <!--
  import standard XML definitions
 -->

 <xs:import namespace="http://www.w3.org/XML/1998/namespace"
  schemaLocation="http://www.w3.org/2001/xml.xsd">
 </xs:import>

 <!--
 import base netconf definitions
 -->

 <xs:import namespace="urn:ietf:params:xml:ns:netconf:
  base:1.0"
    schemaLocation=
```

Service Automation and Dynamic Provisioning Techniques in IP/MPLS Environments
C. Jacquenet, G. Bourdon and M. Boucadair

```
"http://www.iana.org/assignments/xml-registry/schema/netconf.
  xsd" />

<!- -
Symmetrical Operations
- ->

<!- -
<create-subscription> operation
- ->

<xs:complexType name="createSubscriptionType" >
    <xs:complexContent>
        <xs:extension base="netconf:rpcOperationType" >
        <xs:sequence>
            <xs:element name="stream"
                type="streamNameType minOccurs="0" >
                <xs:annotation>
                    <xs:documentation>
                        An optional parameter that indicates
                        which stream of events is of interest. If
                        not present, then events in the default
                        NETCONF stream will be
                        sent.
                    </xs:documentation>
                </xs:annotation>
        </xs:element>
            <xs:element name="filter"
                type="netconf:filterInlineType"
                minOccurs="0" >
                <xs:annotation>
                    <xs:documentation>
                        An optional parameter that indicates
                        which subset of all possible events
                        are of interest. The format of this
                        parameter is the same as that of the
                        filter parameter in the
                        NETCONF
                        protocol operations. If not present,
                        all events not precluded by other
                        parameters will be sent.
                    </xs:documentation>
                </xs:annotation>
            </xs:element>
        <xs:element name="startTime" type="xs:dateTime"
            minOccurs="0" >
```

```
            <xs:annotation>
                <xs:documentation>
                    A parameter used to trigger the replay
                    feature that indicates that the replay
                    should start at the time specified. If
                    start time is not present, this is not a
                    replay subscription.
                </xs:documentation>
            </xs:annotation>
        </xs:element>
        <xs:element name="stopTime" type="xs:dateTime"
            minOccurs="0" >
            <xs:annotation>
                <xs:documentation>
                    An optional parameter used with the
                    optional replay feature to indicate the
                    newest notifications of interest. If
                    stop time is not present, the
                    notifications will continue until the
                    subscription is terminated. Must be used
                    with "startTime".
                </xs:documentation>
            </xs:annotation>
        </xs:element>
      </xs:sequence>
    </xs:extension>
  </xs:complexContent>
</xs:complexType>

<xs:simpleType name="streamNameType" >
    <xs:annotation>
        <xs:documentation>
            The name of an event stream.
        </xs:documentation>
    </xs:annotation>
    <xs:restriction base="xs:string" />
</xs:simpleType>

<xs:element name="create-subscription"
     type="createSubscriptionType"
     substitutionGroup="netconf:rpcOperation" >
     <xs:annotation>
        <xs:documentation>
            The command to create a notification subscription. It
            takes as argument the name of the notification stream
            and filter or profile information. All of those options
```

```
            limit the content of the subscription. In addition,
            there are two time-related parameters ''startTime''
            and ''stopTime'' which can be used to select the time
            interval of interest.
        </xs:documentation>
    </xs:annotation>
</xs:element>

<!--
  One-way operations
-->

<!--
  <Notification> operation
-->

 <xs:complexType name=''NotificationContentType'' />

 <xs:element name=''notificationContent''
    type=''NotificationContentType'' abstract=''true'' />

 <xs:complexType name=''NotificationType'' >
    <xs:sequence>
        <xs:element ref=''notificationContent'' />
    </xs:sequence>
 </xs:complexType>

 <xs:element name=''notification'' type=''NotificationType'' />
</xs:schema>
```

APPENDIX 3

Example of an IP Traffic Engineering Policy Information Base (IP TE PIB)

```
IP-TE-PIB PIB-DEFINITIONS: := BEGIN

  IMPORTS
       Unsigned32, Integer32, MODULE-IDENTITY,
       MODULE-COMPLIANCE, OBJECT-TYPE, OBJECT-GROUP
              FROM COPS-PR-SPPI
       InstanceId, ReferenceId, Prid, TagId
              FROM COPS-PR-SPPI-TC
       InetAddress, InetAddressType
              FROM INET-ADDRESS-MIB
       Count, TEXTUAL-CONVENTION
              FROM ACCT-FR-PIB-TC
       TruthValue, TEXTUAL-CONVENTION
              FROM SNMPv2-TC
       RoleCombination, PrcIdentifier
              FROM FRAMEWORK-ROLE-PIB
       SnmpAdminString
              FROM SNMP-FRAMEWORK-MIB;

  ipTePib  MODULE-IDENTITY

       SUBJECT-CATEGORIES { tbd } - - IP TE client-type to be
                                  - - assigned by IANA
```

Service Automation and Dynamic Provisioning Techniques in IP/MPLS Environments C. Jacquenet, G. Bourdon and M. Boucadair

```
      LAST-UPDATED   "200709280900Z"
      ORGANIZATION   "Your_Organization"
      CONTACT-INFO   "
                      Christian Jacquenet
                      Address
                      Phone:
                      E-Mail: christian.jacquenet@gmail.com"
      DESCRIPTION
          "The PIB module containing a set of policy rule classes
           that describe IP traffic engineering policies to be
           enforced within and between domains."
      REVISION     "200710011600Z"
      DESCRIPTION
            "Initial version."

      ::={ pib tbd} - tbd to be assigned by IANA

ipTeFwdClasses        OBJECT IDENTIFIER  ::=  { ipTePib 1}
ipTeMetricsClasses    OBJECT IDENTIFIER  ::=  { ipTePib 2}
ipTeStatsClasses      OBJECT IDENTIFIER  ::=  { ipTePib 3}

--
-- Forwarding classes. The information contained in these
    classes
-- is meant to provide a detailed description of the traffic-
-- engineered routes. One table has been specified so far, but
   there
-- is room for depicting specific kinds of routes, like MPLS LSP
-- paths, for example.
--
--
--
-- The ipTeRouteTable
--

ipTeRouteTable      OBJECT-TYPE

      SYNTAX         SEQUENCE OF ipTeRouteEntry
      PIB-ACCESS     notify
      STATUS         current
      DESCRIPTION
            "This table describes the traffic-engineered routes
            that are installed in the forwarding tables of the
            routers."
```

```
      ::={ ipTeFwdClasses 1 }

 ipTeRouteEntry      OBJECT-TYPE

      SYNTAX         ipTeRouteEntry
      STATUS         current
      DESCRIPTION
           "A particular traffic-engineered route to a particular
            destination."

      PIB-INDEX    { ipTeRoutePrid }
      UNIQUENESS   { ipTeRouteDest,
                       ipTeRouteMask,
                       ipTeRoutePhbId,
                       ipTeRouteNextHopAddress
                     ipTeRouteNextHopMask
                       ipTeRouteIfIndex }

      ::={ ipTeRouteTable 1 }

 ipTeRouteEntry::= SEQUENCE {
            ipTeRoutePrid                   InstanceId,
            ipTeRouteDestAddrType           InetAddressType,
            ipTeRouteDest                   InetAddress,
            ipTeRouteMask                   Unsigned32,
            ipTeRouteNextHopAddrType        InetAddressType,
            ipTeRouteNextHopAddress         InetAddress,
            ipTeRouteNextHopMask            Unsigned32,
            ipTeRoutePhbId                  Integer32,
            ipTeRouteOrigin                 Integer32,
            ipTeRouteIfIndex                Unsigned32
}

ipTeRoutePrid         OBJECT-TYPE

      SYNTAX        InstanceId
      STATUS        current
      DESCRIPTION
            "An integer index that uniquely identifies this
            route entry among all the route entries."

      ::={ ipTeRouteEntry 1 }

 ipTeRouteDestAddrType    OBJECT-TYPE
```

```
      SYNTAX                 InetAddressType
      STATUS                 current
      DESCRIPTION
            "The address type enumeration value used to specify
            the type of a route's destination IP address."

      ::={ ipTeRouteEntry 2 }

ipTeRouteDest       OBJECT-TYPE

      SYNTAX         InetAddress
      STATUS         current
      DESCRIPTION
            "The IP address to match against the packet's
            destination address."

      ::={ ipTeRouteEntry 3 }

ipTeRouteMask        OBJECT-TYPE

      SYNTAX         Unsigned32 (0..128)
      STATUS         current
      DESCRIPTION
         "Indicates the length of a mask for the matching of the
         destination IP address. Masks are constructed by setting
         bits in sequence from the most significant bit
         downwards for ipTeRouteMask bits length. All other bits
         in the mask, up to the number needed to fill the length
         of the address ipTeRouteDest, are cleared to zero. A
         zero bit in the mask then means that the corresponding
         bit in the address always matches."

      ::={ ipTeRouteEntry 4 }

ipTeRouteNextHopAddrType     OBJECT-TYPE

      SYNTAX                 InetAddressType
      STATUS                 current
      DESCRIPTION
         "The address type enumeration value used to specify the
         type of the next hop's IP address."

      ::={ ipTeRouteEntry 5 }

ipTeRouteNextHopAddress       OBJECT-TYPE
```

```
        SYNTAX                 InetAddress
        STATUS                 current
        DESCRIPTION
              "On remote routes, the address of the next router en
              route; Otherwise, 0.0.0.0."

        ::={ ipTeRouteEntry 6 }

ipTeRouteNextHopMask      OBJECT-TYPE

        SYNTAX                 Unsigned32 (0..128)
        STATUS          current
        DESCRIPTION
            "Indicates the length of a mask for the matching of the
            next hop's IP address. Masks are constructed by setting
            bits in sequence from the most significant bit
            downwards for ipTeRouteNextHopMask bits length. All
            other bits in the mask, up to the number needed to fill
            the length of the address ipTeRouteNextHop, are
            cleared to zero. A zero bit in the mask then means that
            the corresponding bit in the address always matches."

        ::={ ipTeRouteEntry 7 }

ipTeRoutePhbId      OBJECT-TYPE

        SYNTAX          Integer32 (-1|0..63)
        STATUS          current
        DESCRIPTION
          "The binary encoding that uniquely identifies a per-hop
           behaviour (PHB) or a set of PHBs associated with the
           DiffServ Code Point (DSCP) marking of the IP
           datagrams that will be conveyed along this traffic-
           engineered route. A value of -1 indicates that a
           specific PHB ID value has not been defined, and thus
           all PHB ID values are considered a match."

        ::={ ipTeRouteEntry 8 }

ipTeRouteOrigin      OBJECT-TYPE

        SYNTAX INTEGER {
                        OSPF (0)
                        IS-IS (1)
                        BGP (2)
                        STATIC (3)
```

```
                OTHER (4)
            }
    STATUS          current
    DESCRIPTION
        "The value indicates the origin of the route. Either
        the route has been computed by OSPF, by IS-IS,
        announced by BGP-4, is static, or else."

    ::= { ipTeRouteEntry 9 }

ipTeRouteIfIndex        OBJECT-TYPE
    SYNTAX          Unsigned32 (0..65535)
    STATUS          current
    DESCRIPTION
        "The ifIndex value that identifies the local interface
        through which the next hop of this route is
        accessible."

    ::= { ipTeRouteEntry 10 }

--
--
-- Traffic engineering metrics classes.
--
-- The information stored in the following tables is meant to
   provide
-- the description of the metric values that will be taken into
-- account by intra- and interdomain routing protocols for the
-- computation and the selection of traffic-engineered routes. So
-- far, two tables have been identified: one that is based upon the
   traffic engineering extensions of OSPF, another that is based
-- upon the contents of a specific BGP-4 attribute.
--
--
igpTeGroup  OBJECT  IDENTIFIER  ::= { ipTeMetricsClasses 1 }
bgpTeGroup  OBJECT  IDENTIFIER  ::= { ipTeMetricsClasses 2 }

--
-- The ospfTeMetricsTable
--

ospfTeMetricsTable      OBJECT-TYPE

    SYNTAX          SEQUENCE OF ospfTeMetricsEntry
    PIB-ACCESS      install-notify
    STATUS          current
```

```
    DESCRIPTION
         "This class describes the link and traffic engineering
         metrics that will be used by OSPF for TE route
         calculation purposes."

    ::={ igpTeGroup 1 }

ospfTeMetricsEntry      OBJECT-TYPE

    SYNTAX          ospfTeMetricsEntry
    STATUS          current
    DESCRIPTION
         "The collection of OSPF metrics assigned to the router
         on a per interface and per DSCP basis."

    PIB-INDEX           { ospfTeMetricsPrid }
    UNIQUENESS          { ospfTeMetricsLinkMetricValue,
                          ospfTeMetricsDscpValue,
                          ospfTeMetricSubTlvLinkType,
                          ospfTeMetricSubTlvLinkId,
                          ospfTeMetricSubTlvLocalIfAddress,
                          ospfTeMetricSubTlvRemoteIfAddress,
                          ospfTeMetricSubTlvTeMetric,
                          ospfTeMetricSubTlvMaxBandwidth,
                          ospfTeMetricSubTlvMaxRsvBandwidth,
                          ospfTeMetricSubTlvUnRsvBandwidth,
                          ospfTeMetricIfIndex }

    ::={ ospfTeMetricsTable 1 }

ospfTeMetricsEntry::= SEQUENCE {

    ospfTeMetricsPrid                          InstanceId,
    ospfTeMetricsIfMetricValue                 Unsigned32,
    ospfTeMetricsDscpValue                     Integer32,
    ospfTeMetricsTopTlvAddressType             InetAddressType,
    ospfTeMetricsTopTlvRouterAddress           InetAddress,
    ospfTeMetricsTopTlvRouterAddrMask          Unsigned32,
    ospfTeMetricsSubTlvLinkType                Unsigned32,
    ospfTeMetricsSubTlvLinkIdAddressType       InetAddressType,
    ospfTeMetricsSubTlvLinkId                  InetAddress,
    ospfTeMetricsSubTlvLinkIdMask              Unsigned32,
    ospfTeMetricsSubTlvLocalIfAddressType      InetAddressType,
    ospfTeMetricsSubTlvLocalIfAddress          InetAddress,
    ospfTeMetricsSubTlvLocalIfAddrMask         Unsigned32,
    ospfTeMetricsSubTlvRemoteIfAddressType     InetAddressType,
```

```
    ospfTeMetricsSubTlvRemoteIfAddress        InetAddress,
    ospfTeMetricsSubTlvRemoteIfAddrMask       Unsigned32,
    ospfTeMetricsSubTlvTeMetric              Unsigned32,
    ospfTeMetricsSubTlvMaxBandwidth           Unsigned32,
    ospfTeMetricsSubTlvMaxRsvBandwidth        Unsigned32,
    ospfTeMetricsSubTlvUnrsvBandwidth         Unsigned32,
    ospfTeMetricsSubTlvResourceClass          Unsigned32,
    ospfTeMetricsIfIndex                     Unsigned32
}
ospfTeMetricsPrid          OBJECT-TYPE
      SYNTAX               InstanceId
      STATUS               current
      DESCRIPTION
            "An integer index that uniquely identifies this
            instance of the ospfTeMetrics class."

      ::={ ospfTeMetricsEntry 1 }

ospfTeMetricsIfMetricValue         OBJECT-TYPE

      SYNTAX         Unsigned32 (1..65535)
      STATUS         current
      DESCRIPTION
            "The link metric assigned on a per-DSCP and per-
            interface basis, as defined in this instance of the
            ospfTeMetricsTable."

      ::={ ospfTeMetricsEntry 2 }

ospfTeMetricsDscpValue         OBJECT-TYPE

      SYNTAX         Integer32 (-1|0..63)
      STATUS         current
      DESCRIPTION
            "The DSCP value associated with the link metric
            value, as defined in the ospfTeMetricsIfMetricValue
            object. A value of -1 indicates that a specific DSCP
            value has not been defined and thus all DSCP values are
            considered a match."

      ::={ ospfTeMetricsEntry 3 }

ospfTeMetricsTopTlvAddressType         OBJECT-TYPE

      SYNTAX         InetAddressType
      STATUS         current
```

```
        DESCRIPTION
                "The address type enumeration value used to specify
                the IP address of the advertizing router. This IP
                address is always reachable, and is typically imple-
                mented as a "loopback" address."

        ::={ ospfTeMetricsEntry 4 }

ospfTeMetricsTopTlvRouterAddress    OBJECT-TYPE
        SYNTAX          InetAddress
        STATUS          current
        DESCRIPTION
                "The IP address (typically a "loopback" address)
                of the advertising router."

        ::={ ospfTeMetricsEntry 5 }

ospfTeMetricsTopTlvRouterAddrMask    OBJECT-TYPE

        SYNTAX          Unsigned32 (0..128)
        STATUS          current
        DESCRIPTION
                "Indicates the length of a mask for the matching of
                the advertizing router's IP address. Masks are
                constructed by setting bits in sequence from the most
                significant bit downwards for ospfTeMetricsTopTlv-
                RouterAddrMask bits length. All other bits in the
                mask, up to the number needed to fill the length of the
                address ospfTeMetricsTopTlvRouterAddress,are cle-
                ared to zero. A zero bit in the mask then means that the
                corresponding bit in the address always matches."

        ::={ ospfTeMetricsEntry 6 }

ospfTeMetricsSubTlvLinkType    OBJECT-TYPE

        SYNTAX          INTEGER{
                                Point-to-Point (1)
                                Multiaccess (2)
                        }
        STATUS          current
        DESCRIPTION
                "The type of the link, either point-to-point or
                multi-access."

        ::={ ospfTeMetricsEntry 7 }
```

```
ospfTeMetricsSubTlvLinkIdAddressType     OBJECT-TYPE

     SYNTAX          InetAddressType
     STATUS          current
     DESCRIPTION
          "The address type enumeration value used to identify
          the other end of the link, described as an IP address."
     ::={ ospfTeMetricsEntry 8 }

ospfTeMetricsSubTlvLinkId       OBJECT-TYPE

     SYNTAX          InetAddress
     STATUS          current
     DESCRIPTION
          "The identification of the other end of the link,
          described as an IP address."

     ::={ ospfTeMetricsEntry 9 }

ospfTeMetricsSubTlvLinkMask       OBJECT-TYPE

     SYNTAX          Unsigned32 (0..128)
     STATUS          current
     DESCRIPTION
          "Indicates the length of a mask for the matching of the
          other end of the link, described as an IP address. Masks
          are constructed by setting bits in sequence from the
          most significant bit downwards for ospfTeMetrics-
          SubTlvLinkMask bits length. All other bits in the mask,
          up to the number needed to fill the length of the address
          ospfTeMetricsSubTlvLinkId, are cleared to zero. A zero
          but in the mask then means that the corresponding bit in
          the address always matches."

     ::={ ospfTeMetricsEntry 10 }

ospfTeMetricsSubTlvLocalIfAddressType     OBJECT-TYPE

     SYNTAX          InetAddressType
     STATUS          current
     DESCRIPTION
          "The address type enumeration value used to specify
          the IP address of the interface corresponding to this
          instance of the ospfTeMetricsSubTlvLinkType object."

     ::={ ospfTeMetricsEntry 11 }
ospfTeMetricsSubTlvLocalIfAddress       OBJECT-TYPE
```

```
    SYNTAX        InetAddress
    STATUS        current
    DESCRIPTION
            "Specifies the IP address of the interface of the
            advertizing router that is connected to the link
            described as an instance of the ospfTeMetricsSubTlv-
            LinkType object."

    ::={ ospfTeMetricsEntry 12 }

ospfTeMetricsSubTlvLocalIfAddrMask    OBJECT-TYPE

    SYNTAX        Unsigned32 (0..128)
    STATUS        current
    DESCRIPTION
            "Indicates the length of a mask for the matching of
            the IP address of the interface corresponding to this
            instance of the ospfTeMetricsSubTlvLinkType object.
            Masks are constructed by setting bits in sequence
            from the most significant bit downwards for
            ospfTeMetricsSubTlvLocalIfAddrMask bits length.
            All other bits in the mask, up to the number needed to
            fill the lenth of the address ospfTeMetricsSubTlvLo-
            calIfAddress, are cleared to zero. A zero bit in the
            mask then means that the corresponding bit in the
            address always matches."

    ::={ ospfTeMetricsEntry 13 }

ospfTeMetricsSubTlvRemoteIfAddressType    OBJECT-TYPE

    SYNTAX        InetAddressType
    STATUS        current
    DESCRIPTION
            "The address type enumeration value used to specify
            the IP address(es) of the neighbor's interface cor-
            responding to this instance of the ospfTeMetrics-
            SubTlvLinkType object."

    ::={ ospfTeMetricsEntry 14 }

ospfTeMetricSubTlvRemoteIfAddress    OBJECT-TYPE

    SYNTAX        InetAddress
    STATUS        current
```

```
    DESCRIPTION
         "Specifies the IP address of the neighbor's interface
         that is attached to this instance of the
         ospfTeMetricsSubTlvLinkType object."
      ::={ ospfTeMetricsEntry 15 }

ospfTeMetricSubTlvRemoteIfAddrMask       OBJECT-TYPE

    SYNTAX        Unsigned32 (0..128)
    STATUS        current
    DESCRIPTION
      "Indicates the length of a mask for the matching of the IP
      address of the neighbor's interface corresponding to this
      instance of the ospfTeMetricsSubTlvLinkType object. Masks
      are constructed by setting bits in sequence from the most
      significant bit downwards for
      ospfTeMetricSubTlvRemoteIfAddrMask bits length. All other
      bits in the mask, up to the number needed to fill the length
      of the address ospfTeMetricSubTlvRemoteIfAddress, are
      cleared to zero. A zero bit in the mask then means that the
      corresponding bit in the address always matches."

    ::={ ospfTeMetricsEntry 16 }

ospfTeMetricSubTlvTeMetric       OBJECT-TYPE

    SYNTAX        Unsigned32 (1..65535)
    STATUS        current
    DESCRIPTION
      "The link metric that has been assigned for traffic
      engineering purposes. This metric may be different from the
      ospfTeMetricsLinkMetricValue object of the ospfTeMetrics
      class."

    ::={ ospfTeMetricsEntry 17 }

ospfTeMetricSubTlvBandwidthType       OBJECT-TYPE

    SYNTAX        Unsigned32 (0..4294967295)
    UNITS        "bytes per second"
    STATUS        current
    DESCRIPTION
      "Specifies the maximum bandwidth that can be used on this
      instance of the ospfTeMetricsSubTlvLinkType object in this
      direction (from the advertizing router), expressed in bytes
      per second."
```

```
    ::={ ospfpTeMetricsEntry 18 }

ospfTeMetricSubTlvMaxRsvBandwidth      OBJECT-TYPE

    SYNTAX         Unsigned32 (0..4294967295)
    UNITS         "bytes per second"
    STATUS         current
    DESCRIPTION
      "Specifies the maximum bandwidth that may be reserved on
      this instance of the ospfTeMetricsSubTlvLinkType object in
      this direction (from the advertizing router), expressed in
      bytes per second."

    ::={ ospfTeMetricsEntry 19 }

ospfTeMetricSubTlvUnrsvBandwidth      OBJECT-TYPE

    SYNTAX         Unsigned32 (0..4294967295)
    UNITS         "bytes per second"
    STATUS         current
    DESCRIPTION
      "Specifies the amount of bandwidth that has not been
      reserved on this instance of the ospfTeMetricsSubTlvLink-
      Type object in this direction yet (from the advertizing
      router), expressed in bytes per second."

    ::={ ospfTeMetricsEntry 20 }

ospfTeMetricSubTlvResourceClass      OBJECT-TYPE

    SYNTAX         Unsigned32 (0..4294967295)
    STATUS         current
    DESCRIPTION
      "Specifies administrative group membership for the link in
      terms of a bit mask."

    ::={ ospfTeMetricsEntry 21 }

ospfTeMetricIfIndex      OBJECT-TYPE

    SYNTAX         Unsigned32 (0..65535)
    STATUS         current
    DESCRIPTION
      "The ifIndex value that identifies the local interface that
      has been assigned a (set of) metrics."
```

```
    ::={ ospfTeMetricsEntry 22 }

--
-- The isisTeMetricsTable
--

 isisTeMetricsTable      OBJECT-TYPE

    SYNTAX             SEQUENCE OF isisTeMetricsEntry
    PIB-ACCESS         install-notify
    STATUS             current
    DESCRIPTION
      "This class describes the link and traffic engineering
      metrics that will be used by IS-IS for TE route computation
      purposes."

    ::={ igpTeGroup 2 }

 isisTeMetricsEntry      OBJECT-TYPE

    SYNTAX        isisTeMetricsEntry
    STATUS        current
    DESCRIPTION
      "The collection of IS-IS metrics assigned to the router on a
      per interface basis."

    PIB-INDEX        { isisTeMetricsPrid }
    UNIQUENESS       {
                       isisTeMetricsSubTlvIfAddr,
                       isisTeMetricsSubTlvNbrAddr,
                     isisTeMetricSubTlvTeMetric,
                       isisTeMetricsSubTlvMaxLinkBwth,
                       isisTeMetricsSubTlvMaxRsvLinkBwth,
                       isisTeMetricsPriority,
                       isisTeMetricsSubTlvUnRsvBwth,
                     isisTeMetricsIfIndex }

    ::={ isisTeMetricsTable 1 }

 isisTeMetricsEntry::= SEQUENCE {

           isisTeMetricsPrid                  InstanceId,
          isisTeMetricsTlvTeRouterID         InetAddress,
        isisTeMetricsSubTlvIfAddrType         InetAddressType,
         isisTeMetricsSubTlvIfAddr            InetAddress,
```

```
        isisTeMetricsSubTlvIfAddrMask       Unsigned32,
        isisTeMetricsSubTlvNbrAddType       InetAddressType,
        isisTeMetricsSubTlvNbrAddr          InetAddress,
        isisTeMetricsSubTlvNbrMask          Unsigned32,
        isisTeMetricsSubTlvTeMetric         Unsigned32,
        isisTeMetricsSubTlvMaxLinkBwth      Unsigned32,
        isisTeMetricsSubTlvMaxRsvLinkBwth   Unsigned32,
        isisTeMetricsPriority               Integer32,
        isisTeMetricsSubTlvUnRsvBwth        Unsigned32,
        isisTeMetricsIfIndex                Unsigned32
   }

isisTeMetricsPrid OBJECT-TYPE

   SYNTAX        InstanceId
   STATUS        current
   DESCRIPTION
     "An integer index that uniquely identifies this instance of
     the isisTeMetrics class."

   : :={ isisTeMetricsEntry 1 }

isisTeMetricsTlvTeRouterID  OBJECT-TYPE

   SYNTAX        InetAddress
   STATUS        current
   DESCRIPTION
     "Specifies the router ID."

   : :={ isisTeMetricsEntry 2 }

isisTeMetricsSubTlvIfAddrType      OBJECT-TYPE

   SYNTAX        InetAddressType
   STATUS        current
   DESCRIPTION
     "The address type enumeration value used to specify the
     type of the interface IP address."

   : :={ isisTeMetricsEntry 3 }

isisTeMetricsSubTlvIfAddr      OBJECT-TYPE

   SYNTAX        InetAddress
   STATUS        current
   DESCRIPTION
```

```
        "Specifies the IP address of the interface."

    ::={ isisTeMetricsEntry 4 }

isisTeMetricsSubTlvIfAddrMask      OBJECT-TYPE

    SYNTAX        Unsigned32 (0..128)
    STATUS        current
    DESCRIPTION
      "Indicates the length of a mask for the matching of the IP
      address of the neighboring router. Masks are constructed by
      setting bits in sequence from the most significant bit
      downwards for isisTeMetricsSubTlvIfAddrMask bits length.
      All other bits in the mask, up to the number needed to fill the
      length of the address isisTeMetricsSubTlvIfAddr, are cle-
      ared to zero. A zero bit in the mask then means that the
      corresponding bit in the address always matches."

    ::={ isisTeMetricsEntry 5 }

isisTeMetricsSubTlvNbrAddrType      OBJECT-TYPE

    SYNTAX        InetAddressType
    STATUS        current
    DESCRIPTION
      "The address type enumeration value used to specify the
      type of the neighboring router's IP address."

    ::={ isisTeMetricsEntry 6 }

isisTeMetricsSubTlvNbrAddr      OBJECT-TYPE

    SYNTAX        InetAddress
    STATUS        current
    DESCRIPTION
      "Specifies the IP address of the neighboring router on the
      link to which the corresponding interface (defined by the
      ifIndex) is attached."

    ::={ isisTeMetricsEntry 7 }

isisTeMetricsSubTlvNbrMask      OBJECT-TYPE

    SYNTAX        Unsigned32 (0..128)
    STATUS        current
    DESCRIPTION
```

```
    "Indicates the length of a mask for the matching of the IP
    address of the neighboring router. Masks are constructed by
    setting bits in sequence from the most significant bit
    downwards for isisTeMetricsSubTlvNbrMask bits length. All
    other bits in the mask, up to the number needed to fill the
    length of the address isisTeMetricsSubTlvNbrAddr, are
    cleared to zero. A zero bit in the mask then means that the
    corresponding bit in the address always matches."

  : := { isisTeMetricsEntry 8 }

isisTeMetricsSubTlvTeMetric      OBJECT-TYPE

  SYNTAX        Unsigned32 (1..65535)
  STATUS        current
    DESCRIPTION
    "The traffic engineering default metric is used to present a
     differently weighted topology to TE-based SPF computa-
    tions."

  : := { isisTeMetricsEntry 9 }

isisTeMetricsSubTlvMaxLinkBwth      OBJECT-TYPE

  SYNTAX        Unsigned32 (0..4294967295)
  UNITS        "bytes per second"
  STATUS        current
  DESCRIPTION
    "This metric specifies the maximum bandwidth that can be
    used on this link in this direction."

  : := { isisTeMetricsEntry 10 }

isisTeMetricsSubTlvMaxRsvLinkBwth      OBJECT-TYPE

  SYNTAX        Unsigned32 (0..4294967295)
  UNITS        "bytes per second"
  STATUS        current
  DESCRIPTION
    "Specifies the maximum bandwidth that may be reserved on
     this link in this direction, expressed in bytes per sec-
    ond."

  : := { isisTeMetricsEntry 11 }

isisTeMetricsPriority      OBJECT-TYPE
```

```
    SYNTAX         Integer32 (0..7)
    STATUS         current
    DESCRIPTION
      "Specifies one of the eight priority levels, possible values
      ranging from 0 to 7."

    ::={ isisTeMetricsEntry 12}


isisTeMetricsSubTlvUnRsvBwth      OBJECT-TYPE

    SYNTAX         Unsigned32 (0..4294967295)
    UNITS         "bytes per second"
    STATUS         current
    DESCRIPTION
      "Specifies the amount of bandwidth that has not been
      reserved on this link in this direction and having the
      priority isisTeMetricsPriority, expressed in bytes per
      second."

    ::={ isisTeMetricsEntry 13 }

isisTeMetricsIfIndex      OBJECT-TYPE

    SYNTAX         Unsigned32 (0..65535)
    STATUS         current
    DESCRIPTION
      "The ifIndex value that uniquely identifies the interface
      that has been assigned a (set of) metrics."

    ::={ isisTeMetricsEntry 14 }

--
-- The bgpTeTable
--

bgpTeTable              OBJECT-TYPE
    SYNTAX              SEQUENCE OF bgpTeEntry
    PIB-ACCESS          install-notify
    STATUS              current
    DESCRIPTION
      "This class describes the QoS information that MAY be
      conveyed in BGP-4 UPDATE messages for the purpose of
      enforcing an interdomain traffic engineering policy."
```

```
    ::={ bgpTeGroup 1 }

bgpTeEntry          OBJECT-TYPE
    SYNTAX         bgpTeEntry
    STATUS         current
    DESCRIPTION
      "The collection of QoS information to be exchanged by
      BGP peers, as far as the announcement of traffic-
      engineered routes between domains is concerned."

    PIB-INDEX           { bgpTePrid}
    UNIQUENESS          { bgpTeNlriAddress,
                          bgpTeNextHopAddress,
                          bgpTeReservedRate,
                          bgpTeAvailableRate,
                          bgpTeLossRate,
                          bgpTePhbId,
                          bgpTeMinOneWayDelay,
                          bgpTeMaxOneWayDelay,
                          bgpTeAverageOneWayDelay,
                          bgpTeInterPacketDelay}

    ::={ bgpTeTable 1 }

bgpTeEntry::= SEQUENCE {

            bgpTePrid                    InstanceId,
            bgpTeNlriAddressType         InetAddressType,
            bgpTeNlriAddress             InetAddress,
            bgpTeNlriAddressMask         Unsigned32,
            bgpTeNextHopAddressType      InetAddressType,
            bgpTeNextHopAddress          InetAddress,
            bgpTeNextHopMask             Unsigned32,
            bgpTeReservedRate            Unsigned32,
            bgpTeAvailableRate           Unsigned32,
            bgpTeLossRate                Unsigned32,
            bgpTePhbId                   Integer32,
            bgpTeMinOneWayDelay          Unsigned32,
            bgpTeMaxOneWayDelay          Unsigned32,
            bgpTeAverageOneWayDelay      Unsigned32,
            bgpTeInterPacketDelay        Unsigned32
    }

bgpTePrid                OBJECT-TYPE

    SYNTAX               InstanceId
```

```
    STATUS                current
    DESCRIPTION
      "An integer index that uniquely identifies this instance of
      the bgpTeTable class."

    ::={ bgpTeEntry 1 }

bgpTeNlriAddressType   OBJECT-TYPE

    SYNTAX                InetAddressType
    STATUS                current
    DESCRIPTION
      "The address type enumeration value used to specify the
      type of a route's destination IP address."

    ::={ bgpTeEntry 2 }

bgpTeNlriAddress       OBJECT-TYPE

    SYNTAX                InetAddress
    STATUS                current
    DESCRIPTION
      "The IP address to match against the NLRI field of the
      QOS_NLRI attribute of the BGP-4 UPDATE message."

    ::={ bgpTeEntry 3 }

bgpTeNlriAddressMask     OBJECT-TYPE

    SYNTAX                  Unsigned32 (0..128)
    STATUS                  current
    DESCRIPTION
      "Indicates the length of a mask for the matching of the
      NLRI field of the QOS_NLRI attribute of the BGP-4 UPDATE
      message. Masks are constructed by setting bits in sequence
      from the most significant bit downwards for bgpTeNlriMask
      bits length. All other bits in the mask, up to the number
      needed to fill the length of the address bgpTeNlri, are
      cleared to zero. A zero bit in the mask then means that
      the corresponding bit in the address always matches."

    ::={ bgpTeEntry 4 }

bgpTeNextHopAddressType   OBJECT-TYPE

      SYNTAX                  InetAddressType
```

```
      STATUS                current
      DESCRIPTION
          "The address type enumeration value used to specify the
          type of the next hop's IP address."

      ::={ bgpTeEntry 5 }

bgpTeNextHopAddress         OBJECT-TYPE

      SYNTAX                InetAddress
      STATUS                current
      DESCRIPTION
        "On remote routes, the address of the next router en
          route; Otherwise, 0.0.0.0."

      ::={ bgpTeEntry 6 }

bgpTeNextHopMask            OBJECT-TYPE

      SYNTAX                Unsigned32 (0..128)
      STATUS                current
      DESCRIPTION
          "Indicates the length of a mask for the matching of the
          next hop's IP address. Masks are constructed by setting
          bits in sequence from the most significant bit downwards
          for bgpTeNextHopMask bits length. All other bits in the
          mask, up to the number needed to fill the length of the
          address bgpTeNextHopAddress, are cleared to zero. A zero
          bit in the mask then means that the corresponding bit in
          the address always matches."

      ::={ bgpTeEntry 7 }

bgpTeReservedRate           OBJECT-TYPE

      SYNTAX                Unsigned32 (0..4294967295)
      UNITS                 "kilobits per second"
      STATUS                current
      DESCRIPTION
          "Specifies the reserved rate that cannot be used on this
          instance of the bgpTeNlriAddress object in this direc-
          tion (from the advertizing BGP peer), expressed in kilo-
          bits per second."

      ::={ bgpTeEntry 8 }

bgpTeAvailableRate      OBJECT-TYPE
```

```
      SYNTAX           Unsigned32 (0..4294967295)
      UNITS           ``kilobits per second''
      STATUS           current
      DESCRIPTION
         "Specifies the available rate that may be reserved on
         this instance of the bgpTeNlriAddress object in this
         direction (from the advertizing BGP peer), expressed in
         kilobits per second."

      ::={ bgpTeEntry 9 }

bgpTeLossRate      OBJECT-TYPE

      SYNTAX        Unsigned32 (0..4294967295)
      STATUS        current
      DESCRIPTION
         "Specifies the packet loss ratio that has been observed
         on this route instantiated by the bgpTeNlriAddress
         object."

      ::={ bgpTeEntry 10 }

bgpTePhbId         OBJECT-TYPE

      SYNTAX        Integer32 (-1 | 0..63)
      STATUS        current
      DESCRIPTION
         "The binary encoding that uniquely identifies a per-hop
         behavior (PHB) or a set of PHBs associated with the
         DiffServ code point marking of the IP datagrams that are
         to be conveyed along this traffic-engineered route. A
         value of -1 indicates that a specific PHB ID value has not
         been defined, and thus all PHB ID values are considered a
         match."

      ::={ bgpTeEntry 11 }

bgpTeMinOneWayDelay     OBJECT-TYPE

      SYNTAX              Unsigned32 (0..4294967295)
      UNITS              "milliseconds"
      STATUS current
      DESCRIPTION
         "Specifies the minimum one-way delay that has been
         observed on this route instantiated by the bgpTeNlriAd-
         dress object, expressed in milliseconds."
```

```
        : :={ bgpTeEntry 12 }
  bgpTeMaxOneWayDelay     OBJECT-TYPE

        SYNTAX              Unsigned32 (0..4294967295)
        UNITS              ''milliseconds''
        STATUS              current
        DESCRIPTION
            ''Specifies the maximum one-way delay that has been
            observed on this route instantiated by the bgpTeNlriAd-
            dress object, expressed in milliseconds.''

        : :={ bgpTeEntry 13 }

  bgpTeAverageOneWayDelay   OBJECT-TYPE

        SYNTAX              Unsigned32 (0..4294967295)
        UNITS              ''milliseconds''
        STATUS              current
        DESCRIPTION
            ''Specifies the average one-way delay that has been
            observed on this route instantiated by the bgpTeNlriAd-
            dress object, expressed in milliseconds.''

        : :={ bgpTeEntry 14 }

  bgpTeInterPacketDelay     OBJECT-TYPE

        SYNTAX              Unsigned32 (0..4294967295)
        UNITS              ''milliseconds''
        STATUS              current
        DESCRIPTION
            ''Specifies the interpacket delay variation that has been
            observed on this route instantiated by the bgpTeNlriAd-
            dress object.''

        : :={ bgpTeEntry 15 }

 - -
 - - Traffic engineering statistics classes. The information
     contained
 - - in the yet-to-be defined tables aim to report statistics about
 - - COPS control traffic, engineered traffic and potential errors.
 - -
 - -
END
```

APPENDIX 4

Example of an IP TE Accounting PIB

```
- -
- - The PIB defined within the context of IP traffic engineering
- - for accounting purposes has the goal of completing the whole
- - COPS TE reporting system.
- -
IPTE-ACCOUNTING-PIB PIB-DEFINITIONS: := BEGIN

     IMPORTS
          ExtUTCTime, Unsigned32, Unsigned64,
          Integer32, MODULE-IDENTITY, OBJECT-TYPE
                    FROM COPS-PR-SPPI
          TruthValue, TEXTUAL-CONVENTION
                    FROM SNMPv2-TC
          PolicyInstanceId, PolicyReferenceId
                    FROM COPS-PR-SPPI-TC;
          RoleCombination
                    FROM POLICY-DEVICE-AUX-MIB;
          Counter64
                    FROM SNMPv2-SMI;

     ipTeAccountingPib MODULE-IDENTITY

          SUBJECT-CATEGORIES { tbd}
          LAST-UPDATED      ''200201250900Z''
```

Service Automation and Dynamic Provisioning Techniques in IP/MPLS Environments C. Jacquenet, G. Bourdon and M. Boucadair

```
        ORGANIZATION      ''France Telecom R&D''
        CONTACT-INFO      ''
                       Mohamed Boucadair
                       Adresse: 42, rue des Coutures
                       BP 6243
                       14066 Caen Cedex
                       Email: mohamed.boucadair@francetele-
                         com. com''

        DESCRIPTION
                   ''The PIB module that contains classes
                   describing the parameters to be monitored,
                   recorded and/or reported by the PEP for Traffic
                   Engineering accounting purposes.''

          : := { tbd}

--
-- The ipTe Accounting Class
--

ipTeAccountingClasses
        OBJECT IDENTIFIER: := { ipTeAccountingPib 1 }

--
-- The MPLS TE Accounting Class
--
    -- This class defines tables related to MPLS TE
    -- To be done later

lspTeAccountingClasses
        OBJECT IDENTIFIER: := { ipTeAccountingPib 2 }

--
-- ospfTeRouterUsageTable
--

ospfTeRouterUsageTable   OBJECT-TYPE

        SYNTAX          SEQUENCE OF ospfTeRouterUsageEntry
        PIB-ACCESS      report-only
        STATUS          current
        DESCRIPTION
```

```
                "This class defines the usage attributes to be
                reported, and which are related to the router
                identified by the Router-Id."

        ::={ ipTeAccountingClasses 1}

ospfTeRouterUsageEntry      OBJECT-TYPE

        SYNTAX      ospfTeUsageRouterEntry
        STATUS      current
        DESCRIPTION
                "An entry for the ospfTeRouterUsageTable."
        PIB-INDEX     { ospfTeRouterUsagePrid}
        UNIQUENESS    { ospfTeRouterUsageLinkPrid,
                    ospfTeUsageIfActif}

        ::={ ospfTeRouterUsageTable 1}

ospfTeRouterUsageEntry: := SEQUENCE {
                    ospfTeRouterUsagePrid InstanceID,
                    ospfTeRouterUsageLinkPrid Prid,
                    ospfTeRouterUsageIfActif  Counter64 }

ospfTeRouterUsagePrid       OBJECT-TYPE

        SYNTAX      Prid
        STATUS      current
        DESCRIPTION
                "An integer index that uniquely identifies this
                instance of the ospfTeRouterUsage class."

        ::={ ospfTeRouterUsageEntry 1 }

ospfTeRouterUsageLinkPrid       OBJECT-TYPE
        SYNTAX      Prid
        STATUS      current
        DESCRIPTION
                "The PRID of the linkage policy instance used
                to refer this usage policy instance."

        ::={ ospfTeRouterUsageEntry 2 }

ospfTeRouterUsageIfActif       OBJECT-TYPE

        SYNTAX      Counter64
```

```
        STATUS      current
        DESCRIPTION
                "The number of interfaces that are involved in
                an OSPF-TE route computation in the router
                identified by Router-Id."

        ::={ ospfTeRouterUsageEntry 3 }

 --
 -- ospfTeUsageTable
 --
ospfTeUsageTable OBJECT-TYPE

        SYNTAX          SEQUENCE OF ospfTeUsageEntry
        PIB-ACCESS      report-only
        STATUS          current
        DESCRIPTION
                "This class defines the usage attributes to use
                for OSPF TE purposes."

        ::={ ipTeAccountingClasses 2 }

ospfTeUsageEntry    OBJECT-TYPE

        SYNTAX          ospfTeUsageEntry
        STATUS          current
        DESCRIPTION
                "An entry for the ospfTeUsageTable."

        PIB-INDEX       { ospfTeUsagePrid}
        UNIQUENESS      { ospfTeUsageLinkPrid,
                        OspfTeUsageLinkDelay }

        ::={ ospfTeUsageTable 1 }

ospfTeUsageEntry ::= SEQUENCE { ospfTeUsagePrid InstanceID,
                              ospfTeUsageLinkPrid Prid,
                              ospfTeUsageLinkDelay Unsigned32 }

ospfTeUsagePrid         OBJECT-TYPE

        SYNTAX          Prid
        STATUS          current
        DESCRIPTION
```

```
                  "An integer index that uniquely identifies this
                  instance of the ospfTeUsage class."

        ::={ ospfTeUsageEntry 1 }

ospfTeUsageLinkPrid       OBJECT-TYPE

        SYNTAX          Prid
        STATUS          current
        DESCRIPTION
                  "The PRID of the linkage policy instance used
                  to refer this usage policy instance."
        ::={ ospfTeUsageEntry 2 }

ospfTeUsageLinkDelay    OBJECT-TYPE

        SYNTAX          Unsigned32
        STATUS          current
        DESCRIPTION
                  "The one-way delay that has been observed on
                  this route."

        ::={ ospfTeUsageEntry 3 }

--
-- isisTeUsageTable
--

isisTeUsageTable OBJECT-TYPE

        SYNTAX          SEQUENCE OF isisTeUsageEntry
        PIB-ACCESS      report-only
        STATUS          current
        DESCRIPTION
                  "This class defines the usage attributes to use
                  for IS-IS TE purposes."

        ::={ ipTeAccountingClasses 3 }

isisTeUsageEntry     OBJECT-TYPE

        SYNTAX          isisTeUsageEntry
        STATUS          current
        DESCRIPTION
                  "An entry for the isisTeUsageTable."
```

```
        PIB-INDEX        { isisTeUsagePrid}
        UNIQUENESS       { isisTeUsageLinkPrid,
                         isisTeUsageLinkDelay }

        ::={ isisTeUsageTable 1 }

isisTeUsageEntry: := SEQUENCE {
                          isisTeUsagePrid        InstanceID,
                           isisTeUsageLinkPrid      Prid,
                      isisTeUsageLinkDelay         Unsigned32 }

isisTeUsagePrid          OBJECT-TYPE
        SYNTAX           Prid
        STATUS           current
        DESCRIPTION
                  "An integer index that uniquely identifies this
                  instance of the isisTeUsage class."

        ::={ isisTeUsageEntry 1 }

isisTeUsageLinkPrid      OBJECT-TYPE

        SYNTAX           Prid
        STATUS           current
        DESCRIPTION
                  "The PRID of the linkage policy instance used
                  to refer this usage policy instance."

        ::={ isisTeUsageEntry 2 }

isisTeUsageLinkDelay      OBJECT-TYPE

        SYNTAX           Unsigned32
        STATUS           current
        DESCRIPTION
                  "The one-way delay that has been observed on
                  this route."

        ::={ isisTeUsageEntry 3 }

--
-- bgpTeUsageTable
--

bgpTeTable           OBJECT-TYPE
```

```
        SYNTAX          SEQUENCE OF bgpTeUsageEntry
        PIB-ACCESS      report-only
        STATUS          current
        DESCRIPTION
                "This table contains a set of accounting
                information related to the activation of BGP
                process enabling exchange of QoS information."

        : :={ ipTeAccountingClasses 4 }

  bgpTeUsageEntry       OBJECT-TYPE

        SYNTAX          bgpTeUsageEntry
        STATUS          current
        DESCRIPTION
                "An entry to bgpTeUsage Class."

        PIB-INDEX       { bgpTeUsagePrid }
        UNIQUENESS      { bgpTeUsageLinkPrid,
                            bgpTeUsageActIf,
                               bgpTeUsageOneWayDelay }

        ::={ bgpTeUsageTable 1 }

  bgpTeUsageEntry: := SEQUENCE {
              bgpTeUsagePrid          InstanceId,
              bgpTeUsageLinkPrid          Prid,
                    bgpTeUsageActIf          Counter64,
                    bgpTeUsageOneWayDelay      Unsigned32 }

  bgpTeUsagePrid        OBJECT-TYPE

        SYNTAX          InstanceId
        STATUS          current
        DESCRIPTION
                "An integer index that uniquely identifies this
                instance of the bgpTeUsage class."

        : :={ bgpTeUsageEntry 1 }

  bgpTeUsageLinkPrid    OBJECT-TYPE


        SYNTAX          Prid
        STATUS          current
```

```
        DESCRIPTION
                "The PRID of the linkage policy instance used
                to base this usage policy instance upon."

        ::={ bgpTeUsageEntry 2 }

bgpTeUsageActIf        OBJECT-TYPE

        SYNTAX         Counter64
        STATUS         current
        DESCRIPTION
                "Specifies the number of interfaces that are
                involved in the BGP route computation process."

        ::={ bgpTeUsageEntry 3 }

bgpTeUsageOneWayDelay          OBJECT-TYPE

        SYNTAX         Unsigned32
        STATUS         current
        DESCRIPTION
                "Specifies the one-way delay that has been
                observed on this route."

        ::={ bgpTeUsageEntry 4 }

--
-- The Threshold class that accompanies the OSPF and BGP usage
-- tables
--
    --
    -- OSPF threshold attributes
    --

ospfTeThresholdTable OBJECT-TYPE

        SYNTAX         SEQUENCE OF ospfThresholdEntry
        PIB-ACCESS     Install
        STATUS         current
        DESCRIPTION
                "This class defines the threshold attributes
                corresponding to OSPF TE usage attributes
                specified in ospfTeUsageTable."

        ::={ ipTeAccountingClasses 5 }
```

```
ospfTeThresholdEntry OBJECT-TYPE

        SYNTAX          ospfTeThresholdEntry
        STATUS          current
        DESCRIPTION
             "Defines the attributes to hold threshold values."

        PIB-INDEX { ospfTeThresholdId }

        ::={ ospfTeThresholdId 1 }

ospfTeThresholdEntry::= SEQUENCE {
        ospfTeThresholdId                  InstanceID,
        ospfTeThresholdBwThresholds        Integer64,
        ospfTeThresholdRsvBwThresholds     Integer64 }

ospfTeThresholdId    OBJECT-TYPE

        SYNTAX          InstanceId
        STATUS          current
        DESCRIPTION
                "Arbitrary integer index that uniquely
                identifies an instance of the class."

        ::={ ospfTeThresholdEntry 1 }

ospfTeThresholdBwThresholds   OBJECT-TYPE

        SYNTAX          Integer64
        STATUS          current
        DESCRIPTION
                "The threshold the used bandwidth on the link
                shouldn't exceed."

        ::={ ospfTeThresholdEntry 2 }

ospfTeThresholdRsvBwThresholds OBJECT-TYPE


        SYNTAX          Integer64
        STATUS          current
        DESCRIPTION
                "The threshold the reserved bandwidth on the
                link shouldn't exceed."
```

```
        : :={ ospfTeThresholdEntry 3 }

 - -
 - -  ISIS Threshold attributes
 - -

  isisTeThresholdTable   OBJECT-TYPE

        SYNTAX          SEQUENCE OF isisThresholdEntry
        PIB-ACCESS      Install
        STATUS          current
        DESCRIPTION
                ''This class defines the threshold attributes
                corresponding to ISIS TE usage attributes
                specified in isisTeUsageTable.''

        : :={ ipTeAccountingClasses 6 }

 isisTeThresholdEntry   OBJECT-TYPE

        SYNTAX          isisTeThresholdEntry
        STATUS          current
        DESCRIPTION
                ''Defines the attributes to hold threshold
                values.''

        PIB-INDEX{ isisTeThresholdId }

        : :={ isisTeThresholdId 1 }

 isisTeThresholdEntry::= SEQUENCE{
        isisTeThresholdId                   InstanceID,
        isisTeThresholdBwThresholds         Integer64,
        isisTeThresholdRsvBwThresholds      Integer64 }

 isisTeThresholdId   OBJECT-TYPE

        SYNTAX          InstanceId
        STATUS          current
        DESCRIPTION
                ''Arbitrary integer index that uniquely
                identifies an instance of the class.''

        : :={ isisTeThresholdEntry 1 }
```

```
isisTeThresholdBwThresholds   OBJECT-TYPE

        SYNTAX          Integer64
        STATUS          current
        DESCRIPTION
                ''The threshold the used bandwidth on the link
                shouldn't exceed.''

        ::={ isisTeThresholdEntry 2}
isisTeThresholdRsvBwThresholds   OBJECT-TYPE

        SYNTAX          Integer64
        STATUS          current
        DESCRIPTION
                ''The threshold the reserved bandwidth on the
                link shouldn't exceed.''

        ::={ isisTeThresholdEntry 3}

--
-- BGP threshold attributes
--

bgpTeThresholdTableOBJECT-TYPE

        SYNTAX          SEQUENCE OF bgpThresholdEntry
        PIB-ACCESS      Install
        STATUS          current
        DESCRIPTION
                ''This class defines the threshold attributes
                corresponding to BGP usage attributes specified
                in bgpTeUsageTable.''

        ::={ ipTeAccountingClasses 7}

bgpTeThresholdEntry   OBJECT-TYPE

        SYNTAX          bgpTeThresholdEntry
        STATUS          current
        DESCRIPTION
                ''Defines the attributes to hold threshold
                values.''

        PIB-INDEX{ bgpTeThresholdPrid}
```

```
    : :={ bgpTeThresholdId 1}

bgpTeThresholdEntry: := SEQUENCE{
        bgpTeThresholdId          InstanceID,
            bgpTeThresholdNlriAddress        InetAddress,
      bgpTeThresholdNextHopAddress      InetAddress,
        bgpTeThresholdOneWayDelayThreshold      Integer64,
          bgpTeThresholdInterPacketDelayThreshold
  Integer64,
       bgpTeThresholdLossRateThreshold  Integer64}
bgpTeThresholdId      OBJECT-TYPE

     SYNTAX           InstanceId
     STATUS           current
     DESCRIPTION
            "Arbitrary integer index that uniquely
            identifies an instance of the class."

     : :={ bgpTeThresholdEntry 1}

bgpTeThresholdNlriAddress    OBJECT-TYPE

     SYNTAX           InetAddress
     STATUS           current
     DESCRIPTION
            "The IP address to match against the NLRI field
            of QoS_NLRI attribute of the BGP-4 UPDATE
            message introduced in RFC 1771."

       : :={ bgpTeThresholdEntry 2}

bgpTeThresholdNextHopAddress    OBJECT-TYPE

       SYNTAX         InetAddress
       STATUS         current
       DESCRIPTION
            "The address of the next router."

       : :={ bgpTeThresholdEntry 3}

bgpTeThresholdOneWayDelayThreshold   OBJECT-TYPE

       SYNTAX         Integer64
       STATUS         current
```

```
        DESCRIPTION
                ''The threshold of the one-way delay that will
                trigger a report in the next reporting interval
                when exceeded.''

        : :={ bgpTeThresholdEntry 4 }

  bgpTeThresholdInterPacketDelayThreshold  OBJECT-TYPE

        SYNTAX          Integer64
        STATUS          current
        DESCRIPTION
                ''The threshold of the interpacket delay
                variation that will trigger a report in
                the next reporting interval when exceeded.''

        : :={ bgpTeThresholdEntry 5 }

  bgpTeThresholdLossRateThreshold  OBJECT-TYPE

        SYNTAX          Integer64
        STATUS          current
        DESCRIPTION
                ''The threshold, in terms of loss rate, that
                will trigger a report in the next reporting
                interval when exceeded.''

        : :={ bgpTeThresholdEntry 6 }

END
```

APPENDIX 5

Description of Classes of an IP VPN Information Model

A5.1 Introduction

This appendix is the companion document of Chapter 10 'Automated Production of BGP/MPLS-based VPNs'. It details the IP VPN-specific classes of an IP VPN information model that can be used for the dynamic enforcement of IP VPN-specific policies.

A5.2 Policy Class Definitions

```
The class ipvpnPolicyVFICreationAction

   This class specifies the VFI to be created in order dynamically to
  deploy a BGP/MPLS VPN.

   NAME                 ipvpnPolicyVFICreationAction
   DESCRIPTION
                        The class for specifying the VFI to be created.
   DERIVED FROM         PolicyAction
   ABSTRACT             FALSE
   PROPERTIES           AttachedInterface[ref AccessEndPoint[1..n]]

The reference AttachedInterface
   This is a reference to one or several objects of class
   AccessEndPoint.

The class ipvpnPolicyRouteDistributionAction
   This action represents the route distribution process of an IP VPN
   routing table that is implemented by means of RouteTarget and
   results in the definition of routes with/without RouteDistingu-
   isher.
```

Service Automation and Dynamic Provisioning Techniques in IP/MPLS Environments
C. Jacquenet, G. Bourdon and M. Boucadair

This action is intended to be used to implement RFC4364-compliant IP VPNs.

NAME	ipvpnPolicyRouteDistributionAction
DESCRIPTION	The class for representing the route distribution actions. The distribution actions should support point-to-point, hub-and-spoke, full mesh and partial mesh topology requirements.
DERIVED FROM	PolicyAction
ABSTRACT	FALSE
PROPERTIES	DistributionSource [ref AccessEndPoint[1]] DistributionDestination [ref AccessEndPoint [1]] DistributionMandatoryHops [ref AccessEndPoint[0..n]]

The reference DistributionSource

This is a reference to an object of class AccessEndPoint.

The reference DistributionDestination

This is a reference to an object of class AccessEndPoint.

The reference DistributionMandatoryHops

This is a reference to zero or more objects, which points to mandatory hops to be used for the traffic flowing from the DistributionSource to the DistributionDestination. The objects referenced are instances of the class AccessEndPoint.

The class ipvpnPolicyVPNTopologyDescriptionAction

This class specifies the IP VPN service topology and reachability and is intended to be used for configuring an IP VPN database for implementing IP VPNs.

NAME	ipvpnPolicyVPNTopologyDescriptionAction
DESCRIPTION	The class for representing the topology and reachability description actions. The actions should support point-to-point, hub-and-spoke, full mesh and partial mesh topology requirements.
DERIVED FROM	PolicyAction
ABSTRACT	FALSE
PROPERTIES	RoutingSource [ref AccessEndPoint [1]] RoutingDestination [ref AccessEndPoint [1]] RoutingMandatoryHops [ref EdgeNode [2..n]]

The reference RoutingSource

This is a reference to an object of type AccessEndPoint.

The reference RoutingDestination

This is a reference to an object of type AccessEndPoint.

The reference RoutingMandatoryHops

This is a reference to zero or more objects, which points to mandatory hops to be used for the traffic flowing from the ipvpnPolicyRoutingSource to the ipvpnPolicyRoutingDestination. The objects referenced are instance(s) of EdgeNode.

The class ipvpnPolicyNATAction

This class specifies which source addresses need to be translated and what should be the results of this translation.

NAME	ipvpnPolicyNATAction
DESCRIPTION	The class that represents the network address translation action of the "If Condition then Action" semantics associated with a policy rule.
DERIVED FROM	PolicyAction
ABSTRACT	FALSE
PROPERTIES	TranslateFromIPv4Address TranslateToIPv4Address

The property TranslateFromIPv4Address

Specifies the original set of IPv4 addresses that needs to be translated.

NAME	TranslateFromIPv4Address
DESCRIPTION	The original IPv4 address that needs to be translated.
SYNTAX	PolicyIPv4AddrValue

The property TranslateToIPv4Address

Specifies the final set of IPv4 addresses that needs to be translated to.

NAME	TranslateToIPv4Address
DESCRIPTION	The final IPv4 address that needs to be translated to.
SYNTAX	PolicyIPv4AddrValue

The class ipvpnPolicyTrafficTrunkAction

This class indicates the requirements on the traffic trunk to be used to transport the IP VPN traffic.

NAME	ipvpnPolicyTrafficTrunkAction
DESCRIPTION	The class for representing the requirements of the traffic trunk to be used to transport the VPN traffic
DERIVED FROM	PolicyAction

```
ABSTRACT            FALSE
PROPERTIES          Ingress [ref EdgeNode]
                    Egress [ref EdgeNode]
                    Priority [Integer]
                    Preemption [Integer (1-4)]
                    Resilience [boolean]
                    TrafficProfile [QosPolicyTokenBucketTrfcProf]
```

The reference Ingress

This attribute references the Edge Node, which will be the ingress node for the trunk.

The reference Egress

This attribute references the Edge Node, which will be the egress node for the trunk.

The property Priority

This attribute indicates the priority requirement for the trunk.

```
NAME                Priority
DESCRIPTION         The priority requirement for the trunk.
SYNTAX              Integer
```

The property Preemption

This attribute indicates the preemption requirement for the trunk.
The preemption is related to whether the trunk can be preempted to accommodate a new higher priority trunk.

```
NAME                Preemption
DESCRIPTION         The preemption requirement for the trunk.
SYNTAX              Integer(1-4)
```

The property Resilience

This attribute indicates the resilience requirement for the trunk.

```
NAME                Resilience
DESCRIPTION         The resilience requirement for the trunk.
SYNTAX              boolean
```

The property TrafficProfile

This attribute indicates the traffic profile requirement for the trunk.

```
NAME                TrafficProfile
DESCRIPTION         The Traffic Profile requirement for the trunk.
SYNTAX              QoSPolicyTokenBucketTrfcProf
```

The class ipvpnPolicyFirewallAction

Specifies the firewall action to be enforced such as "drop", "pass", "log", "alert", etc. The list of possible actions is limited by the attributes in the action object.

NAME	ipvpnPolicyFirewallAction
DESCRIPTION	The class for representing the firewall action of the "If Condition then Action" semantics associated with a policy rule.
DERIVED FROM	PolicyAction
ABSTRACT	FALSE
PROPERTIES	Action

The property Action

The action defines the type of firewall action to be enforced.

NAME	Action
DESCRIPTION	The firewall action to be enforced
SYNTAX	INTEGER
VALUES	The action can take the following values 0 = Allow 1 = Allow and Log 2 = Allow and Alarm 3 = Deny 4 = Deny and Log 5 = Deny and Alarm

The class ipvpnEncryptionAction

The encryption standard is assumed to be IPsec. This class provides the IPSec parameters that will be used to set up the security association required to handle the encryption and decryption of packets.

NAME	ipvpnEncryptionAction
DESCRIPTION	The class for representing the encryption action of the "If Condition then Action" semantics associated with a policy rule.
DERIVED FROM	PolicyAction
ABSTRACT	TRUE
PROPERTIES	IkeAuthentication IkeEncryption IkeDHGroup IkeTimeout IkeTrafficBasedExpiry IpsecAuthentication IpsecEncryption IpsecDHGroup

IpsecTimeout
IpsecTrafficBasedExpiry
IkePeerAuthenticationMethod

The property IkeAuthentication

The property specifies the authentication algorithm to be used. The IkeAuthentication parameters can be used to generate the corresponding ISA key management protocol (ISAKMP) parameters in cases where ISAKMP is still being used. This draft does not describe a separate set of parameters for ISAKMP. It is left to the policy servers generating the configuration to handle the corresponding translation.

NAME	IkeAuthentication
DESCRIPTION	The property that specifies the authentication algorithm.
SYNTAX	String

The property IkeEncryption

The property specifies the encryption algorithm to be used.

NAME	IkeEncryption
DESCRIPTION	The property that specifies the encryption algorithm.
SYNTAX	String

The property IkeDHGroup

The property specifies the DHGroup to be used during IKE negotiations.

NAME	IkeDHGroup
DESCRIPTION	The property that specifies the DHGroup to be used during IKE negotiations.
SYNTAX	String

The property IkeTimeout

The property specifies the IKE timeout to be used.

NAME	IkeTimeout
DESCRIPTION	The property that specifies the IKE timeout.
SYNTAX	Integer

The property IkeTrafficBasedExpiry

The property specifies the IKE traffic-based expiry to be used.

NAME	IkeTrafficBasedExpiry
DESCRIPTION	The property that specifies the IKE traffic-based expiry to be used.
SYNTAX	Integer

The property IPSECAuthentication

The property specifies the authentication algorithm to be used.

NAME	IPSECAuthentication
DESCRIPTION	The property that specifies the authentication algorithm.
SYNTAX	String

The property IPSECEncryption

The property specifies the encryption algorithm to be used.

NAME	IPSECEncryption
DESCRIPTION	The property that specifies the encryption algorithm.
SYNTAX	String

The property IPSECDHGroup

The property specifies the DHGroup to be used during IPsec negotiations.

NAME	IPSECDHGroup
DESCRIPTION	The property that specifies the DHGroup to be used during the IKE phase II negotiations.
SYNTAX	String

The property IPSECTimeout

The property specifies the IPsec key timeout to be used.

NAME	IPSECTimeout
DESCRIPTION	The property that specifies the IPSEC key timeout.
SYNTAX	Integer

The property IPSECTrafficBasedExpiry

The property specifies the IPsec traffic-based key expiry to be used.

NAME	IPSECTrafficBasedExpiry
DESCRIPTION	The property that specifies the IPsec traffic-based key expiry to be used.
SYNTAX	Integer

The property IkePeerAuthenticationMethod

The IKE peers are the IKE processes that are at the two ends of a control channel related to encryption of traffic at the data layer. The method used by the Internet key exchange (IKE) peers to authenticate each other. The IKE peers are running on the IP VPN nodes.

NAME	IkePeerAuthenticationMethod

DESCRIPTION	The property that specifies the method used by the IKE peers to authenticate each other.
SYNTAX	Unsigned 16-bit integer
VALUE	The possible values are listed below. 0 = ProposalList is to be used (see below) 1 = Preshared key 2 = DSS (D S S) signatures 3 = RSA (R S A) signatures 4 = Encryption with RSA 5 = Revised encryption with RSA 6 = Kerberos (has this number been assigned???) A value of 0 is a special value that indicates that this particular proposal should be repeated once for each authentication method that corresponds to the credentials installed on the machine. For example, if the system has a preshared key and a certificate, a proposal list could be constructed that includes a proposal that specifies preshared key and proposals for any of the public-key authentication methods. DSS and RSA are encryption algorithms that are explained in several encryption specific books such as "Applied Cryptography".

The class ipvpnApplicaionSignatureValue

Specifies the layer-4 to layer-7 characteristics of the packet, including application level decodes that require stateful inspection of the packet, e.g. HTTP, FTP, SMTP, TELNET, etc. This class enables the policies to capture the application layer requirements of the customer with regards to treatment for specific IP traffic.

NAME	ipvpnApplicationSignatureValue
DESCRIPTION	The class for representing application signature to be matched against the traffic
DERIVED FROM	qoSPolicyValue
ABSTRACT	FALSE
PROPERTIES	applicationSignature

This class can have several subclasses, which reflect the application protocol classification granularity.

The property applicationSignature

NAME	applicationSignature
DESCRIPTION	The property that provides a signature used to identify the application by examining the payload of the protocol data unit (PDU).

SYNTAX	String

Topology class definitions

The abstract class "Node"

The abstract class Node is a representation of a generic network node. The class definition is as follows:

NAME	Node
DESCRIPTION	An abstract class representing a network node entity.
DERIVED FROM	ComputerSystem
ABSTRACT	TRUE
PROPERTIES	PEPID

The PEPID single-valued property corresponds to the node identifier. It is a globally unique identifier. The property definition is as follows:

NAME	PEPID
DESCRIPTION	A user-friendly name (e.g. DNS name or primary IP public address) of a node object.
SYNTAX	String

The class "CoreNode"

The class CoreNode is a representation of a router residing at the network core (with respect to the IP VPN service). The class definition is as follows:

NAME	CoreNode
DESCRIPTION	A class representing a network core router.
DERIVED FROM	Node
ABSTRACT	FALSE
PROPERTIES	NONE

The class "EdgeNode"

The class EdgeNode is a representation of a router residing at the network edge (with respect to the IP VPN service). The class definition is as follows:

NAME	EdgeNode
DESCRIPTION	A class representing a network edge router.
DERIVED FROM	Node
ABSTRACT	FALSE
PROPERTIES	NONE

The class "LogicalNetwork"

The class LogicalNetwork is defined by DMTF. It is reported here for convenience. A LogicalNetwork groups together a set of

ProtocolEndpoints of a given type that are able to communicate with each other directly. A LogicalNetwork represents the ability to send and/or receive data over a network. The class definition is as follows:

NAME	LogicalNetwork
DESCRIPTION	A class representing a logical network.
DERIVED FROM	CollectionOfMSEs
ABSTRACT	FALSE
PROPERTIES	NetworkType

The NetworkType single-valued property provides additional information that can be used to help categorize and classify different instances of this class. The property takes values from an enumeration. Some possible values are "Unknown", "Other", "IPv4", "IPv6", "IPX", etc. The property definition is as follows:

NAME	NetworkType
DESCRIPTION	Specify the network type.
SYNTAX	String

The class "Partition"

The provider network is partitioned into domains called "partitions". A partition is an administrative entity. The class definition is as follows:

NAME	Partition
DESCRIPTION	An class representing a (logical) partition.
DERIVED FROM	LogicalNetwork
ABSTRACT	FALSE
PROPERTIES	PartitionID

The PartitionID single-valued property corresponds to the partition identifier. It is unique within the scope of a provider domain. The property definition is as follows:

NAME	PartitionID
DESCRIPTION	A user-friendly name of a partition object.
SYNTAX	String

The class "IP VPN"

The class IP VPN represents an IP virtual private network deployed within the provider network. The class definition is as follows:

NAME	IP VPN
DESCRIPTION	A class representing an IP VPN.
DERIVED FROM	LogicalNetwork
ABSTRACT	FALSE
PROPERTIES	VPNID

The VPNID single-valued property corresponds to the globally unique VPN identifier as defined by IETF. The property definition is as follows:

NAME	VPNID
DESCRIPTION	The standard VPNID.
SYNTAX	Octet

The class "ProtocolEndPoint"

The class ProtocolEndPoint is defined by DMTF. It is reported here for convenience. The class represents a communication point from which data may be sent or received. ProtocolEndPoints link router interfaces and switch ports to LogicalNetworks. The class definition is as follows:

NAME	ProtocolEndPoint
DESCRIPTION	A communication point.
DERIVED FROM	ServiceAccessPoint
ABSTRACT	FALSE
PROPERTIES	ProtocolType

The ProtocolType single-valued property provides additional information that can be used to help categorize and classify different instances of this class. The property takes values from an enumeration. Some possible values are "Unknown", "Other", "IPv4", "IPv6", "IPX", etc. The property definition is as follows:

NAME	ProtocolType
DESCRIPTION	Specify the protocol of endpoint.
SYNTAX	String

The class "AccessEndPoint"

The class AccessEndPoint represents an access IP interface. The class definition is as follows:

NAME	AccessEndPoint
DESCRIPTION	A class representing an access interface.
DERIVED FROM	ProtocolEndPoint
ABSTRACT	FALSE
PROPERTIES	NONE

The class "CoreEndPoint"

The class CoreEndPoint represents a core IP interface. The class definition is as follows:

NAME	CoreEndPoint
DESCRIPTION	A class representing a core interface.
DERIVED FROM	ProtocolEndPoint
ABSTRACT	FALSE

PROPERTIES	IPAddress

The class "VirtualEndPoint"

The class VirtualEndPoint represents a virtual interface (e.g. a tunnel endpoint). The class definition is as follows:

NAME	VirtualEndPoint
DESCRIPTION	A class representing a virtual interface.
DERIVED FROM	ProtocolEndPoint
ABSTRACT	FALSE
PROPERTIES	NONE

The abstract class "NetworkService"

The class NetworkService is defined by DMTF. It is reported here for convenience. This is an abstract base class. It serves as the root of the network hierarchy. Network services represent generic functions that are available from the network that configure and/or modify the traffic being sent. The class definition is as follows:

NAME	NetworkService
DESCRIPTION	A class representing a base network service.
DERIVED FROM	Service
ABSTRACT	TRUE

PROPERTIES	NONE
	//string StartupConditions[ ]
	//string StartupParameters[ ]

The class "VirtualForwardingInstance"

This class represents a VFI. A VFI is a dedicated forwarding process that runs on a border router (i.e. a PE or a CE). VFI forwards customer traffic of a given IP VPN to the virtual links, and vice versa. Hence a VFI is associated with a subset of the access interfaces and virtual interfaces of a border node. The class definition is as follows:

NAME	VirtualForwardingInstance
DESCRIPTION	A class representing a VFI.
DERIVED FROM	NetworkService
ABSTRACT	FALSE
PROPERTIES	VPNID

The following classes define the "associations" that belong to the topology model.

The abstract association "Link"

This abstract association is used to represent a bidirectional link. The class definition for the association is as follows:

NAME	Link
DESCRIPTION	A generic association used to establish a one-to-one bidirectional relationship between the subclasses of ProtocolEndPoint.
DERIVED FROM	Dependency
ABSTRACT	TRUE
PROPERTIES	Antecedent [ref ProtocolEndPoint [1..1]] Dependent [ref ProtocolEndPoint [1..1]]

This abstract association inherits two object references from a higher-level CIM association class, Dependency. It overrides these object references to make them references to instances of the class ProtocolEndPoint. Subclasses of Link then override these object references again, to make them references to concrete "interface" classes.
Note that the semantic of dependent and antecedent properties is changed. These properties just represent a pair of unordered association ends. The [1..1] cardinality indicates that a pair of ProtocolEndpoints can be connected by exactly one Link.

The association "CoreLink"

This association is used to represent a direct reachability between two core interfaces. Interfaces can belong to either ENs or CNs. The class definition for the association is as follows:

NAME	CoreLink
DESCRIPTION	A logical representation of a one-hop reachability between two nodes.
DERIVED FROM	Link
ABSTRACT	FALSE
PROPERTIES	Antecedent [ref CoreEndPoint[1..1]] Dependent [ref CoreEndPoint [1..1]]

This association is a concrete class and can be instantiated. It inherits two object references from the Link class and overrides these object references to make them references to instances of the class CoreEndPoint.

The association "AccessLink"

This association is used to represent a direct reachability between two access interfaces. The class definition for the association is as follows:

NAME	AccessLink
DESCRIPTION	A logical representation of a one-hop reachability between a border node and a customer node.
DERIVED FROM	Link

ABSTRACT	FALSE
PROPERTIES	Antecedent [ref AccessEndPoint [1..1]] Dependent [ref AccessEndPoint [1..1]]

This association is a concrete class. It inherits two object references from the Link class and overrides these object references to make them references to instances of the class AccessEndPoint.

The association "VirtualLink"

This association is used to represent a virtual one-hop reachability (e.g. a tunnel or a MPLS LSP) between two virtual interfaces. The class definition for the association is as follows:

NAME	VirtualLink
DESCRIPTION	A logical representation of a virtual connection traversing the core network.
DERIVED FROM	Link
ABSTRACT	FALSE
PROPERTIES	Antecedent [ref VirtualEndPoint [1..1]] Dependent [ref VirtualEndPoint [1..1]]

This association inherits two object references from the Link class. It overrides these object references to make them references to instances of the class VirtualEndPoint.

The abstract association "NodeInPartition"

The class definition for the association is as follows:

NAME	NodeInPartition
DESCRIPTION	A generic association used to establish a relationship between a generic node and its pertaining partition.
DERIVED FROM	Dependency
ABSTRACT	TRUE
PROPERTIES	Antecedent [ref Node [0..*]] Dependent [ref Partition [1..1]]

This abstract association inherits two object references from a higher-level CIM association class, Dependency. It overrides these object references to make them references to instances of the class Node and Partition. Subclasses of NodeInPartition then override the antecedent references again, to make them references to concrete subclasses of Node.

The association "EdgeNodeInPartition"

The class definition for the association is as follows:

NAME	EdgeNodeInPartition
DESCRIPTION	The association represents the relationship

	between an EdgeNode and its pertaining Partition.
DERIVED FROM	NodeInPartition
ABSTRACT	FALSE
PROPERTIES	Antecedent [ref EdgeNode [2..*]]

The association "CoreNodeInPartition"

The class definition for the association is as follows:

NAME	CoreNodeInPartition
DESCRIPTION	The association represents the relationship between a CoreNode and its pertaining Partition.
DERIVED FROM	NodeInPartition
ABSTRACT	FALSE
PROPERTIES	Antecedent [ref CoreNode [0..*]]

The association "AccessEndPointInVFI"

The class definition for the association is as follows:

NAME	AccessEndPointInVFI
DESCRIPTION	An association used to establish a relationship between a VFI and the access interfaces it serves.
DERIVED FROM	Dependency
ABSTRACT	FALSE
PROPERTIES	Antecedent [ref AccessEndPoint [1..*]] Dependent [ref VirtualForwardingInstance [1..1]]

This association inherits two object references from a higher-level CIM association class, Dependency. It overrides these object references to make them references to instances of the classes AccessEndPoint and VirtualForwardingInstance.

The association "VirtualEndPointInVFI"

The class definition for the association is as follows:

NAME	VirtualEndPointInVFI
DESCRIPTION	A generic association used to establish a relationship between a VFI and the virtual interfaces it works on.
DERIVED FROM	Dependency
ABSTRACT	FALSE
PROPERTIES	Antecedent [ref VirtualEndPoint [1..*]] Dependent [ref VirtualForwardingInstance [1..1]]

This association inherits two object references from a higher-level CIM association class, Dependency. It overrides these object references to make them references to instances of the classes

VirtualEndPoint and VirtualForwardingInstance.

The abstract aggregation "ProtocolEndPointInNode"

This abstract aggregation defines two object references that will be overridden in each of five subclasses, to become references to the subclasses of Node and ProtocolEndPoint. From a general viewpoint, this aggregation expresses what interfaces (physical or virtual) belong to a given node. The class definition for the aggregation is as follows:

NAME	ProtocolEndPointInNode
DESCRIPTION	A generic association used to establish a relationship between a generic node and its interfaces.
DERIVED FROM	Component
ABSTRACT	TRUE
PROPERTIES	GroupComponent [ref Node [0..*]] PartComponent [ref ProtocolEndPoint [0..*]]

The aggregation "AccessEndPointInEdgeNode"

The AccessEndPointInEdgeNode aggregation enables access interfaces to be assigned to a given EN. The class definition for the aggregation is as follows:

NAME	AccessEndPointInEdgeNode
DESCRIPTION	A class representing the aggregation of access interfaces by ENs.
DERIVED FROM	ProtocolEndPointInNode
ABSTRACT	FALSE
PROPERTIES	GroupComponent [ref EdgeNode [1..1]] PartComponent [ref AccessEndPoint [1..*]]

The aggregation "CoreEndPointInEdgeNode"

The CoreEndPointInEdgeNode aggregation enables core interfaces to be assigned to a given EN. The class definition for the aggregation is as follows:

NAME	CoreEndPointInEdgeNode
DESCRIPTION	A class representing the aggregation of core interfaces by ENs.
DERIVED FROM	ProtocolEndPointInNode
ABSTRACT	FALSE
PROPERTIES	GroupComponent [ref EdgeNode [1..1]] PartComponent [ref CoreEndPoint [1..*]]

The aggregation "CoreEndPointInCoreNode"

The CoreEndPointInCoreNode aggregation enables core interfaces to be assigned to a given core router. The class definition for the aggregation is as follows:

NAME	CoreEndPointInCoreNode
DESCRIPTION	A class representing the aggregation of core interfaces by CNs.
DERIVED FROM	ProtocolEndPointInNode
ABSTRACT	FALSE
PROPERTIES	GroupComponent [ref CoreNode [1..1]] PartComponent [ref CoreEndPoint [2..*]]

The aggregation "VirtualEndPointInEdgeNode"

The VirtualEndPointInEdgeNode aggregation enables virtual interfaces to be assigned to a given EN. The class definition for the aggregation is as follows:

NAME	VirtualEndPointInEdgeNode
DESCRIPTION	A class representing the aggregation of virtual interfaces by PEs.
DERIVED FROM	ProtocolEndPointInNode
ABSTRACT	FALSE
PROPERTIES	GroupComponent [ref EdgeNode [1..1]] PartComponent [ref VirtualEndPoint [0..*]]

The aggregation "VFIInEdgeNode"

Each VFI works in an EN. This class associates VFIs with corresponding border routers. The class definition for the aggregation is as follows:

NAME	VFIInEdgeNode
DESCRIPTION	Aggregation between a VFI and an EN.
DERIVED FROM	Component
ABSTRACT	FALSE
PROPERTIES	GroupComponent [ref EdgeNode [1..1]] PartComponent [ref VirtualForwardingInstance [0..*]]

The aggregation "EdgeNodeInIPVPN"

This association identifies which border routers are serving an IP VPN. The class definition for the aggregation is as follows:

NAME	EdgeNodeInIPVPN
DESCRIPTION	Aggregation between an EN and an IP VPN.
DERIVED FROM	Component
ABSTRACT	FALSE
PROPERTIES	GroupComponent [ref IP VPN [1..1]] PartComponent [ref EdgeNode [2..*]]

Index